Laser Acceleration of Particles
(Los Alamos, 1982)

AIP Conference Proceedings
Series Editor: Hugh C. Wolfe
Number 91

Laser Acceleration of Particles

(Los Alamos, 1982)

Edited by
Paul J. Channell
Los Alamos National Laboratory

American Institute of Physics
New York 1982

L.C. Catalog Card No. 82-073361
ISBN 0-88318-190-8
DOE CONF- 820241

PREFACE

A workshop on the Laser Acceleration of Particles was held at the National Security and Resources Study Center of the Los Alamos National Laboratory from February 18-23, 1982. The Chairman of the organizing committee was Dr. Andrew M. Sessler.

This book contains most of the lectures presented at that workshop, papers contriubted by attendees, and an executive summary by the ad hoc committee on recommendations for laser acceleration. Both the executive summary and the workshop summary by Lawson provide overviews of the present state of the field of laser acceleration of particles and can serve as introductions for the non-expert. The Working Group summaries by Pellegrini, Palmer, and Sessler provide more detailed reviews of the three principal subfields, namely, far field accelerators, near field accelerators, and media accelerators.

The workshop and these proceedings were supported by the Office of Energy Research of the U.S. Department of Energy, the Office of Naval Research, the Defense Advanced Research Projects Agency, and the National Science Foundation. Organizational support was provided by the Lawrence Berkeley Laboratory and by Los Alamos National Laboratory. Special thanks are due to Dave Sutter of the Department of Energy and Bob Jameson of Los Alamos National Laboratory for support and encouragement, and to Susan Ovuka of Lawrence Berkeley Laboratory and Sue Nicol of Los Alamos National Laboratory for help with the organizational details.

Paul J. Channell
Los Alamos National Laboratory
Los Alamos, New Mexico

TABLE OF CONTENTS

BRIEF REPORT
on
The Laser Acceleration of Particles Workshop*
February 18-23, 1982
Los Alamos, New Mexico

EXECUTIVE SUMMARY

This first Workshop in the new field of laser-driven particle accelerators brought together some 60 physicists and engineers experienced in particle-accelerator and laser technology.

Our major purpose was to examine the various laser-driven accelerator concepts that have been advanced to assess their potential, explore their limitations, and point out areas where further work is needed. In this we were most successful, as the following report attests.

Furthermore, we examined six specific accelerator schemes that deserve further work and identified basic physics and technology problems common to several of the devices. Stimulated by this Workshop, other specific device proposals are bound to emerge.

Based on the results of the Workshop, we judge that the potential of laser-driven accelerator devices justifies devotion of resources for their further study and experimental exploration.

We found this Workshop profoundly useful in bringing some intellectual order to the laser-driven accelerator field as it now exists. In 1- or 2-yr time, a similar Workshop will again be most valuable in assessing progress in the field and in defining new goals and directions.

I. INTRODUCTION

The possibility of accelerating charged particles by making use of the high fields associated with laser beams has been considered at intervals over the past 20 yr. A number of proposals, some more credible than others, have been discussed in the literature. They have been put forward as new particle-accelerator concepts that might have advantages of economy or performance over existing designs.

- - - - - - - -

* The Organizing Committee of the Workshop on the Laser Acceleration of Particles met with plenary session speakers and the heads of the working groups, on the afternoon and evening of February 23, to consider the Workshop and conclusions to be drawn from it. The group consisted of P. Channell, P. Csonka, P. Morton, R. Palmer, R. Pantell, C. Pellegrini, A. Sessler, J. Slater, P. Sprangle, L. Teng, M. Tigner, and P. Wilson. Also present were D. Keefe and J. Lawson both of whom feel the estimation of the possibilities of laser accelerators is excessive, but nevertheless support the recommendations of this report. The present document is the result of deliberations by this group.

As a particular example, the realization that we seem to be near the "end of the road" for conventional accelerators has generated renewed interest in the possibility of accelerators with "super-high" accelerating fields. It seemed timely, therefore, to convene a Workshop on "laser accelerators," in which the proponents of the various ideas could meet with other experts on accelerators, lasers, and plasma physics to evaluate the proposals.

We were able to identify three broad classes of accelerator devices that have come to be associated with the subject: media accelerators exemplified by devices using the inverse Cherenkov effect and laser-controlled plasma waves; far-field accelerators exemplified by two types of inverse free electron laser; and near-field accelerators that use a loaded guiding structure such as a grating or cavity array.

After a day of introductory talks, the meeting divided into three groups to consider the three identified classes. A vigorous and fruitful interaction between enthusiasts and critics ensued, and on the last day, summaries were given of the states of the various categories. Despite a certain amount of inevitable and stimulating controversy, a fairly clear picture of basic possibilities and constraints emerged.

In the near-field class, it was estimated that acceleration gradients of as much as 1 GeV/m might be attainable, but that the current accelerated would be small.

In the far-field category, useful conceptual clarification of the relation between the Inverse Free-Electron Laser (FEL) and Two-Wave Accelerator was attained. There appear to be limitations to achieving very high energies, but pulsed currents (perhaps even of kiloamps) at energies of several GeV seem attainable.

For media accelerators, an Inverse Cherenkov Accelerator of perhaps 50 GeV
emerged as a possibility. Less certain, but more exciting, are the potentially very high fields attainable in the Beat Wave Accelerator.

As a result of the study, many issues were clarified but additional problems were forseen. Research programs that might resolve these issues have been outlined, so that in a reasonable time period a consensus can be established wherein some approaches will be discarded, some will be affirmed, and some new ideas will evolve.

The initial emphasis, of the research directions outlined in this report, is on issues of basic physics concerning the interactions of laser fields with particles, plasmas, and materials. This program will further lead to technological innovations and advances.

Application to other areas of interest are already foreseen. For example, because very short electron bunches, rich in harmonics, are produced by laser-driven accelerators, these short bunches might be used in turn to drive a device producing tunable radiation in the soft x-ray range.

With the consolidation of knowledge achieved at this Workshop, we believe that laser-driven particle accelerators have great potential and have emerged as a specific field for investigation and systematic development.

II. SHORT SUMMARY OF ACCELERATION MECHANISMS

Six areas have emerged where research should prove productive. However, there are clearly other directions that might be pursued. Some of them have been suggested during the course of the Workshop, such as the Laser Focus Accelerator or the Two Laser-Wave Accelerator. It is perhaps important to emphasize that, at this early stage in the newly emerging field of laser accelerators, it is wise to seriously consider a variety of acceleration schemes.

Some mechanisms of acceleration viewed as potential candidates for high-gradient accelerators are the following (not in order of priority):

1. <u>Grating Accelerator</u> - a special case of a conventional accelerating waveguide in which the waveguide structure is open sided, for example, a diffraction grating in the presence of a strong optical field.
2. <u>High Gradient Structures and Power Sources at Wavelengths Near 1 cm</u> - a scaledown of conventional waveguide structures to millimeter size with emphasis on avoidance of breakdown problems.
3. <u>Inverse Cherenkov Accelerator</u> - relies on index of refraction of background medium to provide phase matching between optical field and electron. A novel geometry has been proposed to alleviate gas-breakdown limitations.
4. <u>Beat-Wave Accelerator</u> - acceleration is achieved by the fields produced as a result of charge separation generated by a traveling plasma wave. The plasma wave, in turn, is set up by a laser pulse whose amplitude is modulated so that the beat frequency matches the plasma frequency.
5. <u>Inverse Free Electron Laser Accelerator</u> - uses the energy exchange occurring when an undulating electron in a periodic magnetic field has a transverse velocity in phase with the transverse E-field of an intense optical beam.
6. <u>Two-Wave Accelerator</u> - similar to (5) except that the magnetic wiggler is replaced by a slow microwave field whose phase velocity relative to the electron beam may be adjusted by changing the relative angle.

III. DESCRIPTION OF ACCELERATION MECHANISMS

1. <u>Grating Accelerator</u>

A grating linac is a special case of a convention linac where the structure is one sided, that is, is formed of bumps on a conducting surface that can be, but does not have to be, a simple periodic grating. A standing-wave solutions is known that has accelerating fields close to the grating that do not radiate away from that surface. The simplicity of this structure allows it to be built on a scale appropriate to 10 μ, that is, CO_2 laser radiation.

A measure of this solution's efficiency is the root of the shunt impedance that gives the accelerating field, given a fixed energy loss at the surfaces. The known grating solution gives 1/5 the field of a conventional linac, a factor that can hopefully be improved.

Transverse stability is achieved by alternating both the grating phase and an appropriate transverse magnetic field. With such focusing, transverse blow-up is negligible if grating errors are less that 1/20 of λ.

The fraction of the electromagnetic energy that can be given to the particles was shown to be independent of λ and thus similar to that in present linacs (~10%).

The advantages of the use of 10 μ radiation compared to the conventional 10 cm waves are (1) negligible electrical breakdown (λ); (2) somewhat less surface heating ($\lambda^{1/8}$); (3) somewhat less rf power for given gradient (λ); (4) far less rf energy per pulse (λ^4).

The first and second points would seem to allow accelerating gradients of 300-1000 MeV/m without grating damage, whereas the third and fourth imply that a 1-TeV accelerator could be powered by only 60 J delivered in a 30-ps pulse of 2 X 10^{12} watts; a seemingly modest requirement for a high-pressure CO_2 laser.

Higher gradients, even up to ~10 GeV/m, may be obtainable if the grating is regarded as disposable. The grating, though turned into a plasma, appears to maintain its essential periodicity long enough to assure acceleration. Although presumably only suitable for low beam currents, such an accelerator would attain 1 TeV in only 100 m and would require a 30-ps pulse of only 2000 J.

2. High-Gradient Structures and Power Sources at Wavelengths Near 1 cm

Based on modest extensions of present technology, it is possible to foresee accelerating structures of more-or-less conventional design operating at a gradient of 200-500 MV/m, in the wavelength range 0.3-3 cm, powered by rf sources with a peak power output ~10^9 W. Such an accelerator might provide the most immediate path toward a 1 TeV on 1 TeV e^+e^- linear collider with a reasonable total length. In this wavelength range, a structure operating at these gradients has sufficient stored energy per unit length to allow acceleration of a few times 10^{10} particles per bunch, leading to a luminosity of ~10^{32}/cm^2/s with reasonable bunch dimensions.

Several areas for theoretical and experimental work have been identified. First, the structure design must be optimized with respect to such figures of merit as E_a^2/w, where E_a is the accelerating field and w is the stored energy per unit length, and E_a/E_s where E_s is the peak surface field. Computer codes must be developed that can calculate the electromagnetic properties of both longitudinal and transverse (deflecting) modes in potentially useful structures. For highly asymmetric structures it may be difficult to develop such codes, although information can still be gained using various bench-measurement techniques. In passing, it can be noted that these measurement techniques can also be used to investigate the properties of gratings or other structures at optical wavelengths, using models that are scaled to centimeter wavelengths.

Secondly, as stated above, power sources are needed with a peak capability of 10^9 W. An FEL operating at $\lambda \sim 1$ cm is a possibility. Other sources have also been suggested, such as energy storage cavities with fast output switching and a photocathode device illuminated by a modulated laser beam. Once a prototype source has been developed, the important question of breakdown limits in this wavelength range can be investigated. There is some feeling that a substantial increase in breakdown field strength is possible at $\lambda \sim 1$ cm as compared to $\lambda \sim 10$ cm, where the accelerating field as limited by breakdown is probably ~ 100 MV/m.

3 Inverse Cherenkov Accelerator

In a medium, such as hydrogen at 1 atm, the index of refraction, n, is greater than unity so that $n - 1 = 10^{-4}$, and hence a wave impinging on a particle at the Cherenkov angle, θ_c, where

$$\cos \theta_c = \frac{1}{n\beta} ,$$

is resonant with the particle. The accelerating field may be expressed in terms of the laser field, E_0, by $E_0 \sin \theta_c$.

The maximum electric field in the gas is limited either by laser intensity or by breakdown. Typically, the breakdown field is 10^4 MV/m, so that (since $\theta_c \sim 15$ milliradians) the accelerating field is 150 MeV/m. However, if the laser light is radially polarized and introduced on a cone, then the maximum field on the axis is just the accelerating field. The maximum field in the gas is off-axis and somewhat reduced from what it is in planar geometry. In this case, use of a laser, of energy such as Antares, (70 Terrawatts) leads to an accelerating field of 400 MeV/m, with the maximum field in the gas just at the breakdown limit.

Because a medium is necessary to produce an index of refraction, there is (inevitably) beam scattering. This can be minimized by using hydrogen gas. Even so, the accelerator appears to be limited to an effective length of 50 m with a corresponding final $\gamma \simeq 10^5$.

The electron beam must not be too large or the energy spread caused by radial variation of the accelerating field becomes intolerable. Using CO_2 light, and taking $\Delta E/E < 0.3$, demands a beam radius of less than 100 microns.

Radial focusing in this accelerator is naturally present because the medium reduces the electric self-defocusing while not affecting the self-magnetic focusing.

In one example, using a laser of ~ 10^{14} W and an accelerator length of 50 m, electrons were shown to gain more than 20 GeV.

4. Beat-Wave Accelerator

If two different laser beams, at frequencies ω_1 and ω_2, are fired into a plasma and if $\omega_1 - \omega_2 = \omega_p$, where ω_p is the plasma frequency, then a strong nonlinear bunching of the plasma takes place. This bunching produces an electrostatic field that is very large and will accelerate the plasma particles or an injected group of particles.

One may think of the process as the nonlinear, or saturated, limit of Raman forward scattering. This last process produces plasma waves whose phase velocity is approximately the velocity of light. The saturation electric field is

$$E_L = \frac{\omega_p cm}{e} ,$$

or

$$E_L = (2\pi \, n r_o^3)^{1/2} \; \frac{mc^2}{r_o} ,$$

where $r_o = e^2/mc^2$ is the classical electron radius so $mc^2/r_o = 1.8 \times 10^{14}$ MeV/m. This accelerating field is very much higher than in other collective accelerators (This device really is a collective accelerator.) because, first, the bunching is over the small distance ω_p/c and, second, because the particle density is much higher than in other collective accelerators (an IREB has $n \leq 10^{14}$ cm^{-3}).

An experiment has shown the basic validity of the concept by observing over 1-MeV electrons in the forward direction. There is excellent agreement between these observations and (extensive) particle simulations. The theory, on the other hand, suggests that a multi-GeV accelerator can be built with each stage (of perhaps a meter in length), producing an energy gain of 2 GeV.

Although extensive particle simulation has been done, there is the need for further 1-D work and, most particularly, 2-D studies. There are a great many questions that need to be answered before one can feel confident of the beat-wave accelerator (especially questions related to 2-D phenomena).

5. Inverse Free-Electron Laser Accelerator

The basic physical principle involved in the FEL accelerator, which has been proved in recent experiments at Los Alamos National Laboratory, TRW, and MSNW, is to produce a transverse oscillation of the particle around its main direction of motion. This provides the mechanism of coupling between the transverse particle motion and the transverse electric field of the laser wave, which also travels in the main direction of the particle motion.

The transverse oscillation of the particle is produced by a static transverse magnetic wiggler field, and is such that the transverse velocity of the particle remains in phase with the transverse laser electric field. This condition requires a change in the parameters of the wiggler magnet as the particle energy increases.

This mechanism can produce accelerating fields of ~0.5 GeV/m, using laser fields of 5×10^{10} to 5×10^{11} V/m.

There are limitations in how far it is possible to achieve this necessary matching between the wiggler magnet parameters and the particle energy. There are also limitations in the maximum achievable laser field, depending upon the distance over which one wants to keep the laser beam focused. Without refocusing the laser beam, the limitation on the maximum energy is approximately a few GeV.

By staging the acceleration and refocusing the laser beam between stages, it is possible to increase the maximum energy. An alternative to refocusing the laser beam is to use multiple, phase-coherent, laser beams.

6. Two-Wave Accelerator

The basic physical principle of the two-wave accelerator (TWA) is similar to that of the FEL, that is, to produce a transverse oscillation of the particle around its main direction of motion. This provides the mechanism of coupling to the transverse electric field of the laser wave, which also travels in the main direction of the particle motion.

In the TWA, the particle oscillation is produced by a microwave field propagating at a small angle with respect to the particle beam, such that its velocity along the direction of the particle beam is nearly equal to the particle velocity. This slow wave greatly increases the transverse particle velocity, leading to a good coupling with the laser wave. An accelerating field of ~0.5 GeV/m can be obtained in this scheme, for a laser field of 10^{10} to 10^{11} V/m and a microwave field of 10^{6} V/m.

The transverse oscillation frequency of the particle must be such as to keep its transverse velocity in phase with the laser electric field. This condition requires a change in the phase velocity of the microwave or of its angle with respect to the particle beam, as the particle energy changes. As in the FEL the particle energy obtainable is limited by the maximum achievable laser field, depending upon the distance over which one wants to keep the laser beam focused. Without refocusing the laser beam, the limitation on the maximum energy is of the order of a few GeV. By staging the acceleration and refocusing the laser beam between stages, it is possible to increase the maxi mum energy. An alternative to refocusing the laser beam is to use multiple, phase-coherent, laser beams.

IV. SUGGESTED R and D PROGRAM

Research and development work thought to be appropriate for furthering the development of the field of laser acceleration of particles is summarized below.

1. Theoretical Investigations
   Much more analytical and numerical work is required. Computer simulations are particularly needed for the Beat-Wave Accelerator and the Grating Accelerator. These studies could be carried out by individuals and groups in their home institutions.
2. Materials Breakdown and Damage Studies
   In all these schemes, material structures made of metal, dielectric or gas come into interaction with high-field and high-power laser beams. The surfaces of these structures are necessary to reflect, transport, focus, diffract, and shape the laser field. The breakdown and damage limits of these materials must be known to determine the maximum attainable field for acceleration. In some cases, new mate- rials may be found or developed to support higher power laser fields. Most of these studies can be performed on existing lasers.
3. Development of High-Quality Radiation Source
   Before effective demonstration projects can proceed, we need to investigate the availability of moderate power lasers with good spatial and temporal coherence. It may be necessary to extend the maximum distance over whichcoherence can be maintained. In addition, high-energy accelerators may require very short-pulse lasers, and these need development.
   For some applications, a good source of high peak power millimeter waves is needed.
4. Focusing, Transporting, and Manipulating High-Power Laser Beams
   In many of the accelerator schemes, the laser beam has to be refocused periodically to maintain a high accelerating field. This is but one facet of the general technology of transporting and manipulating high-power laser beams without loss of quality. A sizable effort should be devoted to understanding, and if necessary developing, this very important and basic technology.
5. Nonlinear Effects in Media
   Nonlinearities must also be investigated if the feasibility of certain acceleration schemes is to be established. This is particularly true for the plasma approaches, in which intense laser beams are focused for short time intervals. The time constants for self-focusing and stimulated Raman scattering are often in the subpicosecond regime, and these nonlinearities may help or hinder the proposed accelerator.

6. Modeling, Testing, and Demonstrating Specific Accelerator Schemes
At an appropriate stage of technological development, experimental demonstrations will be required. Some demonstrations have already been successful on a small scale, and the time has come for some of the schemes--such as the Inverse Cherenkov or the Inverse FEL--to proceed to larger scale tests. For others, more study is required. Many of these tests can be performed using the same basic components: a high-resolution electron beam and a laser beam of high optical quality. It may be desirable at some time in the future to establish a single central facility.

V. CONCLUSIONS AND RECOMMENDATIONS

As a result of this Workshop, it seems clear to us that the areas of study that have come to be grouped under the rubric of laser-driven accelerators are worthy of serious attention by the accelerator community. In the foregoing parts of this report, we have attempted to determine important components of this somewhat amorphous and growing field. We have identified six specific types of accelerating devices that seem attractive to pursue as well as some basic physics and laser development areas that need attention. There are no doubt other devices and areas of basic physics, together with their associated technology, whose importance is yet to emerge.

In addition to accelerators that are clearly tied to the laser as a source, work on high-gradient structures that may be effective over a wide range of wavelengths has come to be associated with the general line of thinking about laser-driven accelerators, and rightly so.

We believe that the potential of the class of devices described above merits the devotion of significant resources to their theoretical and experimental exploration.

This first Workshop on the Laser Acceleration of Particles has manifestly been of great benefit in clarifying the important issues in the field and in understanding some of the basic limitations and advantages of the various accelerator schemes. This clarification and understanding have defined new lines of inquiry for the future. In 1- to 2-yr time, a second workshop to assess progress and reevaluate the position of the field will be a necessity.

REPORT OF THE WORKING GROUP ON MEDIA ACCELERATORS*

Andrew M. Sessler

Lawrence Berkeley Laboratory
University of California
Berkeley, CA 94720

April 12, 1982

ABSTRACT

A summary is given of the activities of those in the Media Accelerator Group. Attention was focused on the Inverse Cherenkov Acceleraotr, the Laser Focus Accelerator, and the Beat Wave Accelerator. For each of these the ultimate capability of the concept was examined as well as the next series of experiments which needs to be performed in order to advance the concept.

I. Introduction

The Media Accelerator Group found itself in the enviable position that for three different accelerators there already existed theories of how they operated and, furthermore, experiments had already been performed which were in accord with these theories. Given this information, it was quickly decided that since only a few days were available to us, we would focus attention upon these three schemes and forego the examination of other proposals. Thus, we only considered the Inverse Cherenkov Accelerator,[1] the Laser Focus Accelerator,[2,3] and the Beat Wave Accelerator.[4,5,6] For each of these we reviewed the theory of its operation; considered the ultimate capability of an accelerator of this type; discussed the various technical, theoretical, and experimental problems which need to be addressed; and outlined theoretical and experimental work which could be undertaken so as to advance our understanding of this particular accelerator.

As you will see, these three devices are quite different in the degree to which they are understood, in the technical problems which must be overcome in order to have them work, in their ultimate promise, in the form which they would take, and in the uses to which they might be put.

* This work was supported by the Director, Office of Basic Energy Sciences Division, of the U. S. Department of Energy Under Contract No. DE-AC03-76SF00098.

Members of the Working Group, whose efforts are summarized here were W. Bostick, A. Chao, F. Cole, G. Fontana, H. Hora, C. Joshi, V. Nardi, D. Neuffer, M. Piestrup, L. Rivkin, R. Ruth, A. Sessler, R. Sudan, D. Sullivan, T. Tijima, and W. Willis.

What the three devices do have in common is that they all require a medium through which the particles to be accelerated must move. For the Inverse Cherenkov Accelerator (ICA) this medium is a gas which has an index of refraction greater than unity and, hence, slows down the laser wave so that it can resonate with the material particle being accelerated. In other regards the medium is not active.

In the Laser Focus Accelerator (LFA) and the Beat Wave Accelerator (BWA) the medium is a plasma and active. The LFA works by employing the non-linear effect in which the index of refraction depends upon the density so that there is self-focussing of the laser beam, while the BWA depends crucially upon media -- i.e., plasma -- motion. The BWA is really a form of collective accelerator in which the medium is organized by the laser light, and the acceleration is done by the electrostatic forces which result from this organization.

Thus the three devices employ the medium in very different ways. The three accelerators may not include the "best" media accelerator, but they are sufficiently different that the study of them even if it is not sufficient to span the range of possibilities, is, at least, suggestive of the range which is possible in media accelerators. Study of these three concepts is bound to be productive; in this report we review the deliberations which a small number of scientists gave, in only a few days, to three fascinating -- and stimulating -- concepts.

II. Inverse Cherenkov Accelerator

This accelerator employs the Cherenkov Mechanism: if a particle moves faster through a medium than light travels in the same medium, then it will radiate (that is form a "wake"). The ICA simply runs this effect backwards; in other words it uses very intense light (from a laser) travelling in a medium in order to accelerate particles.

A. Physical Principles

Imagine a photon of wave number $\underset{\sim}{k}$ and frequency ω impinging upon an electron of momentum $\underset{\sim}{p}_1$ and E_1. If the photon is absorbed then in the final state there is only an electron with momentum $\underset{\sim}{p}_2$ and energy E_2. Conservation of energy and momentum yields:

$$\underset{\sim}{p}_1 + \hbar \underset{\sim}{k} - \underset{\sim}{p}_2, \qquad \text{(II.1)}$$

$$E_1 + \hbar\, \omega = E_2.$$

Because all of this takes place in a medium of index of refraction, n, we have the relation between wave number and frequency

$$|\underset{\sim}{k}| = \frac{\omega n}{c}\,. \qquad \text{(II.1)}$$

In practice, the photon energy and momentum are very small and one quickly deduces that if θ_c (the "Cherenkov angle") is the angle between the photon and the electron then

$$\cos \theta_c = \frac{1}{n\beta}\ , \qquad \text{(II.3)}$$

where β is the relativistic factor of the electron; i.e., $\beta = v/c$ with v the speed of the electron.

In practice, the index of refraction is very close to unity; in fact for H_2 of 1 Atmosphere

$$n-1 \approx 10^{-4}\ . \qquad \text{(II.4)}$$

Since β is also very close to unity, and it is often most convenient to re-write (II.3) in the form:

$$\theta_c^2 + \frac{1}{\gamma^2} \approx 2(n-1) \qquad \text{(II.5)}$$

where γ is the relativistic factor

$$\gamma^2 = \frac{1}{1-\beta^2}\ , \qquad \text{(II.6)}$$

and all three terms in (II.5) are small.

The energy gain per unit length, due to a laser field of electric field strength E, is:

$$W = eE \sin \theta_c \sin \phi, \qquad \text{(II.7)}$$

where ϕ is the phase angle of a particle in the electromagnetic wave. In practical units, the energy gain of electrons, ΔE, in a length, L, subject to Cherenkov light of wavelength, λ, is

$$\Delta E = 68.8 \sin \theta_c \left(\frac{PL}{\lambda}\right)^{1/2} \sin \phi, \qquad \text{(II.8)}$$

where ΔE is in GeV, P is the laser power in terrawatts, L is in meters, and λ is in microns.

Typically, $\theta_c \approx 15$ millirad and E is limited either by laser power or by breakdown in the gas which constitutes the accelerating medium. The breakdown field strength is not precisely known, but we take

$$E_{\text{Breakdown}} \equiv E_b \approx 10^4 \frac{\text{MV}}{\text{m}}\ . \qquad \text{(II.9)}$$

Combining these facts we obtain, from II.7 that W $\approx$ 150 MeV/m. This is a considerable field, but not extraordinary.

One can do better by arranging the light in a geometry which makes all field components, except the accelerating field, vanish on an axis which becomes the line along which one accelerates. A cone of properly polarized (clearly radially) light will produce fields of the form[7]:

$$E_z(r,z) = E_0 J_0\left(\frac{k\ r\ \tan\theta_c}{\beta}\right) e^{-}\ \frac{ikz}{\beta} \quad ,$$

$$E_r(r,z) = i\ \cot\theta\ E_0 J_1\left(\frac{k\ r\ \tan\theta_c}{\beta}\right) e^{-}\ \frac{ikz}{\beta} \quad ,$$

$$H_\theta(r,z) = \frac{i\ n\ E_0}{\sin\theta}\sqrt{\frac{\varepsilon_0}{\mu_0}}\ J_1\left(\frac{k\ r\ \tan\theta_c}{\beta}\right) e^{-}\ \frac{ikz}{\beta} \qquad \text{(II.10)}$$

In these formulas, all symbols have already been defined with the exception of the Bessel functions J_0 and J_1.

As can be seen, from (II.10) only the accelerating field E_z is non-zero at r=0. However, for r≠0 the radial field quickly becomes larger than $E_z(o,z)$(mathematically, because of its large coefficient and, physically, because the cancellation can only be made to occur along a line and must be very exact).

Also, one should note that $E_z(r,z)$ goes to zero as one moves away from the accelerating axis r=0. Thus the accelerated beam must have a radius, r_b, less than the first zero of J_0;

$$r_b < \frac{2.4\ \lambda}{2\pi\ \tan\theta_c} \quad , \qquad \text{(II.11)}$$

It should be noted that the applied fields actually produce a focusing of the accelerated electron beam. Expressions for this were derived[7] and could be employed to quantitatively balance this focusing force against the space-charge defocusing and the beam emittance. The Group, however, did not have the time to persue this further.

B. Full-Scale Machines

We shall give two examples:

Example 1 -- For electrons with energies of tens of Gev's the multiple scattering leading to beam spreading in energy and angle is small enough so that one can make an accelerator of 50 m length, taking:

$$L = 50\ \text{m},\quad \lambda = 10\mu\text{m},\quad P = 70\ \text{TW},\quad \theta_C = 20\ \text{mrad} \qquad \text{(II.12)}$$

One finds an acclelerating gradient of 500 MeV/m and a total energy gain, ΔE, of 25.8 GeV. •

If one considers a beam radius, r_b, of 50 μm, then E_z at the edge of the beam is 90% of E_z on axis, and the radial field within the beam is less than 7.7×10^3 MV/m (which is below the breakdown field of II.9).

However, the radial field peaks at 146 μm and there attains 1.5×10^4 MV/m which is above the spark breakdown value. Thus, one can expect sparks outside the beam, which may be acceptable.

Example 2 --In this example, we considered simply de-rating the laser power, but otherwise keeping the same parameters as before, i.e.,

$$\begin{aligned} L &= 50 \text{ m}, \\ P &= 30 \text{ TW}, \\ \lambda &= 10 \ \mu\text{m}, \\ \Theta_c &= 20 \text{ mrad}. \end{aligned} \tag{II.13}$$

Now the accelerating gradient is 340 MeV/m, and the total energy gain, ΔE, is 17 GeV.

We can now consider a beam radius, r_b, of (say) 100 μm. The field variation of E_z across the beam is now 30%. The radial field still peaks at 146 μm, but now attains the value of 10^4 MV/m, which is just the breakdown field and therefore (presumably) will be just low enough not to cause sparks anyplace.

As you see, these two examples are examples of interesting accelerators. One can imagine other machines such as one in which gas is confined to a narrow tube (less than 146 μm) which explodes when it is irradiated, but still the gas is inertially confined and effective during the accelerating pulse. This, and some other schemes, were not examined in the brief time available to the working group. It was felt that the two examples sufficed to demonstrate that a full-scale machine of an ICA would be interesting.

One can, to obtain even more interesting machines, consider increasing Θ_c, or decreasing λ, or increasing P. None of these changes are easy, however, and the examples, given above are probably near the limit (or even beyond!) of an ICA.

C. Experimental Program

As a next step it was felt to be important to employ a high power CO_2 laser in a "cone-geometry." The rationale is to develop optical and electron beam techniques that can be employed on large scale systems.

The ingredients of the proposed experimental program are:

1. Use the SLAC or SCA electron beams.
2. Use a CO_2 laser with 1 ns pulse and a power rating of (say) 10^{10} W. The CO_2 laser (as contrasted with the Niodinium laser of Ref. 1), decreases the effect of multiple scattering.
3. Use a radially polarized, plane wave cone for injecting the laser beam, while keeping to a minumum the number of optical components.

A sketch of the experimental lay-out is given in Fig. 1.

III. LASER FOCUS ACCELERATOR

A sufficiently intense beam of light, upon entering a medium, will self-focus itself and, consequently, produce a very large gradient or electric field strength. This large gradient will, by non-linear effects, accelerate particles. This is the basic concept of the Laser Focus Accelerator (LFA).

A. Physical Principles

A laser beam of intensity (Power/Unit Area), I, and diameter, d, when $I \gtrsim I_0$ will self-focus in a medium with the focus distance $\approx$ d. The threshold intensity, I_0, a function of wavelength (and not very well known) is roughly:

$$I_0 \approx 10^{18}\ \mathrm{W/cm^2}, \quad \lambda = 1\ \mu\mathrm{m};$$

$$I_0 \approx 10^{16}\ \mathrm{W/cm^2}, \quad \lambda = 10\mu\mathrm{m}; \tag{III.1}$$

In fact, I_0 could be less than the values (III.1) by as much as a factor of 50.

In a medium, characterized by an index of refraction, n, there is a ponderomotive force density, $\underset{\sim}{F}$, given by:

$$\underset{\sim}{F} = -(1+n^2)\ \underset{\sim}{\nabla}(E^2/8\pi) \tag{III.2}$$

If the pulse of laser light of duration, τ, is too long, then the medium, which consists (probably) of a plasma, will disperse and hence there will no longer be self-focusing. Thus the pulse must be shorter than a characteristic time, $_0$, wich depends upon frequency. One finds:

$$\tau_0 \approx 5\ \mathrm{ps}, \quad \lambda = 1\ \mu\mathrm{m};$$

$$\tau_0 \approx 50\ \mathrm{ps}, \quad \lambda = 10\ \mu\mathrm{m}; \tag{III.3}$$

Providing $\tau < \tau_0$ and $I > I_0$ then these two effects produce ions of energy, ΔE, with

$$\Delta E \approx 3ZP, \tag{III.4}$$

where Z is the atomic number of the species accelerated, P is the laser power in terrawatts, and the energy, ΔE, is in MeV. This Eq. (III.4) is due to H. Hora[8] and is the result of analytic work and also of computer studies.

C. The Accelerator

In a single laser focus one can expect particles which would be of interest for a number of applications. We have shown, in Fig. 2, the theoretical curves (Eq. (III.4)) as well as the result of experimental observations. The facts that 15 MeV protons, 38 times ionized tungsten, and ions whose total energy is greater than 100 MeV have all been observed and fall close to the appropriate curves suggests that the theoretical explanation is correct.

On this basis one can expect, in one focus (and no one knows how to have repeated focii or how to have the accelerated particles acted upon by a second focus), $\approx 10^{10}$ ions with an energy of $\approx$ 50 MeV/nucleon. These particles would come out in a pulse ranging from 10 ps to 100 ps with an energy spread which is presently unknown but which might be as small as $\Delta E/E \approx 10\%$. For this one would need P 10^{13} W and the repetition rate, which would depend strongly on the interest in the accelerator, is probably at best 1 Hz.

Such an accelerator could be used for nuclear reaction studies, spallation studies, muon generation, and the study of very short half-life nuclides. It was not clear to the Group just what applications -- if any -- would make the LFA competitive with other (non-laser) accelerators.

It was noted, however, that the CO_2 lasers at LANL and the Nd glass lasers at the Australian National University could be employed to check the predicted dependence of energy gain upon laser power, laser pulse length, and laser wavelength. These experiments are modest in cost and time and could, readily, be fitted into the current program schedules.

IV. BEAT WAVE ACCELERATOR

Perhaps of all the laser acceleration ideas which were considered by participants in this workshop, the Beat Wave Accelerator (BWA) has the most promise and the most uncertainty. That is; the accelerator is based upon controlling very complicated non-linear plasma phenomena which, to date, have only been studied in a one-dimensional approximation (but studied rather extensively by means of particle simulation). On the other hand, the accelerator has the potential of producing higher gradients than seem possible with <u>any</u> other scheme.

A. Physical Principles

The basic idea[4] is to shine into a plasma two laser beams, having angular frequencies ω_1 and ω_2, where the plasma frequency, ω_p is just the difference frequency $\omega_1 - \omega_2$. Under these circumstances the plasma will bunch and there will result an electrostatic field which is then employed to accelerate particles.

The plasma density may be high (10^{17} - 10^{18} cm^{-3}) (much higher, for example, than in intense relativistic electron beams where $n \lesssim 10^{14}$) and the bunching occurs over the distance of a plasma wavelength, $2\pi c/\omega_p$, which can be much less than the characteristic distance of bunching in other collective accelerators. Hence, the accelerating field in the BWA can, in principle, be very much greater than in all other collective (or laser) plasma accelerators.

Because the plasma motion is caused by and organized by the laser light, it is believed that the motion will be stable motion, and hence, that the BWA will work as predicted. The BWA employs, actually, very non-linear plasma motion. In fact, there is essentially complete bunching of the plasma. However, it is useful to consider the basic (linear) interaction of a photon with a plasma.

Consider a single photon of frequency, ω_0, and wave vector, $\underset{\sim}{k}_0$, which undergoes Raman forward scattering. The final state will have a photon, of wave vector, $\underset{\sim}{k}$, and frequency, ω, and a plasmon of wave vector, $\underset{\sim}{K}$, and frequency ω_p. Between initial and final state we must conserve energy and momentum and, thus:

$$\underset{\sim}{k}_0 = \underset{\sim}{k} + \underset{\sim}{K},$$

$$\omega_0 = \omega + \omega_p \qquad \text{(IV.1)}$$

These relations are, of course, just the Manley-Rowe conditions.

Now in a plasma the photon is "dressed"; i.e., surrounded with a polarization charge, and hence

$$\omega_0^2 = k_0^2 c^2 + \omega_p^2,$$

$$\omega^2 = k^2 c^2 + \omega_p^2. \qquad \text{(IV.2)}$$

Alternatively, these formulas can, of course, be obtained from the dispersion relation for waves in a plasma.

Combining (IV.1) and IV.2) we have two equations for two unknowns; namely the frequency, ω, of the scattered light and the wave number, K, of the plasma excitation. One finds that

$$K \approx \omega_p/c,$$

$$\omega = \omega_0 - \omega_p \qquad \text{(IV.3)}$$

The plasma excitation has a phase velocity, v_{ph}, where

$$v_{ph} \equiv \omega_p/K \approx c. \qquad \text{(IV.4)}$$

The process can happen again and again. One finds that in a non-linear treatment one has $v_{ph} \approx v_{group} \approx c$, and, thus, a "wake" which moves along at this speed. One must, also, consider the effect of two laser beams. The basic physics is, however, as described here.

At what field strength, E_L, will the effect saturate One estimate is given by the assumption of essentially complete bunching at the wave length c/ω_p. Thus from

$$\nabla \cdot E = 4\pi ne, \tag{IV.5}$$

we have

$$\frac{\omega_p}{c} E_L = 4\pi ne,$$

and hence

$$E_L = \frac{\omega_p \, cm}{e} . \tag{IV.6}$$

A second estimate is given by assuming the trapping potential, $e\phi$, is of the order $1/2\ mv^2$ with $v=c$. Combining this with

$$\phi \approx \frac{E_L c}{\omega_p} ,$$

one obtains exactly the same formula for the saturation field, E_L, as from the first estimate (IV.6).

The formula for E_L, IV.6, may be written

$$e\,E_L = \left(2\pi n r_o^3\right)^{1/2} \frac{mc^2}{r_o} , \tag{IV.7}$$

where $r_o = e^2/mc^2$ is the classical electron radius and thus $mc^2/r_o = 1.8 \times 10^{14}\ \frac{\text{MeV}}{\text{m}}$. For $n = 10^{17}\ \text{cm}^{-3}$ one obtains $eE_L = 2 \times 10^4$ MeV/m.

If the above estimate is roughly correct (even a factor of 10 degradation is performance still gives 2 GeV/m!) will other waves grow and, perhaps, have a serious effect upon E_L There are lots of other waves in a plasma in particular, the backward scattered Raman wave. Going through energy and momentum balance as before, we now find

$$K \approx 2\ k_o . \tag{IV.8}$$

The phase velocity of this wave is

$$v_{ph} = \frac{c}{2} \frac{\omega_p}{\left(\omega_o^2 - \omega_p^2\right)^{1/2}}; \tag{IV.9}$$

and hence $v_{ph} \approx \left(\frac{\omega_p}{2\omega_o}\right) c$. This wave grows very fast (even faster than the forward going wave), but because the backward wave has $v_{ph} \ll c$ it will be much more strongly Landau damped than the forward-going wave. Thus we believe the backward scattered Raman wave will be very different than the forward-going wave.

Extensive numerical simulation has been done on a 1-D version of the BWA.[6] This work tends to confirm the above analytic estimates.

Most importantly, an experiment has been performed[5] which is in agreement with the theory. Of course, the laser employed was modest and hence the accelerating gradient obtained was modest, but the experiment gives creedance to the theory which predicts, in other regimes, quite remarkable behavior.

In the experiment a very thin (130 Å) carbon foil was irradiated with a 700 ps pulse of CO_2 light. (This experiment employed a single sharp-rising pulse rather than a beat wave.) The foil was heated to very high temperatures ($T \approx 20$ keV) and underdense to the laser light. It was found that the highest energy electrons ($E \geq 400$ keV) were peaked forward and that there were (about) 10^{11} of them. The forward electrons could be characterized by a temperature of 90 - 100 keV, and electrons with energy as high as 1.4 MeV were observed although the laser was only large enough to create a "quivering velocity" of 0.3 c.

B. Parameters of a Full-Scale Machine

Very little time was spent by the Working Group, on full-scale machines. Nevertheless, in Fig. 3 we sketch one version of a large 20-50 GeV accelerator. In addition, an alternative high energy accelerator design is presented in the contributed paper by R. Ruth and A. Chao together with some basic physics calculations on the workings of a plasma/laser accelerator.

C. Theoretical Subjects Suggested by of the BWA

The Group spent considerable time outlining -- attempting to be exhaustive in its deliberations -- the problems which have yet to be addressed. The Group came up with the following problems:

1. How large is the longitudinal electric field

   ($E_L = \dfrac{\gamma_\perp^{1/2} m c\omega_p}{e}$ seems a reasonably accurate estimate.)

2. What is the threshold laser strength (The threshold laser electric field probably is

   $E = \sqrt{2}\,\dfrac{mc\omega}{e}$, but perhaps instabilities lower the

   threshold.)

3. What is the optimum frequency separation of two beams $\omega - \omega$ (Is it ω_p or $2\omega_p$ or some definite function of laser amplitude)

4. What is the effect of the electron distribution function on the beam quality (longitudinal) Is a hot distribution better than a cold one (If one wants to mainly accelerate ions, is it useful to trap electrons)

5. Transverse stability of beam and/or plasma:

a. Self-current, self-magnetic field, filamentation instabilities, return current, etc. (If the beam radius r_0 is larger than $c/\omega_p \approx 1/2 \times 10^{-2}$ cm, then the return current runs on the surface of the electron beam.)
b. Self-Channeling
(This time scale is acoustic time scale: therefore, takes place only when the beam duration is verylong.)
c. Laser coherency and focusing
d. Emmitance growth due to side scattering (Amount of side scattered light energy as a function of Z.)

Some of these subjects, it was felt, could be illuminated by analytic work:

a. A similar analytical approach done for the free electron laser should be done for this concept.
b. Linear stability of possible transverse instabilities should be analyzed.
c. Phase relation of ions to the electrostatic field should be analyzed as a function of energy.

Further 1-D particle simulation work needs to be done on:

a. Saturated accelerating field strength size.

   Already know $E_L \geq \frac{\gamma_\perp^{1/2} \, mc\omega_p}{e}$ due to nonlinearity of saturated wave. (Note γ is determined by laser intensity (Vo/c).)
b. Does an absolute laser threshold exist $(E_0 > 2 \frac{mc\omega_0}{e})$

   or will Raman forward scattering instability saturate E_L in a reasonable distance or time
c. At saturation the nonlinear wave steepening in the single packet case results in the optimum packet case being a plasma wavelength ($\lambda_p = 2\pi c/\omega_p$). Is this true of the beam wave acceleration also
d. Can particle beam quality be improved from its presently observed exponential distribution based on:
   1. Injecting low density preaccelerated particle bunches
   2. Using a hot vs. cold temperature plasma
e. Will a relativistic two-stream instability develop
   1. When ions are included in the simulation (ion-electron)
   2. If low emmitance preaccelerated particles are injected into the beat wave packets (ion-electron, electron-electron, ion-ion)
f. Coherency/synchronism of particles and waves

To address the remaining subjects one will need 2D particle simulation work. This should allow one to study:

a. Does the beam pinch or expand
   Possibilities:
   1. Return current ≈ beam current flows inside beam channel. No self-magnetic field, beam expands radially.
   2. Return current flows outside beam channel. If $N_i/N_e > 1/\gamma^2$ (almost certainly true) beam will pinch.

b. Does Raman side scatter increase emittance or can it be suppressed by Landau damping similar to Raman backscatter

c. Can laser self-channeling (decreased density in beam channel) due to the radial pondermotive force destroy the frequency matching ($\omega_1 - \omega_2 = \omega_p$) condition Can it be overcome where $_{pulse} < r_{spot}/c_{sound\ speed}$.

d. How does diffraction of the laser beam affect the accelerated particles as they transit the focal region

e. Does the particle beam filament 1) 2-d r-θ geometry; 2) 3-D full cylindrical geometry.

D. Two Experiments

The group proposed two experiments which would greatly increase our understanding of the BWA.[9] It was felt that these experiments should be done in the near future.

Laser Requirement:
1. 100 J - 1000 J in ≈ 1 ns. One beam of the Helios laser at LANL would suffice.
2. Multiline CO_2 oscillator going on $P(2\nu)_{10.6\ \mu m}$, $R(1\delta)_{10.27\ \mu m}$ and $P(2\nu)_{9.6\ \mu m}$ bands

Target Requirements:
1. Thin foil targets (C, CH, Au Foils 50-500°); θ pinch plasma source $10^{16} < n_e < 5 \times 10^{18}\ cm^{-3}$)

Diagnostic Requirements:
1. Diagnostic CO_2 beam going on $p(20)_{10.6\ \mu m}$ line
2. IR double grating spectrometer
3. IRMA (infrared multichannel analyzer) or pyroelectric array + data acquisition and handling capability
4. Cu:Ge, Hg:Cd:Te Cold detectors
5. Nuclear emulsion particle detection and Thompson parabolas + CN films
6. X-ray continuum detectors for 10 keV to 300 keV.
7. Usual beam and target diagnostics e.g., - photon drag detection, infared vidicon, calorimeters

Manpower Requirements
1. Two post docs
2. Two graduate students

REFERENCES

1. M. A. Piestrup, G. B. Rothbart, R. N. Fleming, and R. H. Pantell, Jour. Appl. Phys. 46, 132 (1975).

2. H. Hora, E. L. Kane, and J. L. Hughes, Jour. Appl. Phys. 49, 923, (1878).
3. H. Hora, D. A. Jones, E. L. Kane, P. Lalousis, and P. R. Wiles, Proc. of the 4th Intl, Conf. on High-Power Electron and Ion Beam Researcn and Technology, Palalseau, (1981).
4. T. Tajima and J.M. Dawson, Phys. Rev. Lett. 43, 267, 1979.
5. C. Joshi, T. Tajima, J. M. Dawson, H. A. Baldis, and N. A. Ebrahim, Phys. Rev. Lett. 47, 1285, (1981).
6. D. J. Sullivan and B. B. Godfrey, IEEE Trans on Nuclear Science, NS-28, 3395, (1981).
7. G. Fontana, private communication.
8. H. Hora, private communication.
9. C. Joshi, private communication.

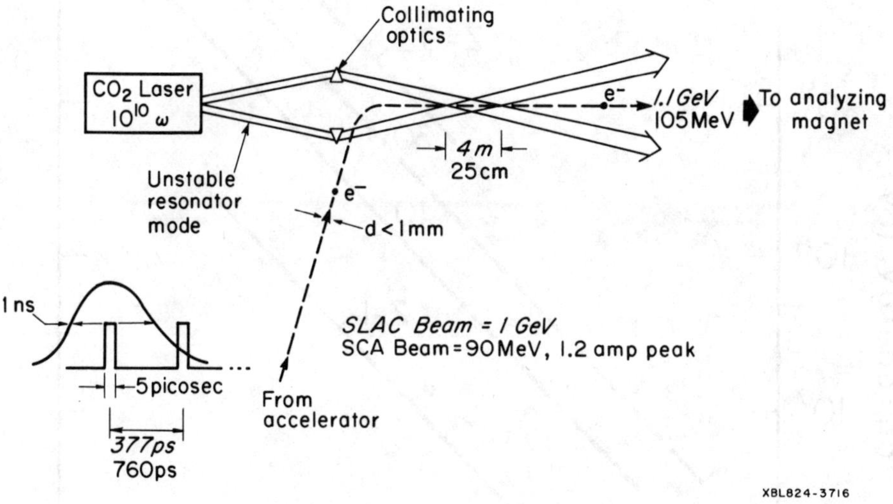

Figure 1. Layout of an Inverse Cherenkov Accelerator experiment.

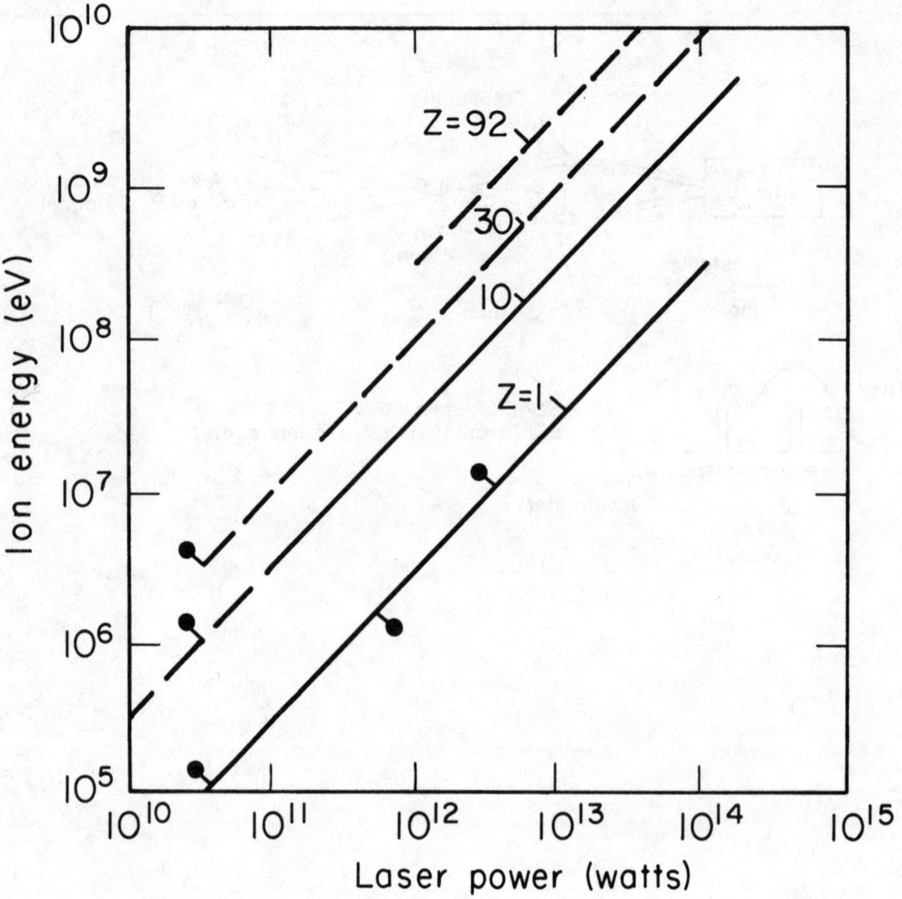

XBL 824-3712

Figure 2. Performance expectations of the Laser Focus Accelerator as a function of laser power. The curves are due to H. Hora[8] and the points indicate experimental observations.

24 Stage Laser Beat Wave Accelerator 20-50 GeV Total Gain

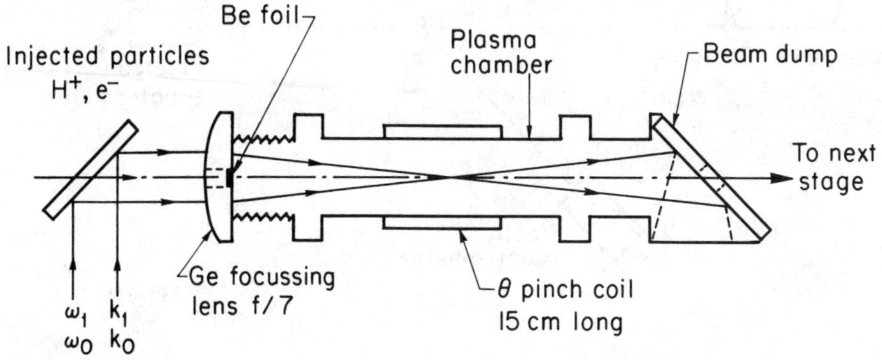

Beam intensity per beam	$\sim 5 \times 10^{15}$ W cm^{-2}, $v_0/c \sim 0.5$
Plasma density	$\sim 10^{17}$ cm^{-3}
Plasma temperature	~ 100 eV
Gain per stage	$\sim 1-2$ GeV

XBL824-3715

Figure 3. A full-scale Beat Wave Accelerator for producing electrons of 20-50 GeV, with 24 stages and each stage giving 1-2 GeV to the electrons.

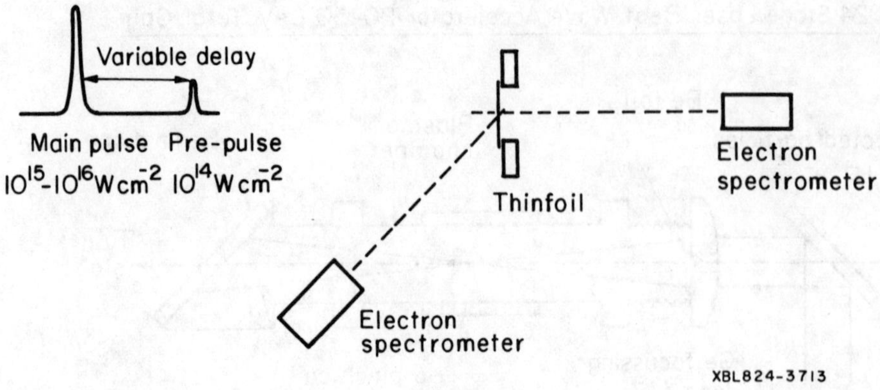

Figure 4. An experiment for the BWA concept. Scaling of maximum electron energy and temperature of the heated electrons with ω_p and v_o/c and T_e

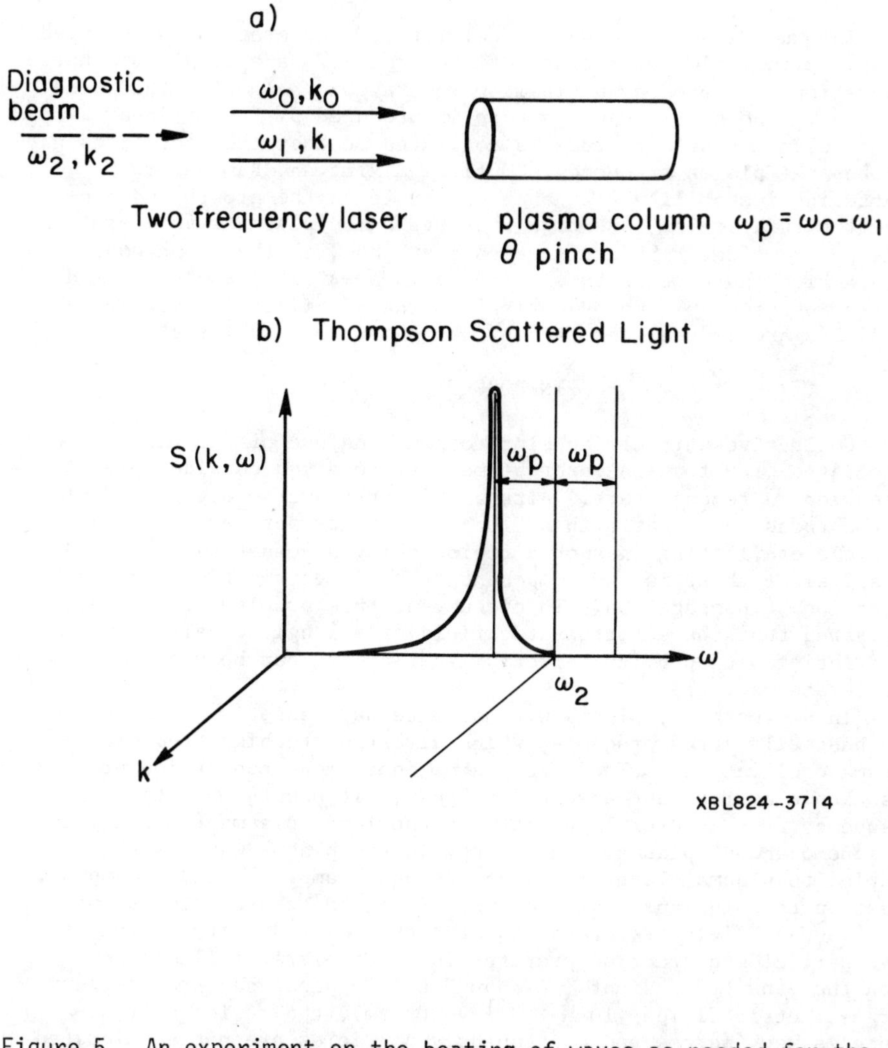

Figure 5. An experiment on the beating of waves as needed for the BWA. From the Fourier transform function, $S(k,\omega)$, of the Thompson scattered light one determines ω, hence $n(k,\omega)/n_o$, and therefore, $E_p(\omega p)$. The experimental set-up is shown in Fig. 5a, and the scattered light is shown in Fig. 5b.

THE PLASMA BEAT WAVE ACCELERATOR - I EXPERIMENTS

C. Joshi
University of California, Los Angeles, CA 90024

ABSTRACT

In the Plasma Beat Wave Accelerator, acceleration is achieved by a longitudinal electrostatic field produced as a result of charge separation generated by a plasma wave traveling close to the speed of light. The plasma wave in turn is produced by two colinear laser pulses whose amplitude is modulated so that the beat frequency matches the plasma frequency. Optical mixing and Raman forward scattering instability play a crucial role in the growth and non-linear saturation of such high phase velocity plasma waves. In this paper we describe experiments that show, i) that long wave-length high phase velocity electron plasma waves can be generated in a homogeneous plasma by optical mixing and ii) relativistic particles are indeed generated by the Raman forward instability.

INTRODUCTION

Collective particle accelerators making use of the high fields associated with focused laser beams have received considerable attention in recent years. Although the transverse electric fields at the focus can be as high as 10^9- 10^{10} volts per meter, a charged particle oscillating in such a field achieves no net acceleration at all since there is no component of the electric field in the direction of propagation. To circumvent this problem, it has been suggested that the electrostatic field of a longitudinal plasma oscillation set up by the electromagnetic wave can be used to accelerate particles[1].

In a laboratory plasma with no external magnetic field, there are basically three processes which give rise to high frequency plasma oscillations. The first is a linear mode conversion process, also known as resonant absorption because it occurs when the frequency of the laser light matches the local plasma frequency in an inhomogeneous plasma. The energy in the plasma wave is then coupled to plasma electrons by collisional damping, Landau damping, electron trapping and wavebreaking. Since the phase velocity of the resonant field propagates towards the lower density region, the wave particle interaction preferentially accelerates electrons down the density gradient. Although this process can produce very large electric fields (10^{10}- 10^{11} volts/cm), the region of resonance is very narrow and consequently particles are not accelerated to ultra-relativistic energies.

The second mechanism is known as high frequency parametric instabilities in which an electromagnetic wave propagating in an underdense plasma undergoes a decay into an electron plasma wave and an electromagnetic (Stimulated Raman Scattering), an ion acoustic (Parametric Decay Instability) or another electron plasma wave (Two-Plasmon Decay). Whenever one of the decay products is an electron plasma wave, very large electric fields in the direction of propagation can be produced. Poisson's equation gives $\nabla \cdot E_p = -4\pi e n_1$ and the maximum amplitude of the electron plasma wave is obtained when the level of density fluctuation $n_1/n_o = 1$. Thus $k_p E_p = -4\pi e n_o$ and using $\omega_p = 4\pi n_o e/m$ we obtain the electric field $E_p = m\omega_p v_p/e$ and the wave potential $e\phi$ as mv_p^2 where v_p is the phase velocity of the wave. For $v_p = c$, $e\phi = mc^2$. In the nonrelativistic case, a particle whose velocity v_e in the direction of propagation of the wave is near the phase velocity can be trapped and gain up to $4mv_p^2\ (v_e/v_p)^{\frac{1}{2}}$. In the relativistic case $v_p \simeq c$ and the maximum energy gained is theoretically up to $(2\omega^2/\omega_p^2)mc^2$.

The third mechanism is known as optical mixing in which two electromagnetic waves beat in a plasma to resonantly drive density fluctuations. The electrostatic field of such a resonantly driven plasma wave can be very large and be used to accelerate the plasma electrons or an injected group of particles.

In this paper we shall discuss the optical mixing process and the stimulated Raman Forward Scattering instability. In particular we shall examine the role of these in the Plasma Beat Wave Accelerator. In the Plasma Beat Wave Accelerator, acceleration is achieved by an electrostatic field produced as a result of charge separation generated by a plasma wave traveling close to speed of light. The plasma wave in turn is produced by colinear two laser pulses whose amplitude is modulated so that the beat frequency matches the plasma frequency. Experiments that show that long wavelength, high phase velocity electron plasma waves can be generated by the optical mixing process in a homogeneous plasma as well as generation of ultrarelativistic particles by the Raman instability are described. In the following paper by D. Sullivan. results of computer simulations using one dimensional electromagnetic particle code to investigate these phenomena in detail are described.

OPTICAL MIXING

The nonlinear excitation of electron plasma waves (EPW) by by beating two electromagnetic (EM) waves has been under considerable investigation lately because of its potential role in the laser electron accelerator[1], cascade plasma heating[2], laser-fusion pellet preheat [3] and as a plasma density diagnostic[4]. Basically, when two coherent EM waves, (ω_o,k_o) and (ω_1,k_1) occupy the same volume the total intensity is modulated in space at $\Delta k = k_o \pm k_1$ and in time $\Delta\omega = \omega_o \pm \omega_1$. In a plasma the ponderomotive force F_{NL} associated with this beat wave can resonantly drive longitudinal electron density fluctuations of wavenumber $k_p = \Delta k$ if $\omega_{EPW} = \Delta\omega$.

ω_{EPW} is the frequency of the electron plasma wave and is related to the plasma frequency via the dispersion relation $\omega_{EPW}^2 = \omega_p^2 + 3k_p^2 v_e^2$. This is the usual optical mixing process. If the two EM waves are colinear as shown in Fig. 1, then the phase velocity $v_p = \omega_{EPW}/k_p$ at which tne density fluctuations propagate is nearly equal to the group velocity of the EM waves $v_g = c(1 - \omega_p^2/\omega_o^2)^{\frac{1}{2}}$ and the three waves are locked into synchronism over thousands of wavelengths if $\omega_p << \omega_o$. Also, since $v_p \simeq c$ there is little Landau damping and the EPW can grow to a very large amplitude.

The behavior of such large amplitude plasma waves driven by beating of two laser beams has been studied by Rosenbluth and Liu[5].

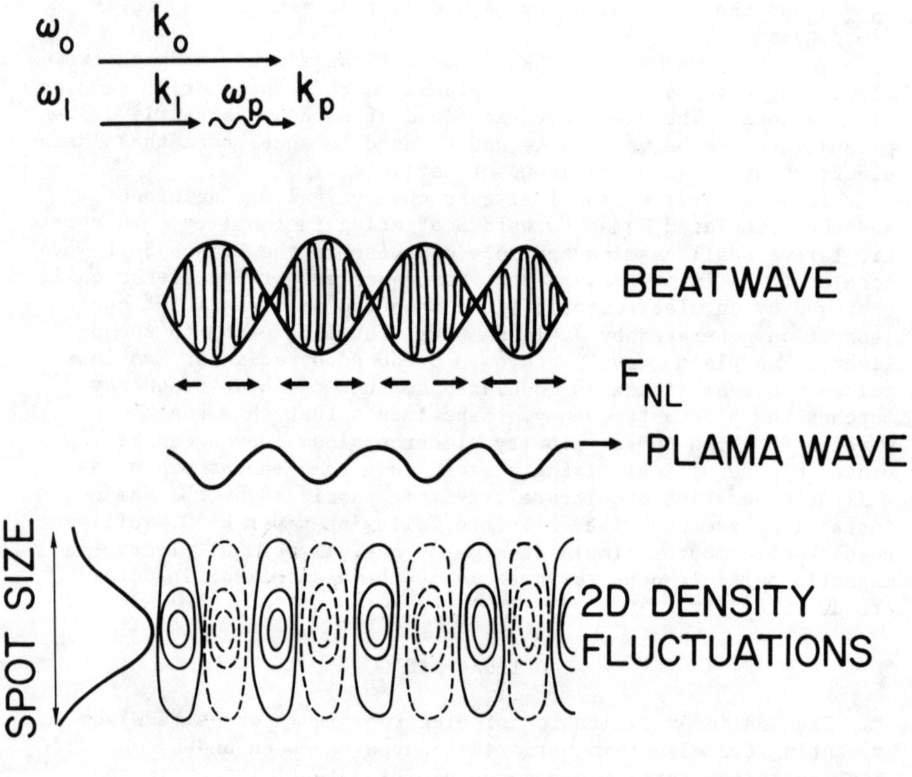

Fig. 1. Resonant excitation of an EPW (ω_{EPW}, k_p) by beating two EM waves (ω_o, k_o) and (ω_1, k_1). Because of the finite spot size the density fluctuations are 2D with solid lines (dotted lines) representing contours of increasing (decreasing) density.

They found that, for $v_o/c \ll 1$, $\tilde{n}/n_o \ll 1$ the density fluctuations grow linearly with time when $\omega_p = \Delta\omega$.

$$\frac{\tilde{n}}{n_o}(t) = \frac{\tilde{n}}{n_o}(t=0) + \frac{1}{4}\,\frac{v_{o(0)}}{c}\,\frac{v_{o(1)}}{c}\,\omega_p t \tag{1}$$

Where $\tilde{n}/n_o$ is the EPW amplitude and $v_{o(0,1)} = eE_{(0,1)}/m\omega_{(0,1)}$ is the electron quiver velocity in the laser fields. Wavebreaking is approached when $\tilde{n}/n_o \to 1$; however, to reach this limit the EPW must be exactly in phase with the beat wave. As $\tilde{n}/n_o \to 1$, relativistic effect on the frequency mismatch becomes important and the EPW saturates at a lower amplitude given by

$$(\tilde{n}/n_o)_{max} \simeq \left(\frac{1}{16}\,\frac{v_{o(0)}}{c}\,\frac{v_{o(1)}}{c}\right)^{1/3} \tag{2}$$

If v_o/c is O(1) for one or both the beams, then the threshold for the stimulated Raman forward scattering (RFS) instability may be exceeded in which case relativistic effects do not provide the saturation mechanism for the EPW.

RAMAN FORWARD SCATTERING INSTABILITY

RFS instability is basically the decay of an incident EM wave into a forward propagating EM wave and an EPW with the usual energy and momentum conservation condition:

$$\begin{aligned} \omega_o &= \omega_1 + \omega_{EPW} \\ k_o &= k_1 + k_p \end{aligned} \tag{3}$$

The threshold for the RFS instability is rather high in an inhomogeneous plasma[6]

$$I\lambda_\mu^2 \gtrsim 5 \times 10^{17} (n_c/n)^2 (\Delta n/n_c)(\lambda/\Delta x) \quad \mathrm{W\,cm^{-2}\mu^2} \tag{4}$$

where the density changes by Δn in a length Δx at an average density n. However, once this threshold is exceeded, the long wavelength high phase velocity EPW can grow rapidly. Relativistic effects do not provide a saturation mechanism because the change in ω_p can be adjusted out by a change in the frequency of the forward scattered light ω_s. Thus in the two beam case, the density fluctuations driven by optical mixing will act as an enhanced noise source to vigouously drive RFS provided that the threshold intensity is exceeded. The other one dimensional instability competing with RFS is the Raman backscatter (RBS). By solving the dispersion relation[7]

$$1 - Z'/(2k^2\lambda_D^2) = Z'v_o^2/[8\lambda_D^2(c^2k^2 \pm 2kk_oc^2 \mp 2\omega\omega_o - \omega_p^2)] \qquad (5)$$

Where Z' is the Fried-Conte function and λ_D is the electron Debye length, the growth rate for the two instabilities can be found[8]. This is shown in Fig. 2. In a cold plasma RBS dominates, however, in a hot underdense plasma the growth rates for the two instabilities become comparable. The longitudinal E field associated with the high phase velocity EPW characteristic of RFS can be very high and is responsible for accelerating either the plasma electrons or externally injected particles to ultrarelativistic energies. This mechanism known as trapping can be more severe than Landau damping. (Fig. 3) Landau damping is strong when v_p is $0(v_e)$ because the slope of the electron distribution function $(\partial f/\partial v)_{v_e}$ has a maximum near the thermal velocity. Landau damping results in the local flattening of the distribution function in the vicinity of v_p thereby producing a heated tail of (nonrelativistic) electrons. When, however, $v_p >> v_e$ Landau damping is small. Furthermore, when the amplitude of such a high phase velocity EPW is small, there are very few electrons in the background thermal distribution which are near enough to the phase velocity to be trapped. However, as pointed out by Dawson and Shanny[9], electron trapping is a nonlinear damping mechanism not local in velocity space. As the wave amplitude increases, the number of particles that can interact with the wave increases rapidly. Consequently, the damping can be much larger than that predicted by the linear theory. The trapping width[10] is given approximately by $v_t = (2eE_pv_p/m\omega_p)^{1/2}$. Electrons or externally injected particles with velocity close to the phase

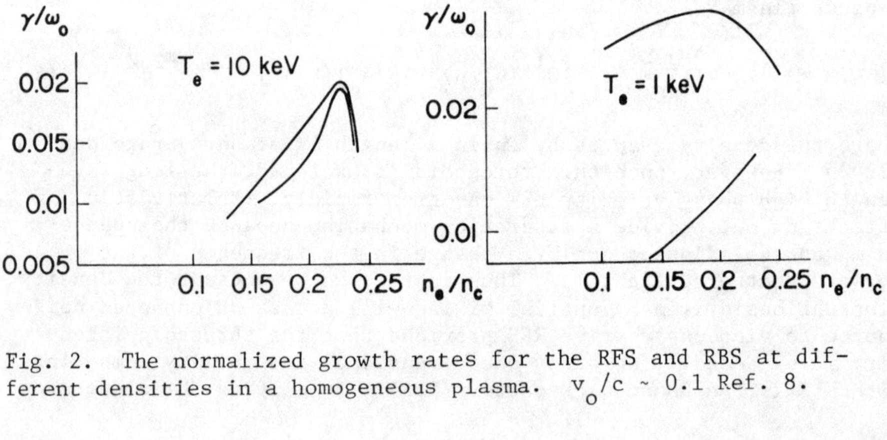

Fig. 2. The normalized growth rates for the RFS and RBS at different densities in a homogeneous plasma. $v_o/c \sim 0.1$ Ref. 8.

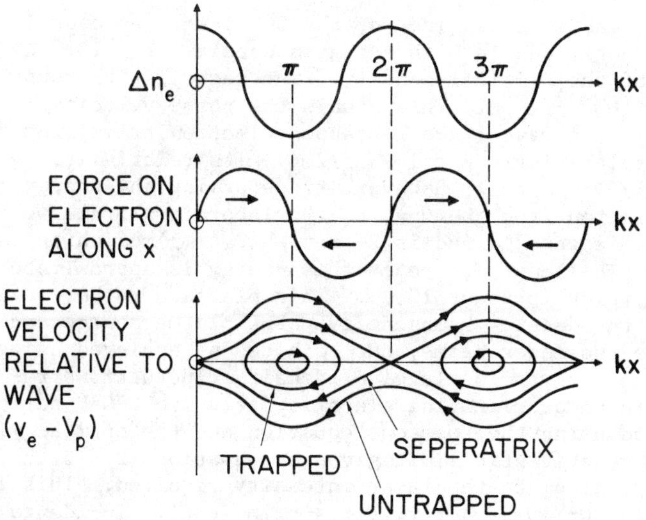

Fig. 3. Trapped particle orbits in an electron plasma wave. The frame of reference is moving at v_p along x.

velocity will be accelerated to $\simeq$ c. Since v cannot exceed c, a small change in v/c greatly increases the relativistic $\gamma = (1-v^2/c^2)^{-\frac{1}{2}}$ so that particles with very large energies can be produced.

The saturation mechanism for the EPW driven by RFS instability is thus particle trapping. Pump depletion should not be a problem since the energy given to the forward scattered EM wave is much greater than that to the EPW. This is dictated by the Manley-Rowe relation or the law of conservation of wave action. Viz:

$$\frac{w_o}{\omega_o} = \frac{w_1}{\omega_1} = \frac{w_{EPW}}{\omega_{EPW}} \qquad (6)$$

where $w = N\hbar\omega$. Since repeated k matching is possible for the RFS instability it can be argued that pump depletion is only important when the original EM wave has cascaded down by multiple RFS to waves near ω_p. The other saturation mechanisms such as harmonic generation, wave-wave coupling, ion dynamics and mainbody heating followed by increased Landau damping may also be significant. In realistic experimental situations Raman sidescattering, filamentation, self-generated B fields due to the electron beam and density inhomogeneities may also influence the RFS instability. Computer simulations using relativistic electromagnetic particle codes[11,12] is the only readily available tool to investigate these competing phenomena.

EXAMPLE

The 10.6 μ and 10.27 μ lines of the CO_2 laser can beat in a plasma to resonantly couple with a plasma density of ~ 10^{16} cm^{-3} which is 0.1% of the critical density. The laser light group velocity is ~ $(1-\frac{1}{2}\,\omega_p^2/\omega_o^2)c$. This equals the phase velocity of the plasma wave. In the wave frame a trapped electron travelling with v_p thus has a relativistic $\beta = 1 - \omega_p^2/2\omega_o^2$. Its relativistic γ is equal to $(1-\beta^2)^{-\frac{1}{2}} = \omega_o/\omega_p$. However, transforming the energy of the trapped electron from the wave to the laboratory frame we find that the maximum energy gained is $2\gamma^2mc^2 = (2\omega^2/\omega_p^2)mc^2$ which is about one GeV. The length to reach this energy is approximately $(2\omega^2/\omega_p^2)(c/\omega_p)$ which is about 10 cm. This places a rather severe requirement on the density homogeneity of the plasma source and the focusing of the two laser beams, but both can be achieved using present day technology. The level of density fluctuations required to obtain an accelerating electric field of 10 GeV per meter can be estimated using the Poissons equation and the optical mixing formula in the relativistic limit given by equation (2) is then used to roughly calculate the laser intensity required. This leads to $I_o = I_1 = 1.4 \times 10^{16}$ W cm^{-2} or $v_{o(1)}/c = v_o(0)/c \simeq 1$. Incidentally, this intensity well exceeds the inhomogeneous RFS instability threshold even if we assume a 10% density ripple per cm.

EXPERIMENTS

Although simple in principle, optical mixing is not straightforward in practice. In fact, until very recently there had been only one experiment[13] in which resonantly driven density fluctuations using two laser frequencies were diagnosed using a probe beam to Thomson scatter of the density fluctuations. In any case, the signal to noise ratio of the Thomson scattered light was only about 3. Diagnosing the EPW driven by two colinear laser beams with the condition $\omega_p \ll \omega_o$ is a tremendously difficult problem because of the short k_p. Also, one must have either a well controlled multiline laser or a tunable laser and a homogeneous, tunable density plasma source. By inserting a 10 cm long SF_6 cell inside a gain-switched TEA CO_2 laser oscillator and varying the SF_6 pressure it is possible to obtain controlled multiline operation mainly on 10.6 μ lines in the P(20) band, 10.27 μ lines in the R(16) band and the 9.6 μ lines in the P(20) band. Each band contains 3 to 5 lines each ~ 40 GHz apart of roughly the same intensity (within a factor of 5). This is shown in Fig. 4. The frequency difference between the 9.6 μ and 10.27 μ gives ω_p = 1.35×10^{13} Hz corresponding to $n_e = 5.7 \times 10^{16}$ cm^{-3}. If the incident laser intensity is well above threshold for RFS, then the growth rate can be larger than the line separation of 40 GHz and the density fluctuations produced by beating of the various lines in the $R(16)_{10.27\,\mu}$ and $P(20)_{9.6\,\mu}$ band are coupled. This relaxes

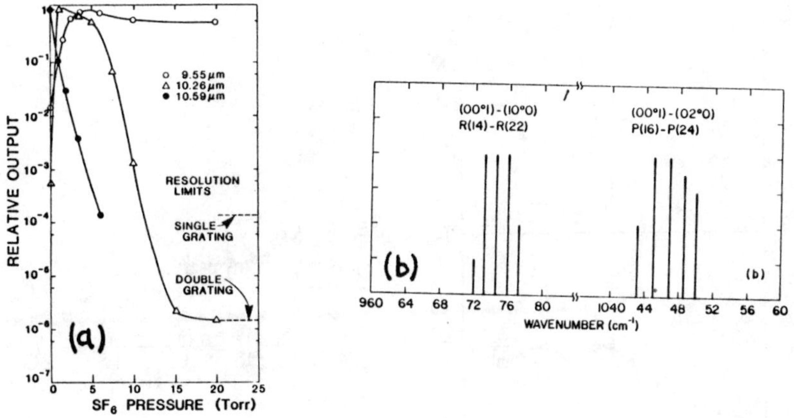

Fig. 4. Multiline operation of a CO_2 oscillator (a) and fine structure of the $R(16)_{10.27\ \mu m}$ and $P(20)_{9.6\ \mu m}$ bands (b).

somewhat the requirement on the density homogeneity along the depth of focus of the laser beams.

For a tunable density source one can either use a θ pinch or a pulsed arc plasma. Density homogeneity required can be obtained in a θ pinch in a fully ionized H_2 or He plasma whereas one has to use a heavy gas such as N_2 or Ar in an arc. Fig. 5 shows the density evolution as a function of time of a 4 Torr, 5 eV Ar plasma produced by a pulsed capacitive arc. The axial density profile measured interferometrically showed that, with the limits of measurement accuracy the condition $\Delta\omega = \omega_p$ can be achieved over the depth of focus of the f/7.5 lens.

The laser beam (75 ns FWHM) containing roughly equal powers in the 9.6 μ and 10.27 μ lines was focused to an intensity of 10^{10} W cm^{-2} by an f/7.5 lens to a 300 μ spot on the plasma axis. The transmitted light plus any forward scattered light was collected by an f/2 lens, analyzed by a double grating infrared spectrometer and detected by a very sensitive Hg:Ge photoconductor. The evidence for optical mixing was obtained by the observation of a new line in the forward scattered light around 11 μm which is produced by Thomson scattering of the 10.27 μm line from the EPW generated by the beat wave. Moreover, this radiation was only generated when $\omega_p = \Delta\omega$ as shown in Fig. 5. Although the FWHM of the input laser pulse was 75 ns the 11 μm line was only about 25 ns wide. Another unusual and at first rather puzzling effect was observed. Whenever $\Delta\omega = \omega_p$ very strong refraction of the beam occurred outside the cone angle of the incident beam. (Fig. 6) This phenomenon has been called resonant self-focusing due to the

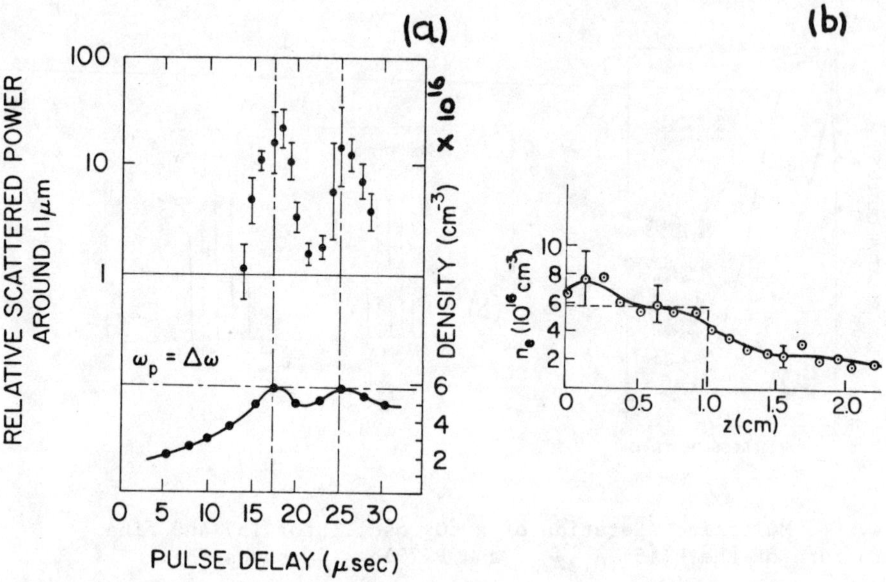

Fig. 5(a). Time evolution of the arc density and scattered power in frequency $\omega_2 = \omega(10.27\ \mu) - \omega_p$ and (b) axial electron density profile at 18 μs.

ponderomotive force of the EPW[14].

The ponderomotive force F_{NL}(plasmon) can be much larger than F_{NL}(light). The amplification factor A is given by

$$A = \frac{F_{NL}(\text{plasmon})}{F_{NL}(\text{light})} = \frac{-(\omega_p^2/\omega_o^2)\ \nabla \langle E_o^2\rangle/8\pi}{-\nabla \langle E_p^2\rangle/8\pi} \tag{7}$$

E_p is the electric field of the EPW. Poisson's equation gives $E_p - 4\pi e n_1/k_p$ and since $v_o = eE_o/m\omega_o$ and $v_p = c$ we obtain the amplification factor

$$A = \left(\frac{v_p}{v_o}\ \frac{n_1}{n_o}\right)^2 = \left(\frac{n_1/n_o}{v_o/c}\right)^2 \tag{8}$$

Using the value of $n_1/n_o \simeq 0.4\%$ estimated from the absolute levels of Thomson scattered light from the EPW and $I_o = 10^{10}\ \text{W cm}^{-2}$ we obtain $A > 20$. Since the wavelength of the EPW is about the same as the diameter of the focal spot, the logitudinal and transverse gradients of the electric fields are comparable and a density depression is created on axis which causes deflection of the beam

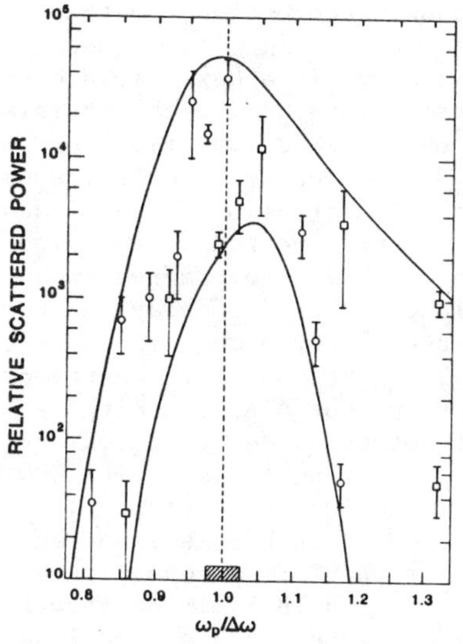

Fig. 6. Forward refracted light in the 10.27 μm line vs. $\omega_p/\Delta\omega$ with density measured by ruby interferometry (□) and by Stark broadening of a seed gas H_α line. The shaded areas indicate the spread of difference frequencies in the incident beam.

by refraction. It is not clear whether the amplitude of the EPW in this experiment was limited by relativistic effects or ion dynamics, however, the time duration of the EPW was almost certainly limited by the ion motion. The wave potential corresponding to $n_1/n_0 \sim 0.4\%$ was about 2.5 kV i.e. $\ll mc^2$ and since T_e was only 5 eV no hot electrons due to trapping were expected. However, this experiment did demonstrate that short k_p EPW can be generated resonantly via the optical mixing process. More experiments are needed to check out the predictions of the optical mixing theory at low laser powers, perhaps with shorter laser pulses.

The role of RFS instability in hot electron generation was investigated in another experiment[15]. 130 Å thick, self-supported carbon foils were irradiated at normal incidence by intense, $v_0/c \sim 0.3$, 700 ps FWHM, CO_2-laser pulses. 1-5% of the incident energy was backscattered and roughly 50% of the incident energy was transmitted by the plasma. Thus it can be assumed that the foil plasma becomes underdense around the peak of the laser pulse. The electron temperature of the bulk distribution wad deduced from the slope of the ion spectra recorded absolutely using Thomson parabolas, to be ~ 20 keV for both front and rear expansions. The angular distribution of the electrons escaping the plasma was measured using two absolutely calibrated electron spectrometers in the range 0.4 - 1.5 MeV.

Fig. 7 shows the absolute electron spectra measured at $\theta \simeq 5^{o}$ in the forward direction and $\theta \simeq 15^{o}$ in the backward direction from the thin carbon foil plasmas. If a Maxwellian distribution is assumed then these distributions can be characterized by temperatures of 90-100 keV in the forward direction and of 40-50 keV in the backward direction. Electrons with energies up to 1.4 MeV were observed in the forward direction. The highest energy electron emission (> 1 MeV) was strongly peaked in the direction of the laser. Electrons up to 400 keV were observed nearly isotropically, however, probably attributable to $2\omega_p$ decay and Raman sidescattering. Integrating over the measured angular distribution, assuming azimuthal symmetry, ~ 10^{11} electrons with energy greater than 400 keV were found to escape the plasma. Although no direct measurements of the target potential due to this loss of electrons were made, we note that target potentials of ~ 200 keV have been measured under similar irradiance conditions[16].

A simple estimate shows that RFS is important in our experiment. The growth rate for the RFS process[17] is given by $\gamma = \frac{1}{2}(v_o/c)\omega_p^2/\omega_o$ and the finite length limit on growth is $\gamma L/(cv_g)^{\frac{1}{2}} > 1$, where $v_g = 3\, k_p v_e^2/\omega_p$ and L is the interaction length. Assuming $v_o/c \sim 0.3$, $T_e \sim 20$ keV, $\omega_p/\omega_o \sim 0.46$ we obtain for $L/\lambda \sim 50$, $\gamma/\omega_o \sim 0.03$ and we have nearly 27 e-folding growths from the initial noise level. For backscatter the growth rates are comparable to those for the forwardscatter but backscatter suffers much more

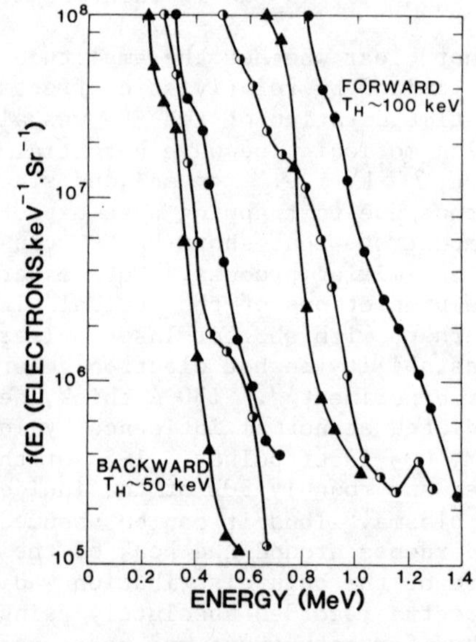

Fig. 7. Experimental electron energy distributions in the forward and backward directions from thin C foil plasmas.

severe Landau damping due to the shorter wavelength and the lower phase velocity of the backward EPW. The assumption of a homogeneous plasma with $L/\lambda \sim 50$ is reasonable, since we expect the instability to occur in the density plateau region, separating the front and the rear expansions. This region has a density scalelength somewhere in between the focal spot diameter (150 μ) and the ion acoustic speed times the FWHM pulse length (1000 μ). In any case the depth of focus of the laser beam was ~ 50 wavelengths.

Simulations were carried out on the 1D relativistic electromagnetic particle code[11] with the periodic boundary conditions where similar wave setups were used as in reference 1. The plasma was initially thermal, $T_e \sim 20$ keV and uniform, $\omega_p/\omega_o \sim 0.46$. The propagating electromagnetic pulse had a $v_o/c \sim 0.3$. The distribution function $f(P_{\parallel})$ as well as the electrostatic wave spectra are displayed in Fig. 8. The temperature and the maximum electron energy observed in the simulation distributions were similar to the experimentally measured values. For instance, simulations show electrons with $(\varepsilon_{max})_F \simeq 1.3$ MeV and $(T_{HOT})_F \sim 100$ keV in the forward direction compared to experimental values $(\varepsilon_{max}) \sim 1.4$ MeV and $(T_{HOT})_F \sim 90$-100 keV. Similarly, simulations show $(\varepsilon_{max})_B \sim 0.9$ MeV and $(T_{HOT})_B \sim 60$ keV in the backward direction compared to experimental values of $(\varepsilon_{max})_B \sim 0.8$ MeV and $(T_{HOT})_B \sim 40$-50 keV. In view of the possible influence of the target potential on the experimentally measured electron distributions, this rather excellent agreement between the experiment and the simulations may be rather fortuitous, particularly for the maximum electron energy unless the target potential was indeed much smaller than ε_{max}. The electrostatic wave spectra (Fig. 8b) shows that the backscatter mode k_b (which grows initially) is swamped by other modes with a smaller wave number, the most intense of which is the plasma wave associated with forwardscatter k_p. In addition, there are some wavenumbers which are less than k_p. Thus the heated electron dis-

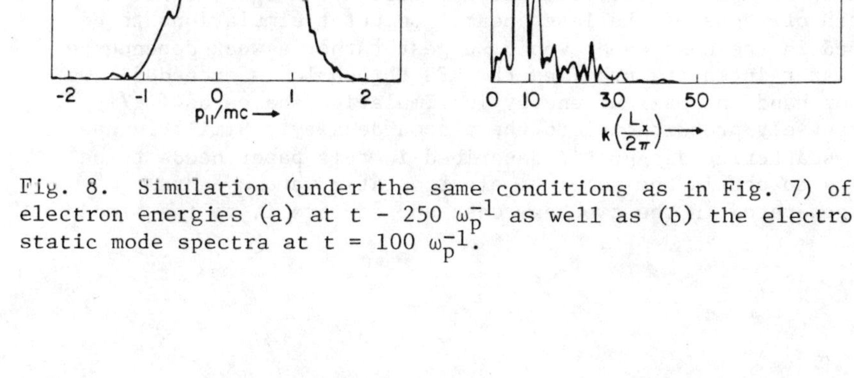

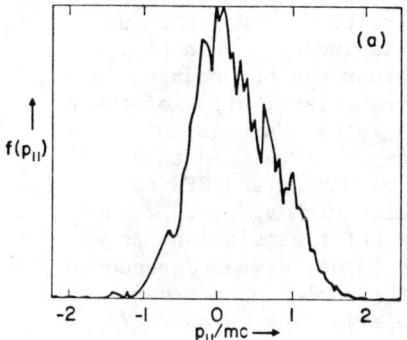

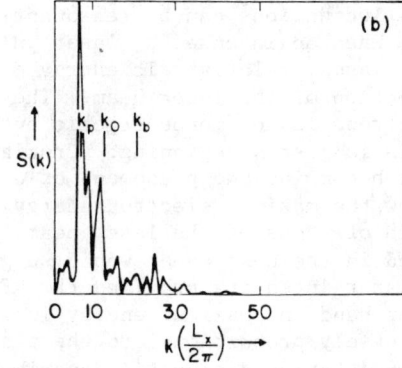

Fig. 8. Simulation (under the same conditions as in Fig. 7) of electron energies (a) at t - 250 ω_p^{-1} as well as (b) the electrostatic mode spectra at t = 100 ω_p^{-1}.

tributions obtained by the experiment and the simulations agree well with most of the electron heating due to the RFS process (and/or multiple RFS processes since repeated k matching is possible only for the RFS process), but is not so much due to the backward process.

The reason why the backscattering is suppressed is the following: When the backscattering EPW is excited, heavy Landau damping or electron trapping by this EPW saturates it at a low level thus limiting the backscattering to a small value. The phase velocity of the backscattering EPW $v_p = \omega_p/k_b$ which for this case is ~ 1.6 v_e. Thus this wave is heavily Landau damped to begin with and as it grows in amplitude, more and more particles will be trapped by it and the damping will grow. The trapping width is given approximately by $\Delta v_t = (2eE_{EPW}v_p/m\omega_p)^{\frac{1}{2}}$, where E_{EPW} is the electric field for the plasma wave. The condition that a large number of electrons are trapped is given[9] by $v_p - \Delta v_t \lesssim 2v_e \equiv 2(T_e/m)^{\frac{1}{2}}$. The maximum electrostatic wave intensity is obtained for a cold plasma by setting $v_e = 0$. This gives for the saturation amplitude E_{EPW} for the longitudinal wave as $(eE_{EPW}/m\omega_p) = v_p/2$. For our case, the phase velocity is $O(v_e)$ and we expect little growth; in any case, the saturation intensity for a cold plasma is less than 6% of the light waves. In addition the two plasmon decay instability would also saturate at a low level even if ω_o were chosen to be $2\omega_p$ because strong Landau damping sets in much earlier for $2\omega_p$ decay than it does for the forward Raman process. The RFS process appears to be the last parametric process to saturate in this hot underdense plasma.

FUTURE EXPERIMENTS

The two experiments described in this paper have demonstrated the following:

a) Using a multiline CO_2 laser of only 10^{10} W cm^{-2}, electron density fluctuations can be reasonantly excited for up to 25 ns.

b) When an intense CO_2 laser pulse is incident on a hot, tenuous plasma, relativistic energy electrons can be produced in the direction of the laser beam. The energy distribution of these hot electrons is not monoenergetic but is rather Maxwellian.

This is just a beginning. Crucial experiments need to be done which combine the phenomena of optical mixing and RFS to check how the maximum electron energy scales with v_o/c, ω_p/ω_o and the depth of focus of the laser beam. Computer simulations to be described in the next paper would suggest rather a weak dependence on the laser intensity provided the RFS threshold is exceeded. On the other hand the maximum energy in simulation scales as $(\omega_o/\omega_p)^2$, i.e. inversely proportional to the plasma density. Similarly the Thomson scattering diagnostic described in this paper needs to be exploited to obtain the complete electrostatic wave spectrum $S(k,\omega)$ generated in the two beam case and compare it to the simulation spectrum. It is also of some importance to investigate to what extent two dimensional effects such as sidescattering and bubble formation are important. Another crucial question that may be more conveniently addressed to in simulations is 'how important is the background temperature of the plasma in determining the relative importance of Landau damping and particle trapping?' Experiments designed to answer these questions should form the next stage of research on the plasma beat wave accelerator. If the results look promising and very large longitudinal electric fields can indeed be produced using existing 1 μ or 10 μ laser facilities, then it would be of enormous interest to see if an externally injected, nearly monoenergetic beam of electrons or protons can be accelerated to GeV(s) while keeping it relatively monoenergetic.

ACKNOWLEDGEMENTS

I wish to acknowledge the contribution my colleagues C. Clayton, Dr. F. F. Chen, Dr. N. A. Ebrahim and Dr. H. A. Baldis in successfully carrying out the experiments described in this paper. This work was supported by DOE contracts DE-AS08-81DP40163, DE-AS08-81DP40135, DE-AS08-81DP40166 and NSF grant ECS-80-03558.

REFERENCES

1. T. Tajima and J. M. Dawson, Phys. Rev. Lett. 43, 267 (1974).
2. B. I. Cohen, A. N. Kaufman, and K. M. Watson, Phys. Rev. Lett. 29, 581 (1972).
3. N. A. Ebrahim, C. Joshi and H. A. Baldis, Phys. Rev. A 25 No. 4 (1982) to be published.
4. N. Krall, A. Ron and N. Rostoker, Phys. Rev. Lett. 13, 83 (1964) G. Weyl, Phys. Fluids 13, 1802 (1970).
5. M. N. Rosenbluth and C. S. Liu, Phys. Rev. Lett. 29, 701 (1972).
6. K. Estabrook, W. L. Kruer and B. F. Lasinski, Phys. Rev. Lett. 45, 1399 (1980).
7. W. L. Kruer, K. Estabrook, B. F. Lasinski and A. B. Langdon, Phys. Fluids 23, 7 (1980).
8. K. Estabrook in Laser Program Annual Report UCRL-50021-80, Vol. 2, p. 3-34 (1980), Lawrence Livermore National Laboratory.
9. J. M. Dawson and R. Shanny, Phys. Fluids 11, 1506 (1968).
10. This formula is nonrelativistic but is illustrative for the present purpose.
11. A. T. Lin, J. M. Dawson and H. Okuda, Phys. Fluids 17, 1995 (1974).
12. M. Ashour-Abdalla, J. N. Laboeuf, T. Tajima, J. M. Dawson and C. F. Kennel, Phys. Rev. A 23, 1406 (1981).
13. B. L. Stansfield, R. Nodwell and J. Meyer, Phys. Rev. Lett. 26, 1219 (1971).
14. C. Joshi, C. E. Clayton and F. F. Chen, Phys. Rev. Lett. 48, 874 (1982).

15. C. Joshi, T. Tajima, J. M. Dawson, H. A. Baldis and N. A. Ebrahim, Phys. Rev. Lett. 47, 1285 (1981).
16. R. F. Benjamin, G. H. McCall and A. W. Ehler, Phys. Rev. Lett. 42, 890 (1979).
17. J. J. Thomson, Phys. Fluids, 21, 2082 (1978).

THE PLASMA BEATWAVE ACCELERATOR - II SIMULATIONS

D. J. Sullivan and B. B. Godfrey
Mission Research Corporation, Albuquerque, New Mexico 87106

ABSTRACT

One-dimensional relativistic and electromagnetic simulations show the Plasma Beatwave Accelerator to be a viable approach to obtaining large accelerating fields. The plasma wave excited by Raman Forward Scattering has a longitudinal electric field on the order of 100 MeV/cm to 10 GeV/cm based on plasma densities of 10^{16} cm^{-3} to 10^{20} cm^{-3}. The use of optical mixing to drive the Raman instability greatly reduces the laser intensity needed to excite large amplitude plasma waves. TeV energies for plasma electrons or preaccelerated charged particles in short distances appear feasible. Wave synchronism is maintained in space and time provided the laser beatwave intensity is properly chosen. Streaming instabilities may be overcome in the same manner. The frequency matching condition appears less strenuous than anticipated. Two- and three-dimensional effects must still be resolved.

INTRODUCTION

The Plasma Beatwave Accelerator[1,2] is one of a number of particle accelerator concepts utilizing the high fields inherent in intense laser beams. In this case, EM waves interact nonlinearly with an underdense plasma through optical mixing or Raman Forward Scattering (RFS) to excite large amplitude plasma waves. The plasma self-fields provide the acceleration. Thus, the concept is a collective effect accelerator. The longitudinal electrostatic field of the plasma depends on the plasma density and is typically on the order of $mc\omega_p/e$, where m is the electron rest mass, e is the electron charge, c is the speed of light and ω_p is the electron plasma frequency. For a 10^{18} cm^{-3} plasma this expression is equal to 1 GeV/cm. It explains the very high accelerating fields obtainable with this mechanism.

The Plasma Beatwave Accelerator can be used in two different regimes. In one case preaccelerated and bunched particles are injected into the plasma. If streaming instabilities can be suppressed, a low emittance TeV beam can be generated in a relatively short distance. This scheme, however, requires many plasma stages in order to obtain this high energy. A second regime relies on a single plasma region. A high intensity laser pulse would result in a large flux of particles with an exponential energy distribution up to several GeV. The maximum energy in the distribution is adjustable by varying the ratio of ω/ω_p as noted below.

Because the previous companion paper,[3] as well as others,[4] covers the theory in detail, we will only briefly describe the acceleration mechanisms before discussing the simulations.[5] Our results address optical mixing, Raman Forward Scattering, wave synchronism, frequency matching and streaming instabilities. In the summary an example of a 1 TeV proton accelerator is given.

ACCELERATION MECHANISM

The process depends upon the interaction between an intense electromagnetic wave and the electrons of an underdense plasma. The nonlinear ponderomotive force associated with the light wave's propagation in the plasma displaces the electrons. This leads to a charge separation and coincident restoring force producing a train of plasma oscillations. The phase velocity of the wake plasma wave is equal to the group velocity of the EM wave, which is derived from the dispersion relation $\omega^2 = k^2c^2 + \omega_p^2$ to be

$$\omega_p/k_p = v_p = v_g = (1 - \omega_p^2/\omega^2)^{1/2}c \quad , \tag{1}$$

where k_p is the plasma wavenumber, v_p the plasma wave phase velocity, v_g the EM wave group velocity, and ω the EM wave frequency. Because of the mobility of electrons, and the fact that large changes in energy for relativistic electrons translate into small velocity changes, the electrons are synchronous with the wave front for long periods.

In the wave frame the electrostatic field associated with the plasma can be viewed as a particle mirror with the maximum electron acceleration taking place when the electron experiences a momentum change of $2\gamma\beta mc$, where β is the wave velocity normalized to c and γ is the usual relativistic factor given by $(1 - \beta^2)^{-1/2}$. Transforming back to the laboratory frame yields a maximum electron energy of $\gamma^{max}mc^2$ where $\gamma^{max} = 2\ \omega^2/\omega_p^2$. The critical plasmon electric field derived from wavebreaking arguments is $E_z = mc\omega_p/e$, which implies an accelerating field on the order of a GeV/cm for a 10^{18} cm^{-3} plasma density. The optimal pulselength was expected to be half of a plasma wavelength or $\pi\ \omega_p^{-1}$. This is a severe constraint on the mechanism, because it implies a 0.056 picosecond laser pulse for a 10^{18} cm^{-3} plasma. This critical flaw can be overcome by using two lasers or two bands of a single laser with a frequency difference of ω_p to produce a beatwave. The beatwave then provides the means to obtain the large intensity gradients necessary for this acceleration mechanism.

One can determine a relationship for the minimum E-field amplitude, E_o^{min}, needed to capture electrons at rest in the lab frame from trapping arguments and the nonlinear ponderomotive force equations. In the wave frame trapping requires that the potential be large enough to stop the electrons moving in the negative direction. Therefore, $e\phi^{wave} > \gamma mc^2$. Transforming back to the laboratory frame yields that $e\phi^{lab} > mc^2$. Using the relation that $E_z = k_p\phi^{lab}$ and $v_p = v_g$, one obtains for the minimum plasma wave electric field

$$E_z^{min} = \frac{mc\omega_p}{e}\ \frac{c}{v_g} > \frac{mc\omega_p}{e} \tag{2}$$

If the plasma is hot, then a smaller E_z can be used to trap those electrons in the tail of the distribution. Likewise, a smaller electric field would be necessary for preaccelerated particles.

If one neglects ion motion due to the high frequency nature of the phenomena, the resotring force on the electrons is due solely to space charge. Thus, the nonlinear ponderomotive force per cm^3 can be equated to the energy density gradient of the resulting plasma wave, or

$$\frac{\nabla |E_z|^2}{8\pi} = \frac{\omega_p^2}{\omega^2} \frac{\nabla E_0^2}{16\pi} \tag{3}$$

where E_0 is the EM wave electric field. Solving for E_0 and substituting E_z^{min} from equation (2) yields

$$E_0^{min} = \sqrt{2} \frac{mc}{e} \frac{\omega^2}{(\omega^2 - \omega_p^2)^{1/2}} > \sqrt{2} \frac{mc\omega}{e} \tag{4}$$

This is valid provided the laser pulse is shorter than the plasma period, $2\pi \omega_p^{-1}$.

The minimum laser intensity is readily calculated from equation (4) to be[5]

$$I^{min} = \frac{m^2 c^3}{4\pi e^2} \frac{\omega^4}{\omega^2 - \omega_p^2} \tag{5}$$

This is an absolute minimum, since the analysis assumed an effectively instantaneous risetime. Nevertheless, the strong dependence on laser frequency must be noted. For example, a 1.06 μm Nd-Glass laser requires an intensity 100 times greater than a 10.6 μm CO_2 laser or 2.4×10^{18} W/cm^2 assuming $\omega/\omega_p \gg 1$.

An absolute minimum power necessary to trap electrons from rest can be obtained assuming $E_0^{min} = \sqrt{2}\, mc\omega/e$, and the focal spot size is at the diffraction limit. Then

$$P^{min} \approx (1.22\, \pi\, f^{\#})^2 \frac{mc^2}{e} \cdot \frac{mc^3}{e} \tag{6}$$

where $f^{\#}$ is the focal number of the laser optics. Since such a tight focus can only be maintained over a few laser wavelengths due to diffraction, this is not the best configuration for electron acceleration. Rather, a more powerful laser beam over a larger area is necesary, which will assure the minimum intensity over the acceleration saturation length of $\gamma^{max}\, c/\omega_p$.

The laser pulse is coupled to plasma motion in the transverse direction. In contrast to acceleration in the longitudinal direction where only a fraction of the electrons at any given position are trapped and accelerated, all electrons experience the same transverse acceleration for a set value of z and time. It is, therefore, appropriate to quantify the motion by a plasma relativistic factor, γ_p. An estimate for γ_p can be obtained from the relativistic equation for a cold fluid

$$\frac{d}{dt}(\gamma - 1)\, mc^2 = e\, \vec{E} \cdot \vec{v} \tag{7}$$

If we assume that the electron velocity approaches the speed of light, and E has the form $E = E_0 \cos \omega t$ then

$$\gamma_p = \sqrt{2}\, \alpha \sin \omega t \, \frac{c}{v_g} \tag{8}$$

where $\alpha = E_0/E_0^{min}$. The effect of the relativistic plasma enters into the wavebreaking limit for the plasma electric field and the maximum electron energy, because

$$\left(\omega_p^{rel}\right)^2 = \frac{4\pi ne^2}{\gamma_p m} = \frac{\omega_p^2}{\gamma_p} \tag{9}$$

where ω_p^{rel} is the relativistic plasma frequency. The critical electric field is then modified to be[5]

$$E_z^{cr} \simeq \gamma_p^{1/2} \frac{mc\omega_p}{e}, \tag{10}$$

which helps explain simulation results where the magnitude of E_z has attained values several times $mc\omega_p/e$. Note, however, that at these amplitudes the plasma waves are extremely nonlinear. The nonlinearity of the plasma probably contributes more than relativistic effects to this saturation amplitude. Similarly, the maximum energy becomes

$$\gamma^{max} = 2\, \gamma_p^{max}\, (\omega/\omega_p)^2 \tag{11}$$

where

$$\gamma_p^{max} = \sqrt{2}\, \alpha\, c/v_g\ . \tag{12}$$

By increasing γ_p and, therefore, ω/ω_p^{rel} the value of v_g is effectively increased. The enhancement in the maximum energy can be viewed as a larger momentum change at the particle mirror. Note, however, that γ_p is not constant in space or time, and particles can be decelerated if the value of γ_p is not maintained synchronously with the particle motion. From equation (5) it is also evident that the relativistic effect decreases the minimum wave intensity necessary for electron trapping.

SIMULATIONS

Simulations were carried out in conjunction with this analysis using a two-dimensional, fully relativistic and electromagnetic, particle-in-cell code. The code can solve self-consistently for the time dependent trajectories of tens of thousands of plasma particles over thousands of plasma periods. All variables are expressed in dimensionless terms. Therefore, length is in units of c/ω_p; time is measured in units of ω_p^{-1}; and particle velocity is given by $v_i = \beta_i\gamma$ ($i = 1,2,3$), where ω_p is the initial electron plasma frequency.

Initially, short EM wavepackets were modelled to observe the acceleration. In the laser simulations a plane-polarized electromagnetic wave is launched into a Cartesian geometry. Periodic boundary conditions in the transverse (y) direction make configuration space effectively one-dimensional. In general, the simulation has 1250 cells in the longitudinal (z) direction modelling a length of 100 c/ω_p. The cells in the y direction appeared uniform because of periodicity. Each of these macrocells initially contained 24 particles. The ions were taken to be an infinitely massive neutralizing background. A vacuum region 10π c/ω_p long was left between the left hand boundary and the plasma in order to accurately determine the dynamics of laser injection into the plasma. Different runs were made in which the values of ω/ω_p, E_0, plasma gradient length, $\nabla_n z$, electron temperature, T_e, and pulse length, τ_p, were varied. The canonical simulation has values: $\omega = 3\omega_p$, $E_0 = E^{min}$, $\nabla_n z = 0.01$ c/ω_p, $T_e = 0$ and $\tau_p = \pi\omega_p^{-1}$.

When the laser pulse encounters the plasma, the nonlinear ponderomotive force resulting from the intensity gradient causes the electrons to snowplow. This continues until the force arising from charge separation is greater than the ponderomotive force, and the electrons attempt to restore the charge imbalance by moving in the negative z direction. This motion initiates a train of large amplitude plasma waves. The electrons are trapped by the waves and accelerated. In several code runs at low values of ω/ω_p, the electron density momentarily exceeded the critical density for the laser pulse and part of the wave was reflected. The plasma wavetrain and resultant electron bunching are shown in Figures 1 and 2. The wave steeping evident in the electric field profile indicates the nonlinear nature of the mechanism even at minimum intensity.

If the laser intensity exceeds the minimum required by a factor of two, the plasma becomes so turbulent that a discernible plasma wavetrain cannot be established. Coherent acceleration takes place primarily at the pulse front. This is depicted in Figure 3. More importantly, both the plasma wave accelerating field (Figure 4) and

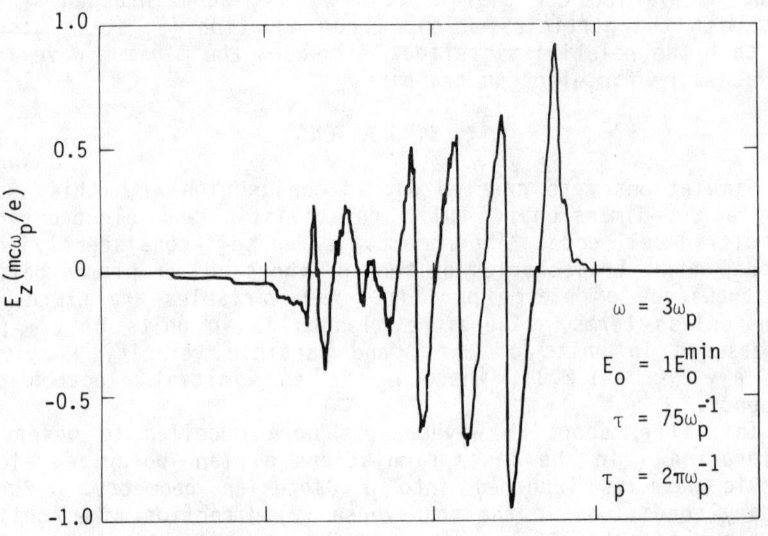

Figure 1. Plot of the plasma wave electric field versus axial distance z.

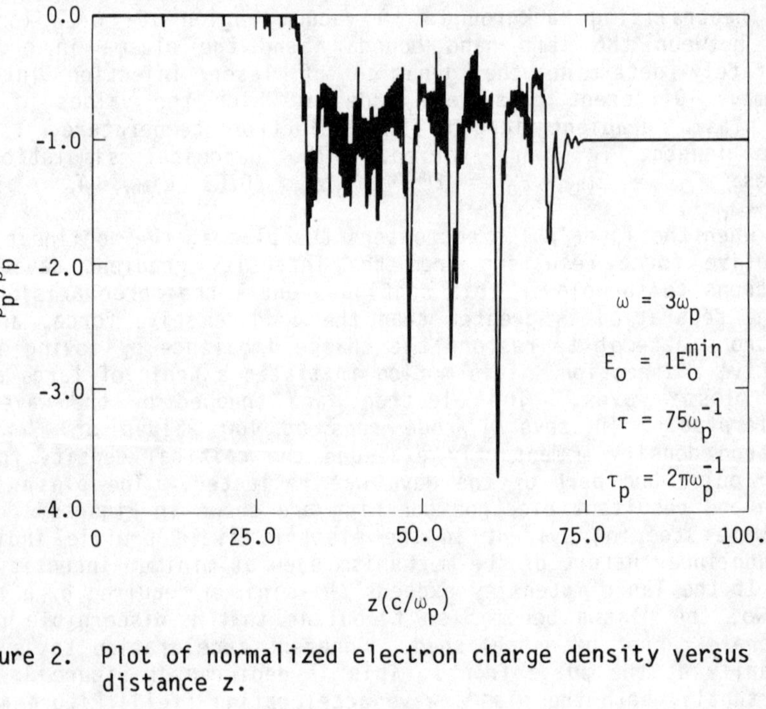

Figure 2. Plot of normalized electron charge density versus axial distance z.

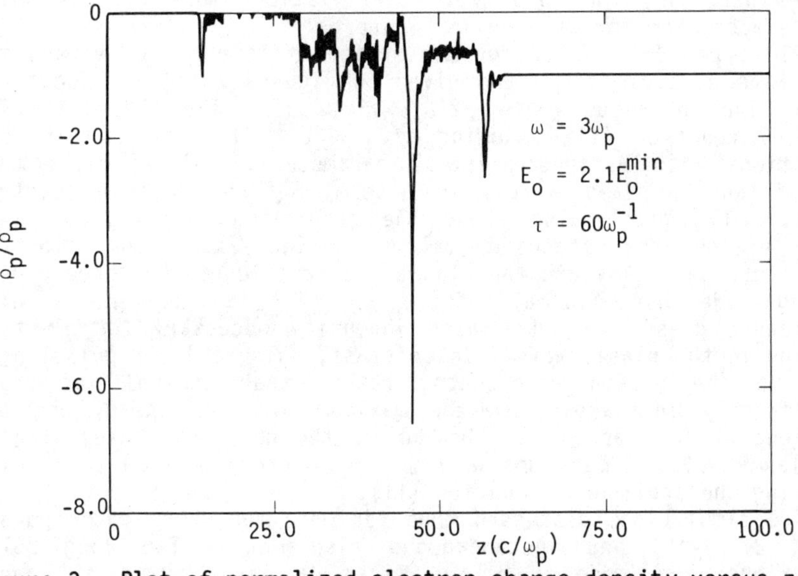

Figure 3. Plot of normalized electron charge density versus z. When the electromagnetic wave intensity is approximately twice the minimum required, the acceleration takes place primarily in one place.

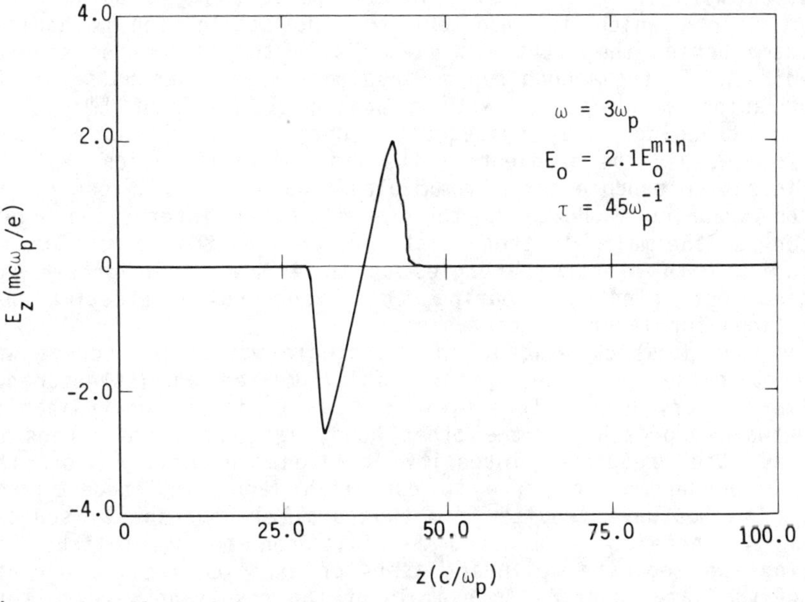

Figure 4. Plot of longitudinal electric field versus axial distance. The amplitude exceeds the anticipated wavebreaking limit.

particle acceleration (Figure 5) exceed their anticipated maximums.[1] This is partially the result of relativistic plasma effects which must be accounted for as seen in Figure 6.

Plots of γ^{max} and energy conversion efficiency, η, versus the laser wave electric field are given in Figures 7 and 8. The horizontal line indicates $\gamma^{max} = 2(\omega/\omega_p)^2$ or 18. The sloped line is given by Equation (11) assuming $c/v_g \simeq 1$. The symbols refer to simulations with various physical parameters. It is clear that laser intensity plays a role in determining the maximum electron energy. The graph also shows the results of varying T_e, $\nabla_n z$, and τ_p from the standard values. Increasing the electron temperature to 10 keV and the plasma gradient length by three orders of magnitude has minimal effect on γ^{max}. However, a high temperature does lower the laser intensity necessary for electron trapping in the plasma wave. In contrast, using a laser pulselength equal to the plasma wavelength, rather than one-half of that, significantly increases both the maximum electron energy and the efficiency. This appears to be due to the nonlinear steepening of the plasma wave. Code runs with $\omega = 5\ \omega_p$ produced similar results verifying the scaling of Equation (11).

The simulations discussed thus far were for short laser pulses ($\tau_p < 2\pi\ \omega_p^{-1}$) and instantaneous risetimes. Two long pulse simulations were made. In the first, a laser pulse of length greater than 100 c/ω_p (the length of the simulation) and an instantaneous risetime was injected into the plasma. The maximum acceleration at the wavefront was the same as the equivalent simulation with $\tau_p = 2\pi\ \omega_p^{-1}$. In long pulses, it is pulse shape and not length which is important for the accelertion mechanism. The plasma behind the front was heated to a temperature of several hundred keV. In the second run a Gaussian shaped laser pulse of $72\pi\ \omega_p^{-1}$ duration was injected with a peak wave field of E_0^{min}. No electron trapping and subsequent coherent acceleration were observed, because the gradient of the laser wave (Equation (3)) was insufficient to produce the minimum plasma wave fields necessary for electron trapping. However, the laser plasma interaction again produced a thermal electron distribution in the direction of propagation with electron energies up to 4 MeV. The process of producing hot electrons during the laser pulse risetime has implications for laser pellet fusion.

The one drawback evident in the single wavepacket scheme was the short pulse ($\tau \leq 2\pi\ \omega_p^{-1}$). This implied an instantaneous risetime to very high intensities. It is obviously unattainable. The beatwave approach, on the other hand, is a practical means of obtaining the required intensity gradient necessary for the nonlinear ponderomotive force to establish large amplitude plasma waves. The beatwave results from two parallel coherent EM sources ω_0 and ω_1 where $\omega_0 - \omega_1 = \omega_p$. This can be accomplished by utilizing two separate colinear lasers or exciting two appropriate bands of the same laser.[3] An example of the resultant wave pattern is given in Figure 9 where $\omega_0 = 10.6\ \omega_p$ and $\omega_1 = 9.6\ \omega_p$. Figure 9 is a probe history of the combined laser electric fields

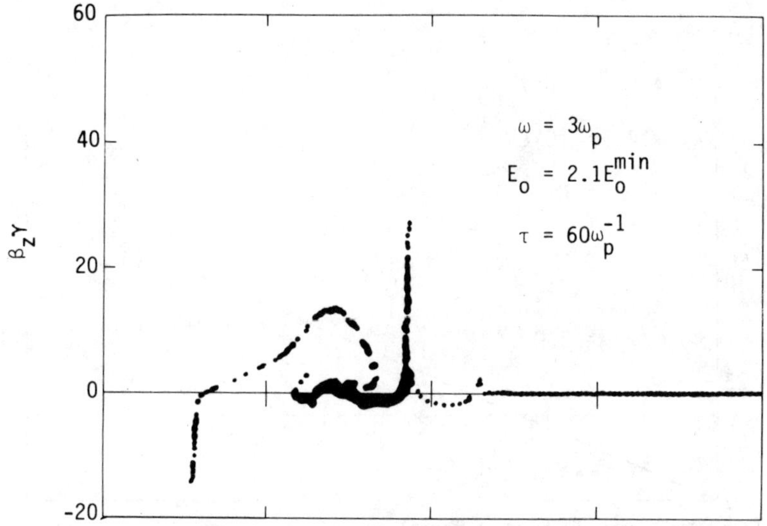

Figure 5. Particle plot of the longitudinal electron acceleration versus z.

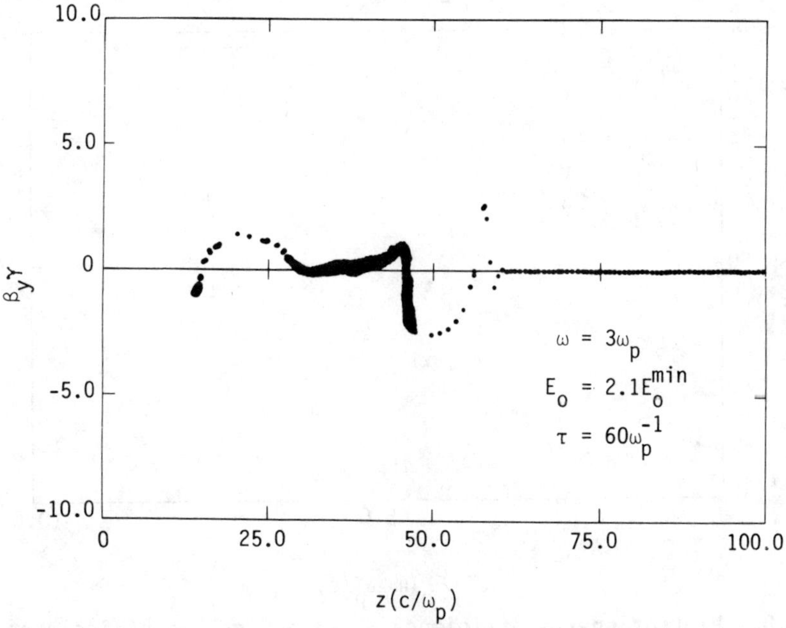

Figure 6. Particle plot of the transverse electron acceleration as a function of z.

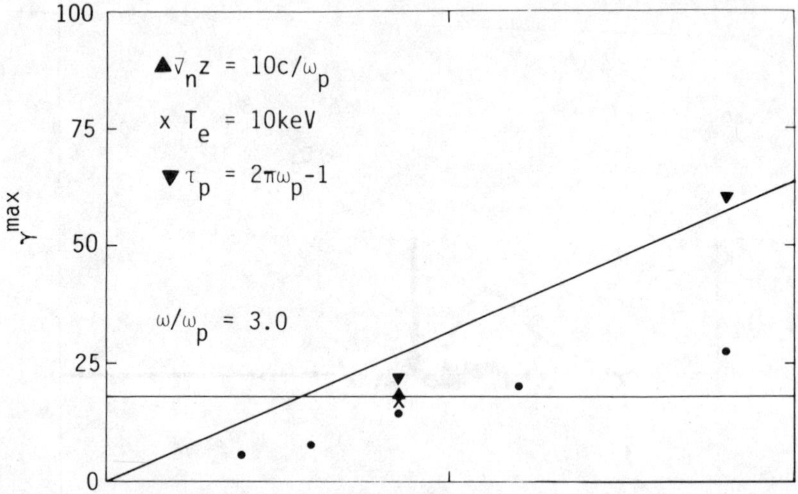

Figure 7. Graph of γ^{max} as a function of incident EM wave electric field amplitude, E_0. The horizontal line is $\gamma^{max} = 2(\omega/\omega_p)^2 = 18$. The sloped line is given by equation 11. The dots indicate simulation results with $\nabla_n z = 0.01\ c/\omega_p$, $T_e = 0$, and $\tau_p = \pi\ \omega_p^{-1}$. The other symbols are as noted.

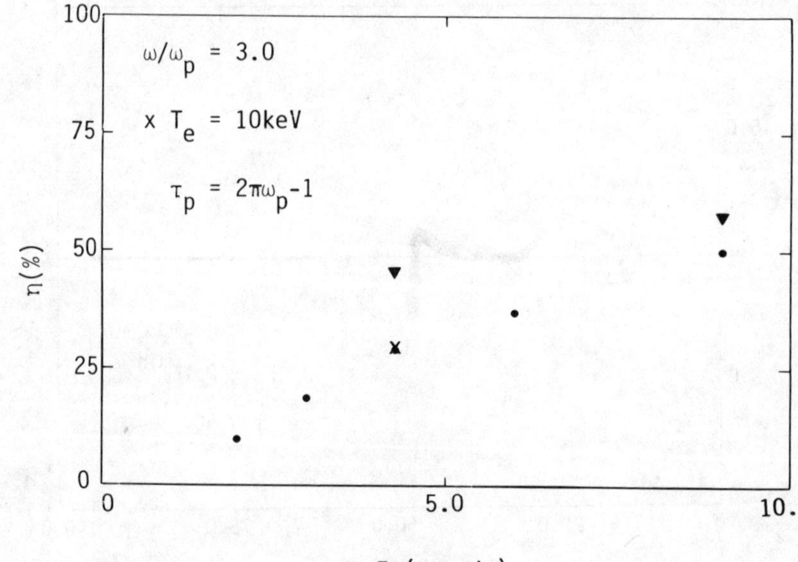

Figure 8. Plot of energy efficiency η, as a function of incident EM wave electric field amplitude, E_0. Symbols are the same as for Figure 7.

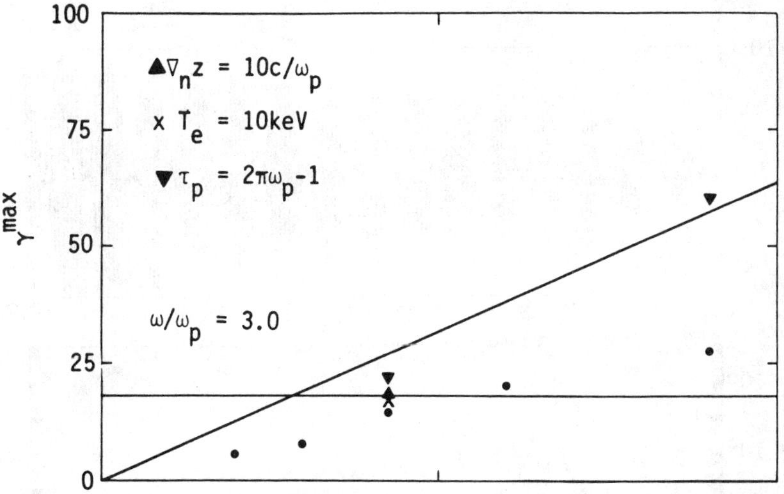

Figure 7. Graph of γ^{max} as a function of incident EM wave electric field amplitude, E_0. The horizontal line is $\gamma^{max} = 2(\omega/\omega p)^2 = 18$. The sloped line is given by equation 11. The dots indicate simulation results with $\nabla nz = 0.01 \cdot c/\omega_p$, $T_e = 0$, and $\tau_p = \pi\ \omega_p^{-1}$. The other symbols are as noted.

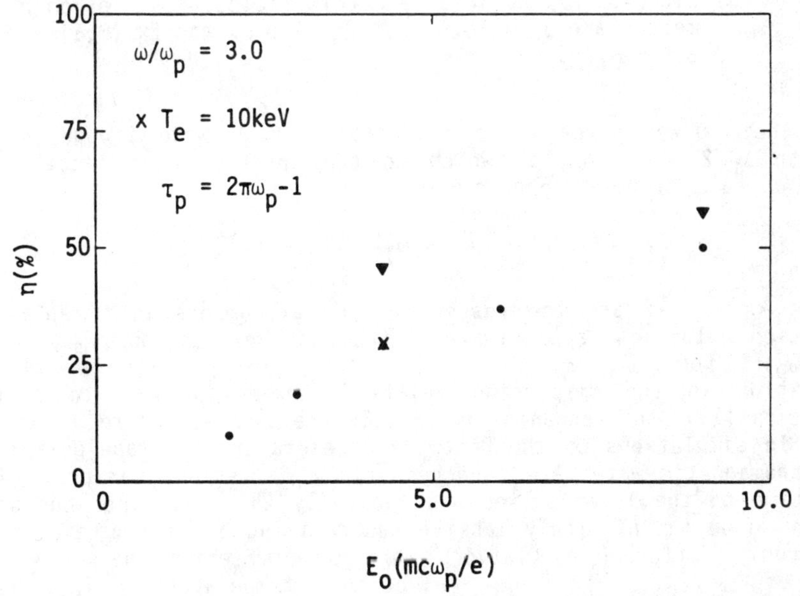

Figure 8. Plot of energy conversion efficiency η, as a function of incident EM wave electric field amplitude, E_0. Symbols are the same as for Figure 7.

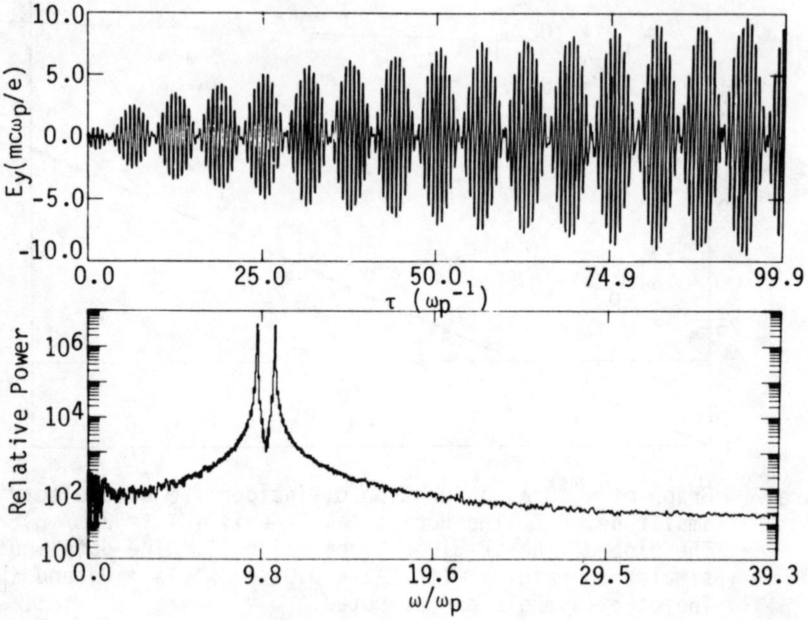

Figure 9. Time history and power spectrum at injection of the two plane-polarized laser electric fields. Simulation parameters are $\omega_0 = 10.6\ \omega_p$, $\omega_1 = 9.6\ \omega_p$ and $E_0\ (0,1) = 5.0\ mc\omega_p/e$.

and their power spectrum at injection. Each wave trough has a length $\lambda_p/2 = \pi\ c/\omega_p$ in which to trap particles and accelerate them up to a maximum energy of[2]

$$\gamma^{max} = 2\gamma^2 = 2[1 - (\omega_0 - \omega_1)^2/(k_0 - k_1)^2c^2]^{-1} \qquad (13)$$

where k_0 and k_1 are the respective EM wavenumbers in the plasma. Equation (13) is approximately the old expression $\gamma_{max} = 2(\omega_0/\omega_p)^2$ for $\omega_0,\ \omega_1 \gg \omega_p$. Thus, except for a slight variation in the dispersion relation, which is used to derive equation (13), the mechanism is exactly the same as before.

In simulations on the Beatwave Accelerator two plane-polarized electromagnetic waves are launched into a Cartesian geometry. The risetime on the laser pulses was typically 25 ω_p^{-1}. The ions were taken to be an infinitely massive neutralizing background except in one run. Different simulations were made in which the values of ω_0/ω_p, ω_1/ω_p, E_0 and electron temperature, T_e were varied. Otherwise the simulations had the same structure as before.

NONLINEAR WAVE-WAVE PROCESSES

As described in the previous paper, an alternative method of depicting the beatwave acceleration mechanism is in terms of two nonlinear wave-wave interactions. At low intensities where $E_0(0,1) \ll mc\omega(0,1)/e$ optical mixing can result in large amplitude plasma waves.[7,8] This occurs provided $\omega_p = \omega_0 \pm \omega_1$. Then, the beat frequency of the electromagnetic waves is in resonance with the plasma, and electron density fluctuations can grow. The result is linear growth of the plasma wave amplitude in time until relativistic plasma effects cause a frequency mismatch and wave saturation.

The second interaction is the Forward Raman Scattering Instability.[1,2,9] As the laser waves transit the plasma, they undergo successive scatterings where $\omega_n = \omega_0 - n\omega_p$. Alternatively, this may be viewed as $k_n = k_0 - nk_p$. This is a parametric three-wave process where a large electromagnetic wave interacts with a plasma to generate another forward moving electromagnetic wave and a plasma wave. Conservation of energy and momentum require the frequency and wavenumber matching conditions noted above. As is usual for parametric processes, the pump wave must exceed a certain intensity threshold for the instability to take place.[10]

The two wave-wave interactions, however, are not mutually exclusive - one limited to low intensities and the other to high. Rather, the plasma wave excited by optical mixing helps drive the parametric instability. To examine this a simulation with $v_{osc}(0,1)/c \equiv eE_0(0,1)/mc\omega(0,1) = .04$ and $.05$ was conducted. The plasma was cold at $\omega_p\tau = 0$. The linear growth of the plasma wave is depicted in Figure 10. At this time ($\omega_p\tau = 150$) the density fluctuations are 10% of the initial density. The analytical equation[8] indicates saturation should occur at a fluctuation level of 5%. Indeed, time histories of the laser fields indicate a cascade to lower frequencies in steps of ω_p indicative of an RFS instability. By $\omega_p\tau = 300$ the plasma field amplitude has saturated at $.2\ mc\omega_p/e$, the density variations are as high as .75 of the initial value, and the RFS instability is fully developed as seen in Figure 11. However, as expected, these field levels are not sufficient to trap plasma electrons at rest.

The importance of this result must be stressed. It indicates that large amplitude plasma waves can be grown from relatively low intensity laser sources, due to RFS excitation by optical mixing. Consider the cold plasma simulation just discussed. If ω_0 is $2.36\times10^{14}\ sec^{-1}$ ($\lambda_0 = 10\ \mu m$), then λ_1 is 8 μm and the plasma density is $7\times10^{17}\ cm^{-3}$. The equivalent laser intensity is only $1.3\times10^{14}\ W/cm^2$, yet the plasma wave accelerating field is 166 MeV/cm.

This lower intensity makes the Plasma Beatwave Accelerator an extremely attractive candidate for application to high energy physics accelerators. The lower intensity implies the focal spot size may be larger, thus avoiding diffraction of the laser pulse in the

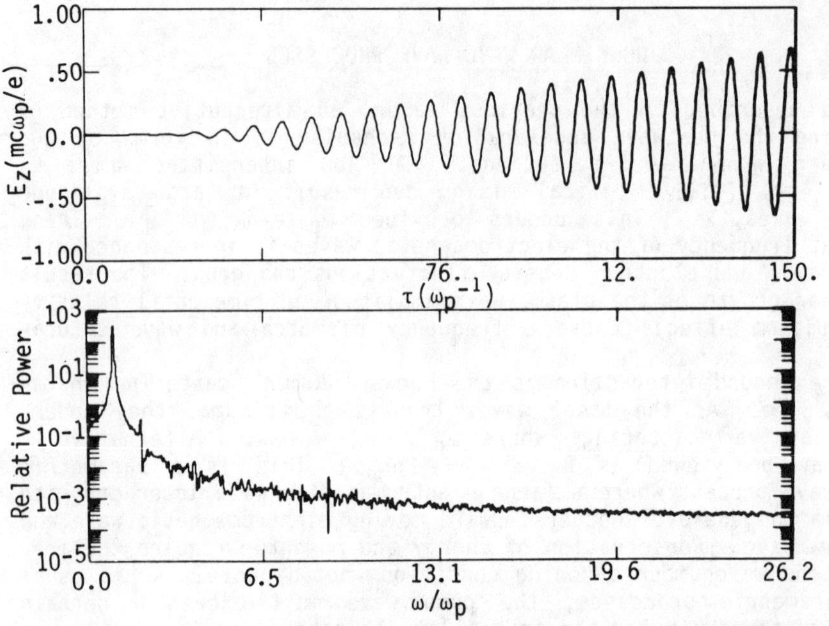

Figure 10. Time history and power spectrum inside the plasma of the longitudinal electric field. Simulation parameters are $\omega_0 = 5.0\ \omega_p$, $\omega_1 = 4.0\ \omega_p$ and $E_0\ (0,1) = 0.2\ mc\omega_p/e$.

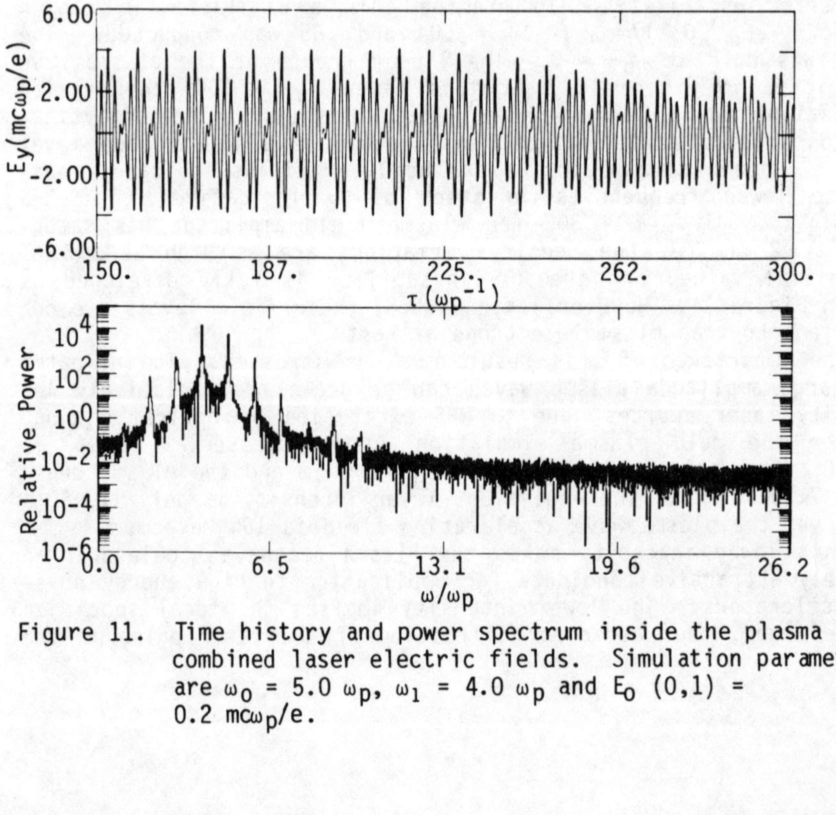

Figure 11. Time history and power spectrum inside the plasma of the combined laser electric fields. Simulation parameters are $\omega_0 = 5.0\ \omega_p$, $\omega_1 = 4.0\ \omega_p$ and $E_0\ (0,1) = 0.2\ mc\omega_p/e$.

plasma. The fact that plasma electrons are not trapped is advantageous for injection of preaccelerated, bunched particles. As will be seen in the next section, large numbers of trapped particles lead to wave damping and loss of synchronism. A disadvantage is the possibility for two-stream instability growth. This will be addressed later.

In a subsequent simulation where E_0 (0,1) were each reduced by a factor of 2 and T_e = 10 keV, neither optical mixing nor RFS was evident. This was apparently due to high frequency density spikes in the hot plasma. Theory predicts that plasma inhomogeneity raises the RFS threshold and inhibits optical mixing.

At higher intensities the RFS instability grows exponentially to a saturation level. Figure 12 depicts the plasma wave amplitude and spectrum for the case of v_{osc} $(0,1)/c \sim 0.5$. Note the presence of spectral components at $\omega = \omega_0 + \omega_1$ as well as $\omega = \omega_0 - \omega_1 = \omega_p$ and higher harmonics. Graphic evidence of this instability is also given in Figure 13, which is the time history and power spectra of the laser waves inside the plasma. The downward cascade of energy into multiples of ω_p is a clear signature of Raman instability. The upward cascade is the result of multiple four-wave interactions in which each wave is coupled to all other waves in a multiwave parametric process.[11] The sidebands are often referred to as Stokes (downshifted) and Anti-Stokes (upshifted) lines.

This simulation is particularly interesting, because the parameters model a CO_2 laser emitting in the 9.6 μm and 10.6 μm bands at a combined intensity of 1×10^{16} W/cm^2. The laser pulse is incident on a hot 10^{17} cm^{-3} plasma. Excitation of these CO_2 lines at this intensity and production of such a plasma is not beyond current technology. The higher saturation level of the electrostatic field is only a factor of three larger in this simulation than in the run where optical mixing was important. This implies a field of 316 MeV/cm based on the lower plasma density. The combined laser intensities here, however, are equivalently 100 times larger. Nevertheless, there are some advantages to this higher intensity.

The high value of v_{osc}/c and its associated relativistic effects evident in Figure 14b lead to particle trapping and acceleration of a small number of electrons indicated in Figure 14a. This also occurs if T_e = 0 for the same parameters. This should be obvious, since the trapped particles originate at wavelength intervals where $\beta_y\gamma$ is a maximum and are not associated with any overall temperature. These results are important if the plasma is to be the source of accelerated particles. The maximum energy will be approximately 100 MeV after .34 cm. In addition, the particle bunching at higher intensity is more pronounced and coherent (see Figure 15). As a reference, assume the focal spot size for the laser pulse described above is 10^{-4} cm^2. Then, based on simulation results, a particle pulse occurs every .18 psec, contains 5×10^{10} electrons, carries a peak current of 60 kA at a current density of 600 Megamps/cm^2. The main reason for belaboring this point is to indicate the potential the plasma has for confining and accelerating high density bunches of preaccelerated ions. Finally, consider the plasma electron distribution functions reproduced in Figure 16. The

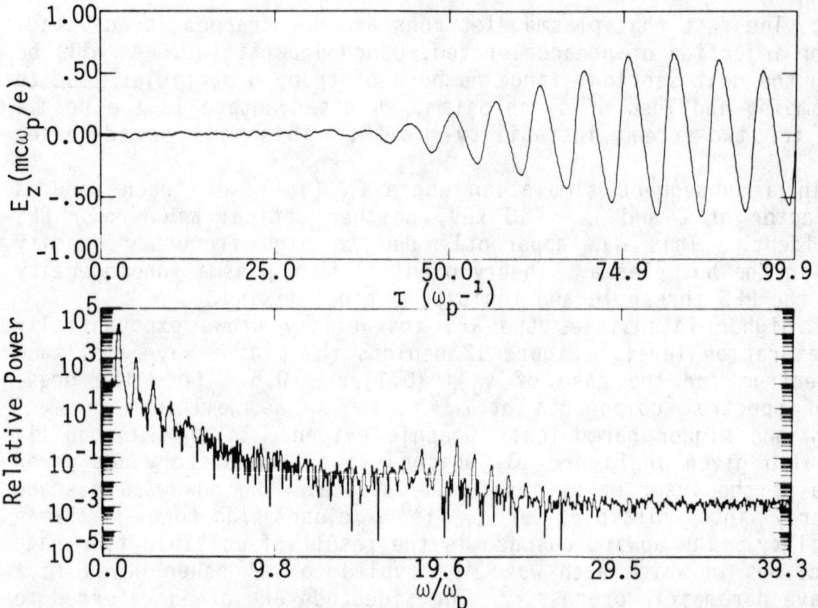

Figure 12. Time history and power spectrum inside the plasma of the longitudinal electric field. Simulation parameters are $\omega_0 = 10.6\ \omega_p$, $\omega_1 = 9.6\ \omega_p$ and $E_0\ (0,1) = 5.0\ mc\omega_p/e$.

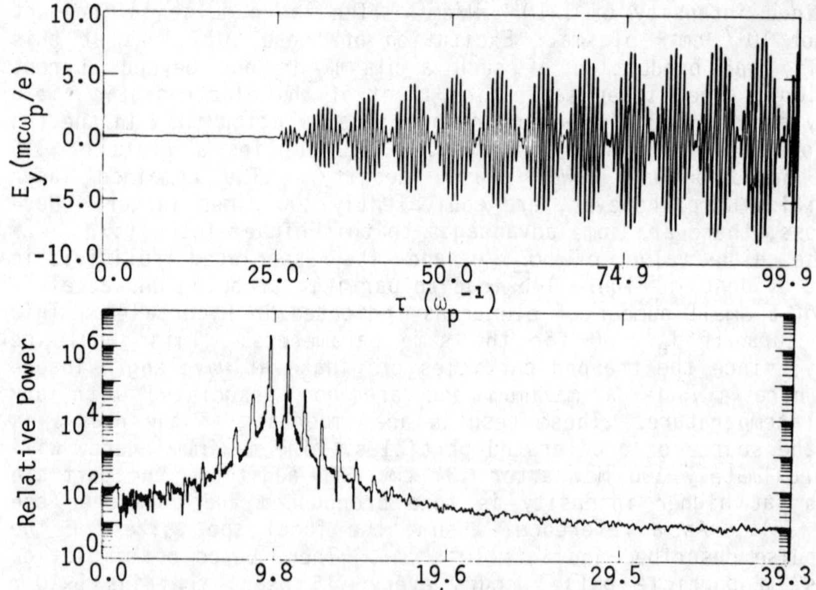

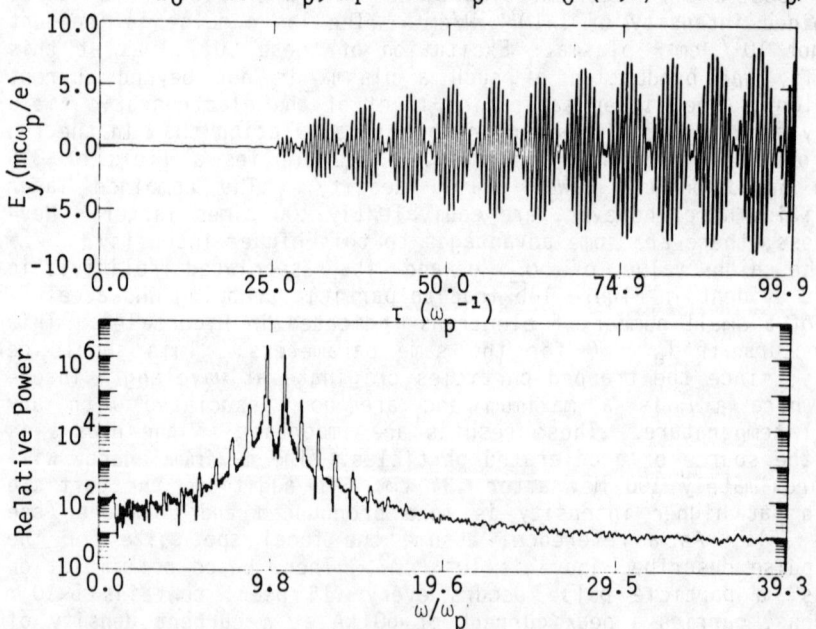

Figure 13. Time history and power spectrum inside the plasma of the two plane polarized laser wave electric fields. Parameters are ω_0 10.6 ω_p, $\omega_1 = 9.6\ \omega_p$ and $E_0\ (0,1) = 5.0\ mc\omega_p/e$.

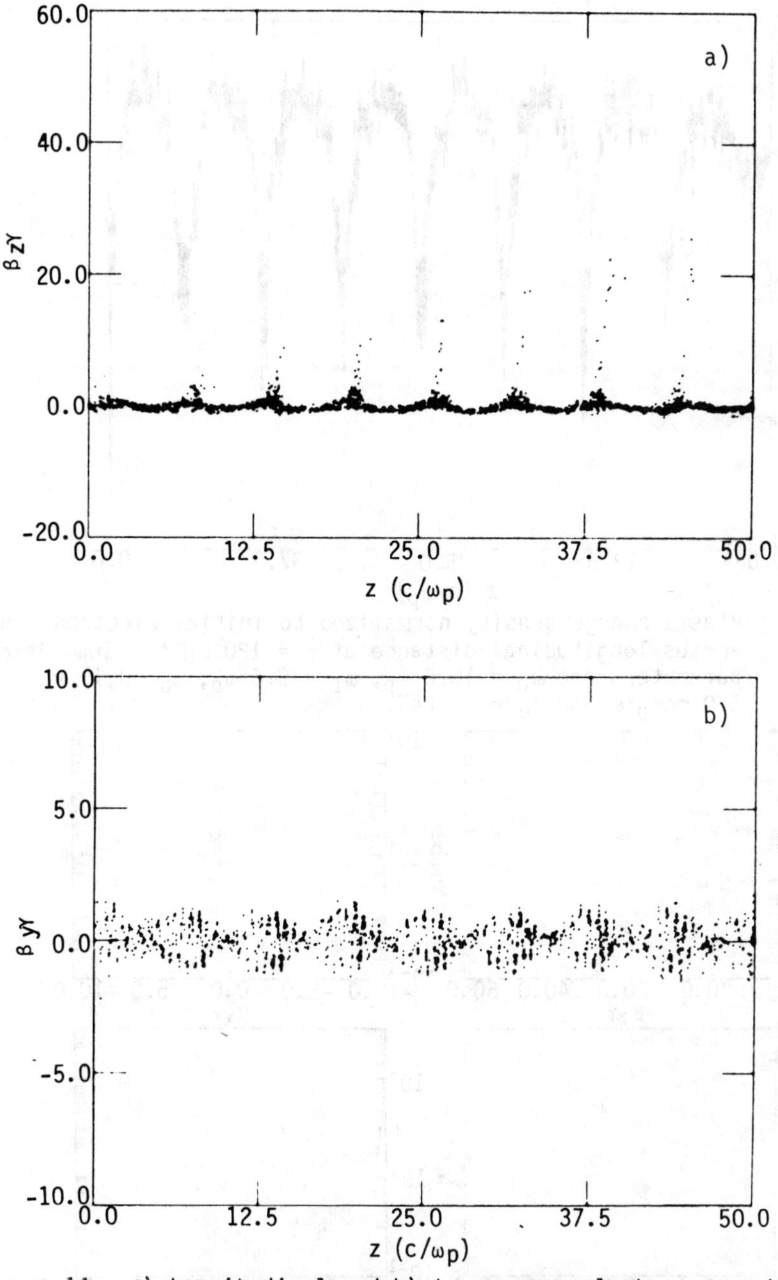

Figure 14. a) Longitudinal and b) transverse electron momentum phase space at $\tau = 120\ \omega_p^{-1}$. Simulation parameters are $\omega_0 = 10.6\ \omega_n$, $\omega_1 = 9.6\ \omega_p$, $E_0\ (0,1) = 5.0\ mc\omega_p/e$ and $T_e = 10.$ keV.

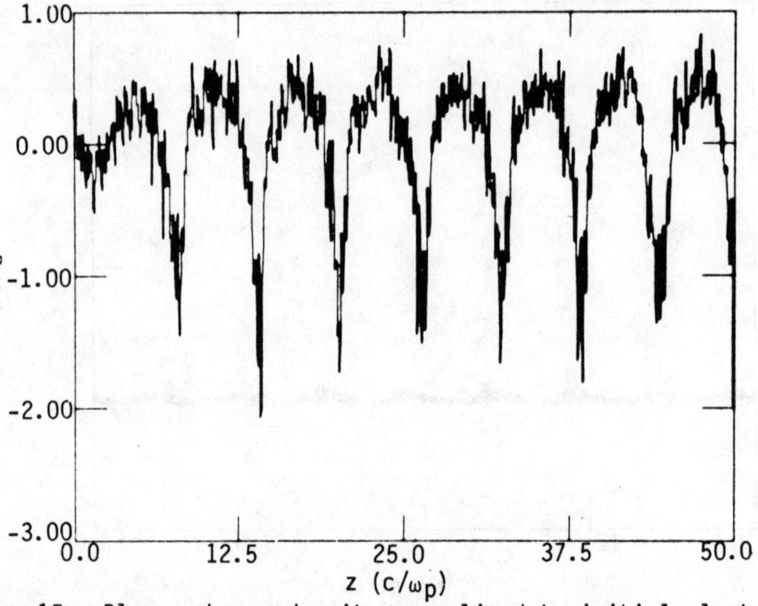

Figure 15. Plasma charge density normalized to initial electron density versus longitudinal distance at $\tau = 120\ \omega_p^{-1}$. Simulation parameters are $\omega_0 = 10.6\ \omega_p$, $\omega_1 = 9.6\ \omega_p$, $E_0\ (0,1) = 5.0\ mc\omega_p/e$ and $T_e = 10$ keV.

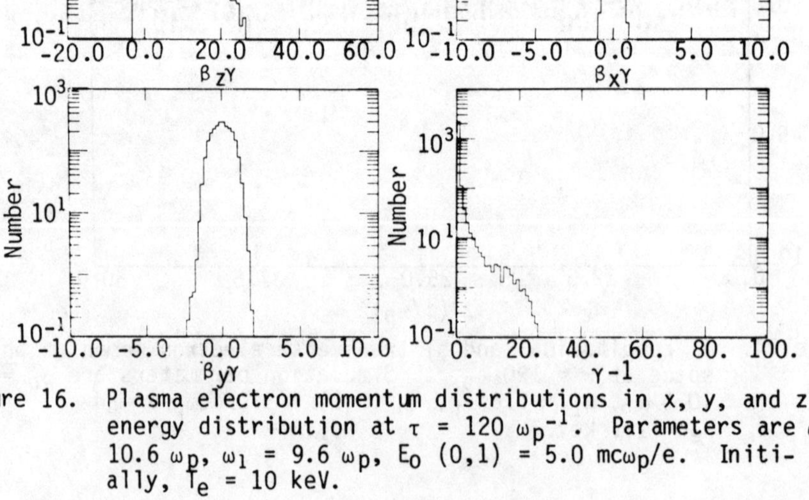

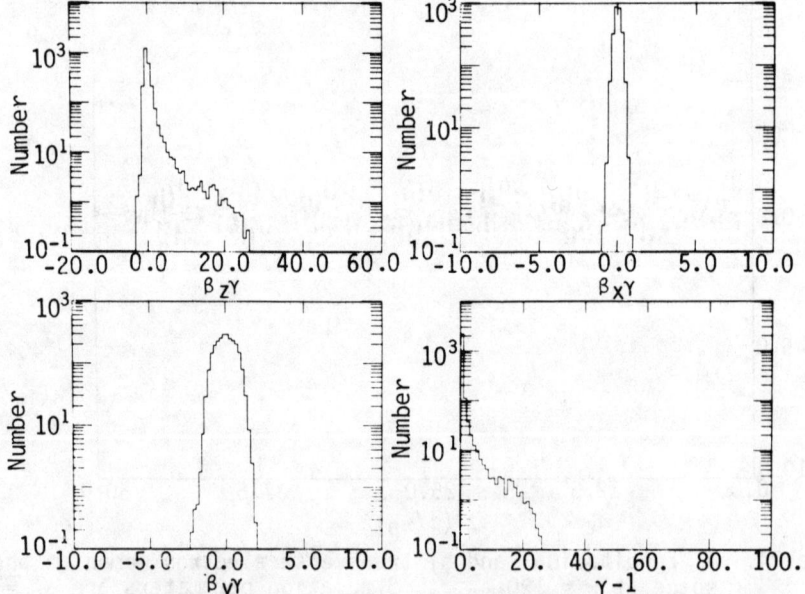

Figure 16. Plasma electron momentum distributions in x, y, and z and energy distribution at $\tau = 120\ \omega_p^{-1}$. Parameters are $\omega_0 = 10.6\ \omega_p$, $\omega_1 = 9.6\ \omega_p$, $E_0\ (0,1) = 5.0\ mc\omega_p/e$. Initially, $T_e = 10$ keV.

monotonically decreasing distribution functions are absolutely stable against streaming instabilities.[12] This will be covered in more detail later.

WAVE SYNCHRONISM

In order for plasma electrons or preaccelerated particles to attain the maximum acceleration given by equation (11), they must maintain synchronism with the wave for a distance on the order of $\gamma^{max}c/\omega_p$. A number of processes may be involved in destroying wave coherence. All result from driving plasmas with very intense laser pulses. The effects are seen in a simulation where the combined $v_{osc}/c > 1$.

By far the main degradation results from too many trapped particles. As the amplitude of the plasma wave increases, more particles can be trapped and nonlinear damping of the wave results as energy is transferred to the particles.[13] A slight increase in the trapping width of the wave[3,4] exponentially increases the number of trapped particles in a thermal distribution. Figure 17a shows large numbers of trapped particles. Simulation movies confirm that they are executing trapping oscillations which damp the plasma waves. This leads to incoherent density fluctuations and turbulent heating which make the plasma nonuniform. If this changes the plasma dispersion relation appreciably, the frequency matching condition will be upset.

High intensity laser pulses ($v_{osc}/c > 1$) result in plasmas where relativistic effects may be of comparable importance. The transverse plasma oscillation is evident in Figure 17b. This results in the forward Raman scattered wave having a frequency, ω_s, not equal to ω_1. Likewise the cascade process to lower frequencies due to multiple Raman scattering results in $\omega_s \neq \Delta\omega \equiv \omega_0 - \omega_1$. Although RFS still takes place, the two laser pulses no longer reinforce each other. Note also that from Figure 17b the γ_p associated with the plasma is a function of z. Thus, a range of scattering frequencies are present again leading to turbulence.

These combined processes lead to rapid loss of synchronism between the laser and plasma waves. Indeed, after a short time ($\tau = 120\ \omega_p^{-1}$) a plasma wave train is no longer discernible (see Figure 18). The deleterious effects, however, can be negated simply by lowering the laser intensity as little as a factor of two so that $E_0 = mc\omega/e$ rather than $\sqrt{2}\ mc\omega/e$. A coherent train of plasma waves in both space and time can be generated under these conditions. This is apparent in Figures 19 and 12. By choosing the proper laser intensity the number of particles being accelerated can be adjusted to limit damping of the plasma waves. Thus, acceleration of a few particles to maximum energy can be maintained.

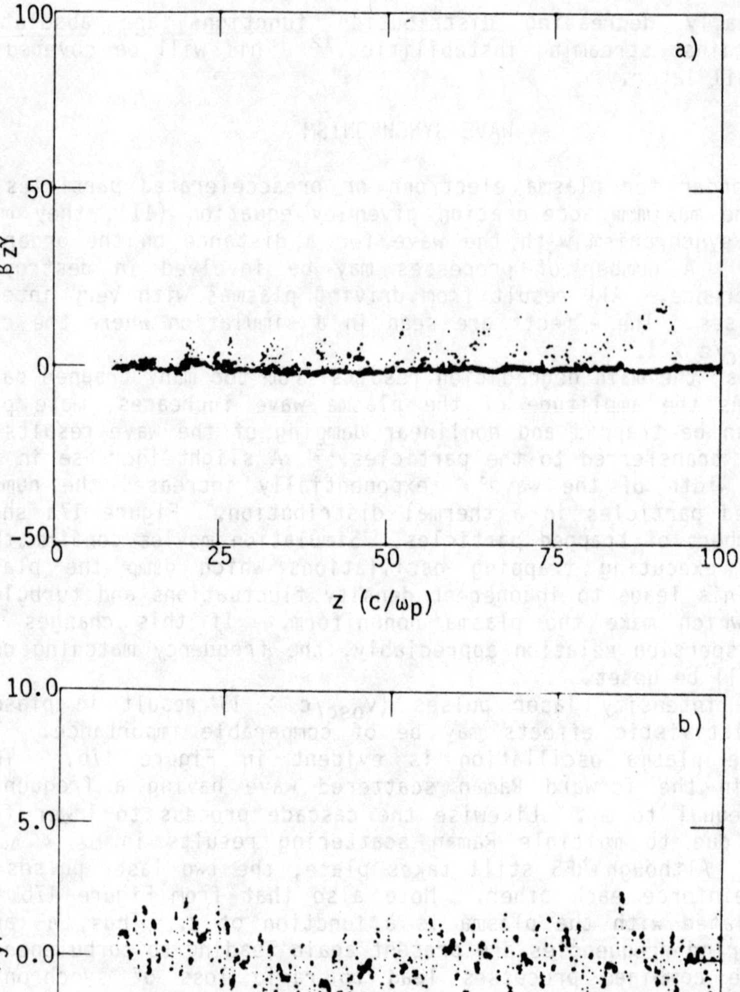

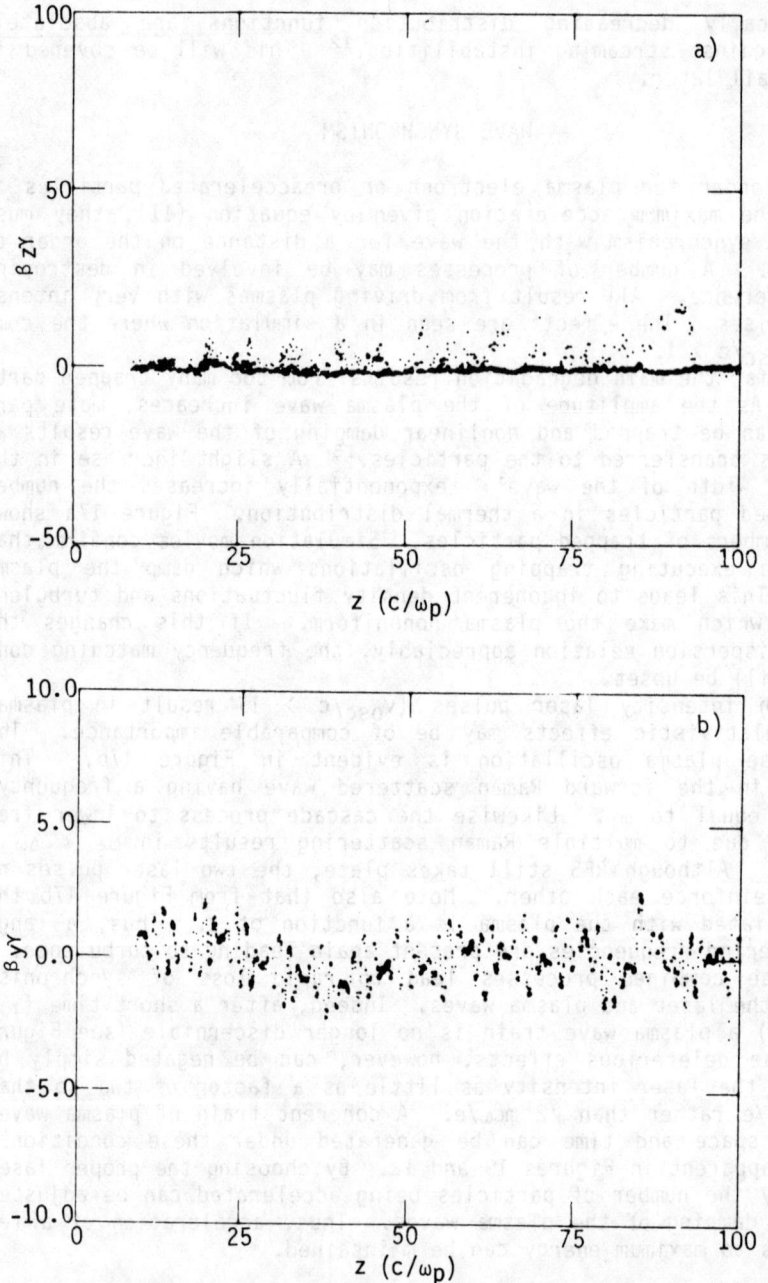

Figure 17. Phase space plots of a) longitudinal and b) transverse electron momentum at $\tau = 120\ \omega_p^{-1}$ for a simulation with $\omega_0 = 5.0\ \omega_p$, $\omega_1 = 4.0\ \omega_p$, $E_0\ (0,1) = 2.8\ mc\omega_p/e$ and $T_e = 0$.

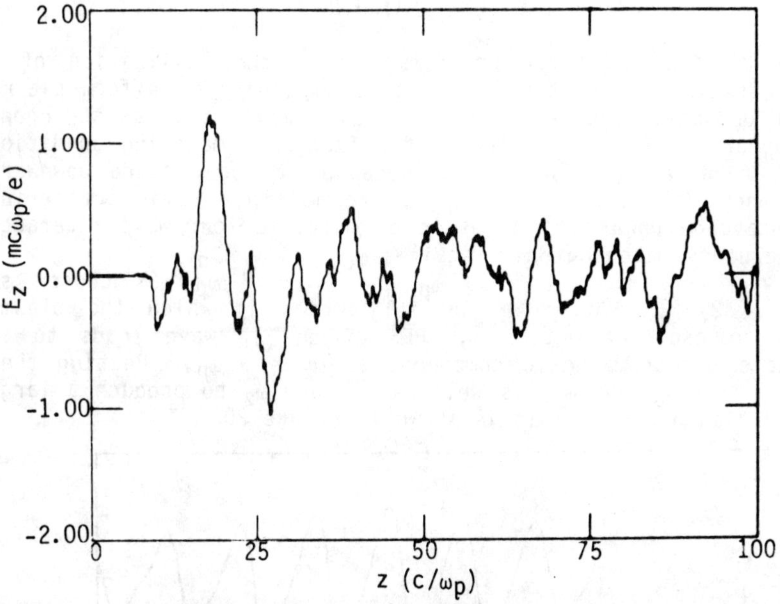

Figure 18. Longitudinal electric field as a function of distance $\tau = 120\ \omega_p^{-1}$. Simulation parameters are $\omega_0 = 5.0\ \omega_p$, $\omega_1 = 4.0\ \omega_p$, $E_0\ (0,1) = 2.8\ mc\omega_p/e$ and $T_e = 0$.

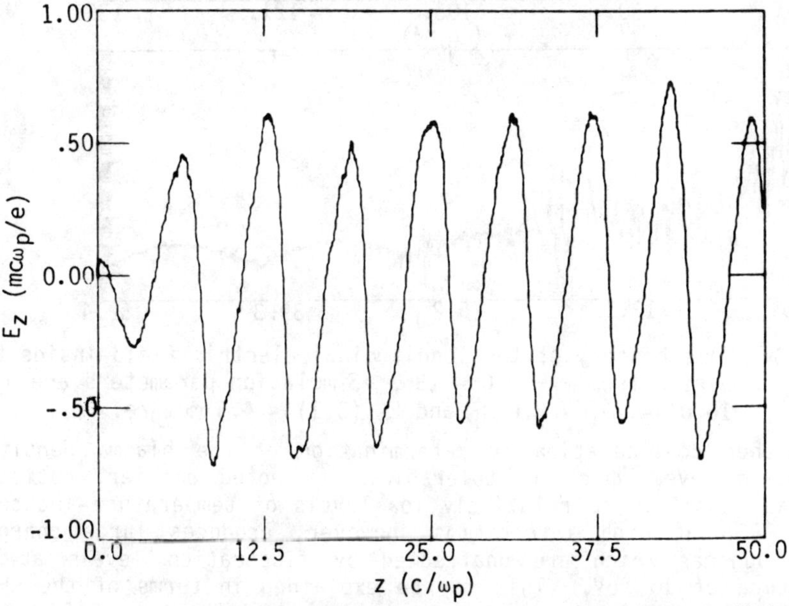

Figure 19. Longitudinal electric field as a function of distance $\tau = 120\ \omega_p^{-1}$. Simulation parameters are $\omega_0 = 10.6\ \omega_p$, $\omega_1 = 9.6\ \omega_p$, $E_0\ (0,1) = 5.0\ mc\omega_p$ and $T_e = 10$ keV.

FREQUENCY MATCHING

One of the practical problems facing the realization of a Plasma Beatwave Accelerator is the need to create a uniform plasma of a set density. However, the constraint may not be as stringent as it appears. In the RFS regime the frequency matching condition does not require $\omega_0 - \omega_1 = \omega_p$. Because of the cascade downward in frequency at intervals of ω_p due to multiple Raman Scattering and the cascade upward as a result of multiple four wave interactions, frequency matching only requires $\omega_0 - \omega_1 = n\omega_p$.

In one code run $\omega_0 = 10\ \omega_p$ and $\omega_1 = 8\ \omega_p$ was chosen so that $\Delta\omega = 2\ \omega_p$. After an initial period in which the plasma attempts to oscillate at $2\ \omega_p$, RFS of the ω_0 wave leads to an appreciable electromagnetic component at $\omega_2 = 9\ \omega_p$. Beating then occurs between ω_0 and ω_2, as well as, ω_1 and ω_2 to produce a large amplitude plasma wave. This is shown in Figure 20.

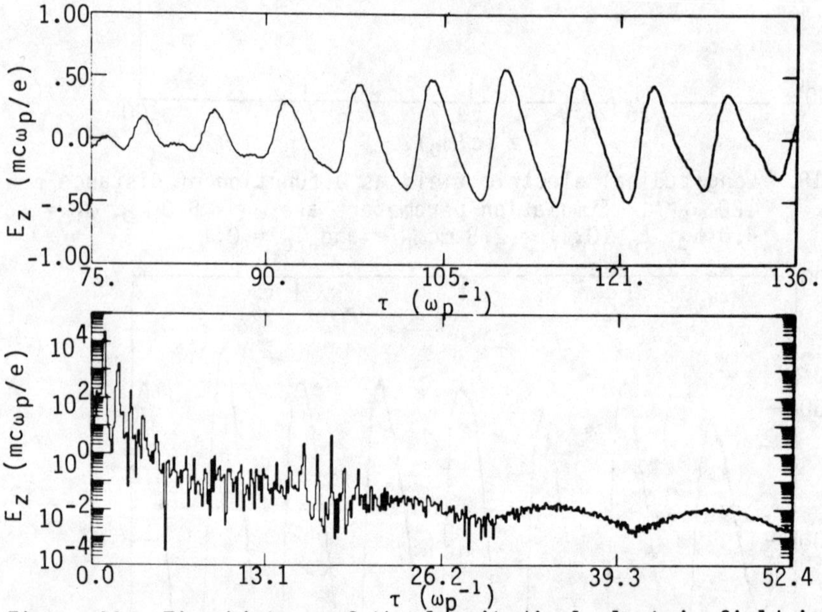

Figure 20. Time history of the longitudinal electric field inside the plasma from $\omega_p\tau = 0 - 136$. Simulation parameters are $\omega_0 = 10.0\ \omega_p$, $\omega_1 = 8.0\ \omega_p$ and $E_0\ (0,1) = 4.5\ mc\omega_p/e$.

Another consideration is determination of the plasma density fluctuation level which is tolerable. As noted earlier, optical mixing is sensitive to relatively low levels of temperature-induced noise. RFS at high intensities, however, produces large charge density bunches which are unaffected by fluctuation levels at a temperature of 10 keV. This may be explained in terms of the RFS intensity threshold for inhomogeneous plasmas derived in Reference 10. However, further correlation studies between instability growth and noise levels would be helpful.

STREAMING INSTABILITIES

One class of virulent instabilities which must be addressed for this accelerator are the various forms of streaming instabilities. As the particles are accelerated through the plasma, waves with a phase velocity equal to the beam velocity will grow. This only occurs, however, if the beam distribution superimposed on the plasma distribution is increasing ($\partial f/\partial v > 0$), where f is the distribution function. If the distribution is monotonically decreasing ($\partial f/\partial v < 0$) everywhere, then the plasma is absolutely stable against streaming instabilities.[12] This is the case where electrons are drawn out of the plasma to be trapped and accelerated. As shown in Figure 16, the momentum and energy distributions of the plasma electrons are monotonically decreasing and stable against the two-stream.

In this same simulation ions with a mass ratio of m_i/m_e = 359 were included. The laser pulse had a risetime of 100 $\omega_p{}^{-1}$. In terms of ion plasma periods this is equivalent to a risetime of 5 picoseconds for a Xe^{+8} plasma of density 10^{17} cm^{-3}. Admittedly, this is an artificially short risetime. However, it does allow a period, before electron trapping can take place, for the ions to react to the laser pulse and create unwanted density fluctuations and a nonthermal ion distribution. This does not occur. The ions do not react to the electromagnetic waves even at this fictiously low mass ratio. The plasma distribution remains stable against ion-electron as well as electron-electron two-stream instabilities. The real concern with regard to streaming instabilities is related to preaccelerated particles injected into the plasma. The distribution will be susceptible to the relativistic two-stream instability. However, its effects may be mitigated by the presence of large amplitude plasma waves at approximately the same phase velocity. Also, the injected particles can be at densities orders of magnitude below the background plasma so that the instability growth rate will be low.

In an attempt to answer some of these questions, a 1 MeV electron beam with a density of 10^{14} cm^{-3} was coinjected with a 10^{16} W/cm CO_2 laser emitting at 10.6 μm and 9.6 μm. Thus, the beam to plasma density ratio was .001. The plasma was cold. As before, a small number of plasma electrons are trapped and accelerated resulting in a monotonically decreasing distribution. Because the electron beam was not bunched at injection, its accelerated emittance is large. Notably the beam particles bunch at a phase in the wave which is different from the plasma electron bunches. Yet no streaming instability results. A more conclusive test using prebunched, monoenergetic ions at a laser intensity which does not allow plasma trapping will be conducted shortly.

SUMMARY

This paper has attempted to convey a number of salient features of the Plasma Beatwave Accelerator utilizing results from one-dimensional, relativistic, electromagnetic simulations. The beatwave, which is constructed by parallel propagation of two laser pulses, is a practical means of creating large amplitude plasma waves. The accelerating plasma field is on the order of $mc\omega_p/e$. Nonlinear wave-wave processes are responsible for plasma wave growth. If $\omega_0 - \omega_1 = \omega_p$, optical mixing may be used to enhance the Raman Forward Scattering Instability and drastically reduce the intensity necessary to excite large electrostatic waves. Wave synchronism is maintained provided particle trapping does not damp the wave and produce a turbulent plasma. This implies a combined $v_{osc}/c \lesssim 1$. In this same intensity range some plasma particle trapping and acceleration appear to be beneficial in suppressing streaming instabilities. Frequency matching in the RFS regime only requires that $\omega_0 - \omega_1 = n\omega_p$. The downward and upward cascade in frequency at ω_p intervals produces the beatwave conditions necessary to create the plasma waves.

A number of practical considerations in applying this concept to high energy accelerators deserve mention,[14] aside from the very high field strengths. The acceleration takes place far from any material structures. The medium it is operating in cannot breakdown, since it is already fully ionized. The particle acceleration takes place in the same direction as the laser pulse is propagating. Finally, extremely high energy (TeV) laser particle accelerators require staging, because of the limited range over the laser beam can maintain a focus. Here, the Plasma Beatwave Accelerator has a significant advantage. The acceleration takes place in a plasma wave rather than in vacuum. Therefore, if $\omega_0, \omega_1 \gg \omega_p$, the problem of phase locking the accelerated particles from one accelerator stage to another is greatly reduced.

Consider the following not-too-rigorous example. The object is to accelerate protons to 1 TeV. A number of 1 kJ, 1 nsec CO_2 lasers are conditioned to emit two beams at either 10.6 μm and 9.6 μm, or 10.6 μm and 10.2 μm. The need for this duality will become obvious. The combined laser intensity is 10^{16} W/cm^2. The focal spot size is 10^{-4} cm^2. The plasma density, which is determined by $\omega_0 - \omega_1 = \omega_p$, is either 10^{17} cm^{-3} or 1.5×10^{16} cm^{-3}. The plasma source will be a low energy gas z-pinch producing a uniform, cold plasma. The plasma would be approximately .5 mm in radius and 10 cm long. The beatwave peak-to-peak distance is $\lambda_p/2 = \pi\ c/\omega_p$. This is 53 μm for the higher density plasma and 137 μm for the lower. Thus, the particle bunch length must be less than 53 μm, which is achievable with current synchrotrons.

A conservative estimate based on simulations for E_z is 15 GeV/m and 6 GeV/m for the higher and lower density plasmas, respectively. The high density system can be utilized to accelerate the 1 GeV injected protons synchronously up to $\gamma^{max} = 2\ (\omega/\omega_p)^2 = 225$, or 210 GeV. This will require a plasma length of 14 meters, or 140 ten cm stages. The γ^{max} of the protons in the high density stage is 1429, or 1.34 TeV. To obtain a 1 TeV beam, therefore, requires an additional 132 meters. The total acceleration length is less than 150 meters. Allowing 1 meter between stages for refocusing of the lasers implies a total acceleration length of 1650 meters. Thus, we have a one mile linear accelerator capable of producing a 1

TeV proton beam. Clever engineering should reduce this size considerably. The number of protons per bunch would approach 10^{10}.

This example does not mean to imply all of the problems are solved. This concept is still in an embryonic stage. Theoretically, the main unanswered questions concern the transverse dynamics and self-fields of the beam. These must be addressed using fully two-dimensional and in some cases three-dimensional simulations. A partial list of these issues is given here.

Does the particle beam pinch or expand radially under the influence of its combined azimuthal-magnetic and radial electric fields? If the plasma return current flows inside the beam channel, the self-magnetic field generated by the net current may be too weak to overcome the beam's radial electric field: The beam will expand radially. On the other hand, if the return current flows outside the beam channel and $n_i/n_e > 1/\gamma^2$ (almost certainly true), the beam will pinch due to its azimuthal magnetic self-field.

How does diffraction of the laser pulse in the focal region affect its intensity, plasma wave amplitude and particle acceleration? Laser self-channeling due to the radial ponderomotive force will decrease the plasma density in the beam channel and destroy the frequency matching condition. Can it be overcome if $\tau_{pulse} < r_{spot}/C_s$, where r_{spot} is the laser focal spot size radius and C_s is the ion acoustic velocity? Raman backscatter is suppressed by Landau damping in a hot plasma.[2-4,9] Will Raman sidescatter be similarly suppressed or will it increase beam emittance and/or grow faster than Raman forward scattering? Will the particle beam break up due to filamentation instabiltiies?

All of these questions are critical for a high energy physics accelerator, due to the multiple plasma stages and low beam emittance necessary for conducting research.

ACKNOWLEDGEMENT

The authors would like to thank Drs. F. Cole, A. Sessler, and W. Willis for many helpful discussions on current accelerator technology and issues. We are particularly indebted to Drs. C. Joshi and T. Tajima for useful talks on optical mixing, nonlinear wave damping and experimental techniques. This work was supported by the Air Force Weapons Laboratory under contract F29601-78-C-0082.

REFERENCES

1. T. Tajima and J. M. Dawson, Phys. Rev. Lett. 43, 267 (1979).
2. T. Tajima and J. M. Dawson, IEEE Trans. Nuc. Sci. NS-28, 3416 (1981).
3. C. Joshi, this proceeding.
4. T. Tajima and J. M. Dawson, this proceeding.
5. Equation (5) and (10) of Reference 6 are incorrect.
6. D. J. Sullivan and B. B. Godfrey, IEEE Trans. Nuc. Sci. NS-28, 3395 (1981).
7. B. I. Cohen, A. N. Kaufman, and K. M. Watson, Phys. Rev. Lett. 29 581 (1972).

REFERENCES (Continued)

8. M. N. Rosenbluth and C. S. Liu, Phys. Rev. Lett. 29, 701 (1972).
9. Joshi, T. Tajima, J. M. Dawson, H. A. Baldis, and N. A. Ebrahim, Phys. Rev. Lett. 47, 1285 (1981).
10. K. Estabrook, W. L. Kruer, and B. F. Lasinski, Phys. Rev. Lett. 45, 1399 (1980).
11. D. Eimerl, R. S. Hargrove, and J. A. Pasiner, Phys. Rev. Lett. 46, 651 (1981).
12. C. S. Gardner, Phys. Fluids 6, 839 (1963).
13. J. M. Dawson and R. Shanny, Phys. Fluids 11, 1506 (1968).
14. These criteria were put forward by R. Pantell at the Laser Particle Accelerator Workshop (LANL, February 1982).

LASER ACCELERATOR BY PLASMA WAVES

T. Tajima
Department of Physics and Institute for Fusion Studies
University of Texas, Austin, Texas 78712

J. M. Dawson
Department of Physics, University of California
Los Angeles, California 90024

ABSTRACT

Parallel intense laser beams ω_o, k_o and ω_1, k_1 shone on a plasma with frequency separation equal to the plasma frequency ω_p is capable of creating a coherent large electrostatic field and accelerating particles to high energies in large flux. The photon beat excites through the forward Raman scattering large amplitude plasmons whose phase velocity is equal to $(\omega_o - \omega_1)/(k_o - k_1)$, close to c in an underdense plasma. The plasmon traps electrons with electrostatic field $E_L = \gamma_\perp^{1/2}\ mc\omega_p/c$, of the order of a few GeV/cm for plasma density $10^{18}cm^{-3}$. Because of the phase velocity of the field close to c this field carries trapped electrons to high energies: $W \simeq 2mc^2(\omega_o/\omega_p)^2$. Preaccelerated particles (ions, for example) coherent with the plasmon fields can also be accelerated. The (multiple) forward Raman instability saturates only when a sizable electron population is trapped and most of the electromagnetic energy is cascaded down to the frequency close to the cut-off (ω_p).

I. INTRODUCTION

A particle in an electromagnetic field of small amplitude is accelerated perpendicular to the direction of this electromagnetic wave $\underset{\sim}{k}$ and executes oscillations along the electric field direction $\hat{\underset{\sim}{E}}$. No net acceleration, therefore, is achieved. This holds true even if the electromagnetic field amplitude is large and there is the magnetic acceleration (see Fig. 1). The magnetic acceleration with the electric acceleration makes the particle execute a figure 8 orbit with no net acceleration. A spatially or temporally localized packet of electromagnetic waves cannot cancel all the oscillatory motion and thus leaves a net acceleration. The amount of acceleration, however, remains small.

In order to gain net acceleration by electromagnetic waves, there have been many attempts, which may be categorized into two: the virtual photon approach and the real photon approach. Most of the conventional accelerators including (proposed) collective accelerators are in the first category. Consider Fig. 2(a). In

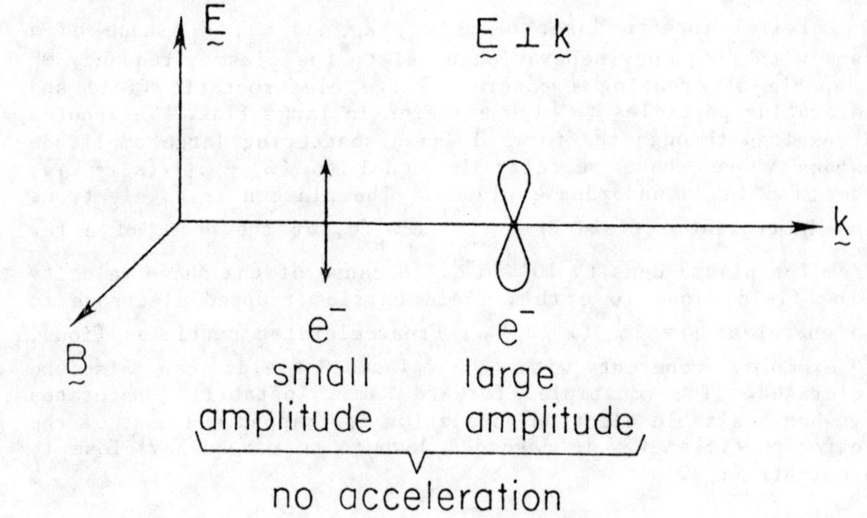

Fig. 1: Particle motion in a propagating plane electromagnetic wave. The wavenumber $\tilde{k}$ is perpendicular to electric field $\tilde{E}$ and magnetic field $\tilde{B}$. No net acceleration along the $\tilde{k}$-direction is achieved.

order to obtain an electric field component parallel to the wave propagation $k_{\parallel}$, some (metal) reflector is placed. The wave has $E_{\parallel}$ component, but unfortunately the phase velocity of the wave is larger than the speed of light $\omega/k_{\parallel} > \omega/k = c$. Thus no coupling. One may confine the electromagnetic field by two conductors in a wave-guide [Fig. 2(b)] instead of a semi-infinite case in Fig. 2(a). The characteristics of wave phase velocity of a wave-guide is well-known [See Fig. 2(c)]. The phase velocity is again always larger than c: $v_{ph} = c/\sqrt{1-(\omega_c/\omega)^2}$ where ω_c is the cut-off frequency for the wave-guide. An accelerator such as SLAC's alleviates this problem by implementing a periodic structure in the wave-guide. A periodically rippled wave-guide introduces the so-called Brillouin effect into the phase velocity characteritics. The Brillouin diagram for the frequency vs. wavenumber in the ripple wave-guide is depicted in Fig. 2(d). Here sections of phase velocity less than c are realized. Particles can now surf on such field crests and obtain net acceleration. The intensity of the fields is limited by materials considerations such as electric breakdowns.

When we utilize the real photon in a plasma, there is no physical limitation due to the materials considerations. The characteristics of the phase velocity of the real photon in a plasma is similar to the one in a wave-guide. See Fig. 2(e). The phase velocity of the electromagnetic wave in a plasma is $v_{ph} = c/\sqrt{1-(\omega_p/\omega)^2}$, always larger than c in an underdense plasma, indicating again the difficulty to directly couple the wave to accelerate electrons. Here ω_p is the plasma frequency. It is possible, however, to nonlinearly couple to the plasma if the amplitude of the electromagnetic waves are sufficiently large. In Refs. 1-3 we discussed a laser electron accelerator scheme by exciting a large amplitude Langmuir wave created either by a strong photon wavepacket with a very short spatial pulse length as a photon wake or by two beating photons. In Ref. 3 we concluded that using two photon beams is much more effective in acceleration than using a very short photon pulse. In the present paper, therefore, we primarily focus on the two beam case although the physics involved in the case of short wavepacket is quite common with the present case. The basic mechanism of particle acceleration is as follows. The two injected laser beam induce plasmons (or a Langmuir wave) through the forward Raman scattering process. This may be regarded as optical mixing. The resultant large amplitude plasma wave with phase velocity very close to the speed of light grows and is sustained by the laser lights. It grows until the amplitude becomes relativistic, i.e. the quivering velocity of the electrostatic field becomes c, so that the wave begins trapping electrons in the tail of distribution and accelerating them. The trapped electrons can be accelerated to high energies since the electrostatic wave is propagating with a phase velocity very close to c.

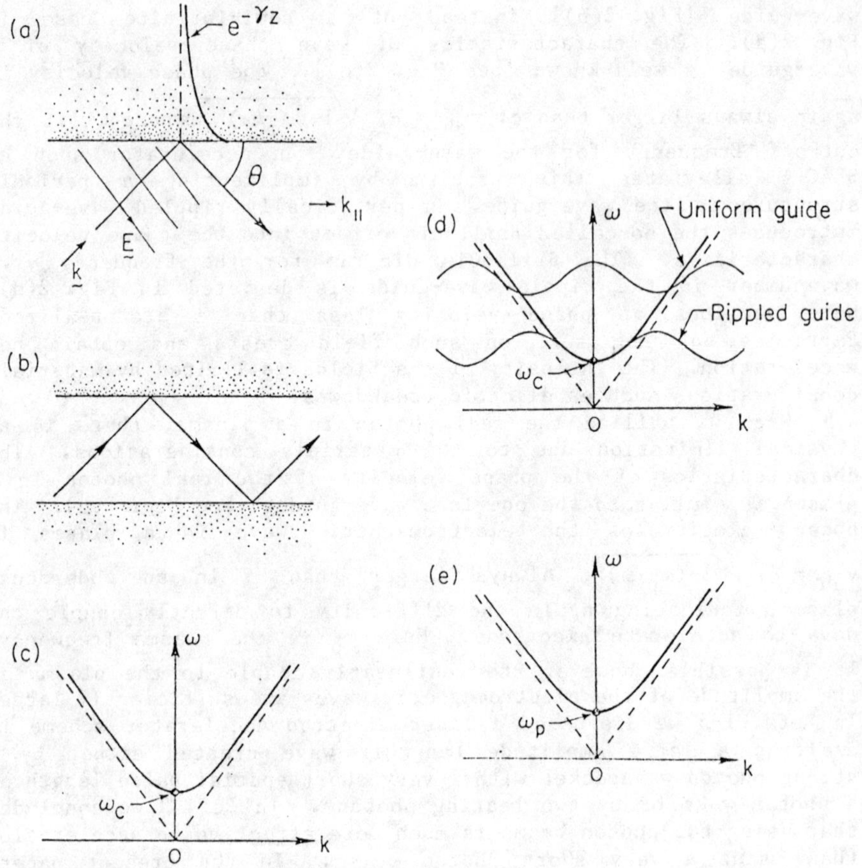

Fig. 2: Virtual photons and photons in a plasma. (a) A plane electromagnetic wave reflects on the metallic surface. There is a field component parallel to $k_\parallel$, $E \sin\Theta$. In the metal the EM field exponentially decays. (b) If we put two metallic plates together, we get a waveguide. Again a field component parallel to $k_\parallel$ exists. (c) The dispersion relation of the EM wave in the waveguide. k is the parallel wavenumber. (d) The dispersion relation of the EM waves in the uniform and ripple waveguides. (e) The dispersion relation of the EM wave in a plasma.

In terms of the available technology, the pulsed intense electron or ion beam technology[4] delivers an electric field of $\sim 10^7$ V/cm and a power density of 10^{13} W/cm^2. On the other hand, the laser technology is capable of delivering a power density of 10^{18} W/cm^2 for a glass laser and 10^{16} W/cm^2 for a CO_2 laser. As we shall see later, the electrostatic field can reach 10^9 V/cm for the glass laser case and a similar but somewhat smaller value for the CO_2 laser case.

II. LASER BEAT ACCELERATOR

Two large amplitude travelling electromagnetic waves, (ω_0,k_0) and (ω_1,k_1), injected in an underdense plasma induce a plasma wave (ω_p, k_0-k_1) through the beating of two electromagnetic waves if the frequency separation of two electromagnetic waves is equal to the plasma frequency:

$$\omega_0 - \omega_1 = \omega_p \quad , \tag{1}$$

$$k_0 - k_1 = k_p \quad , \tag{2}$$

where k_p is the wavenumber of the plasma wave. The beat of two electromagnetic waves gives rise to a nonlinear ponderomotive force which sets off the plasma oscillations.[1,2] This process may also be regarded as a nonlinear optical mixing[5] as well as a forward Raman scattering.[2,6,7] It may be possible to achieve the objective through the forward Raman instability,[6] i.e. the second electromagnetic wave (ω_1,k_1) grows from a thermal noise. It is important that the plasma is sufficiently underdense so that ω_0 is much larger than ω_p (see Fig. 3). This will ensure that the phase velocity of the plasma wave v_p is very close to the speed of light:

$$v_p = \frac{\omega_p}{k_p} = \frac{\omega_0 - \omega_1}{k_0 - k_1} \quad . \tag{3}$$

In the limit of $\omega_p/\omega_0 \ll 1$, Eq. (3) yields

$$v_p = \frac{\omega_p}{k_p} = \lim_{\omega_p/\omega_0 \to 0} \frac{\omega_0 - \omega_1}{k_0 - k_1} = v_g^{EM} = c\left(1 - \frac{\omega_p^2}{\omega_0^2}\right)^{1/2} \quad . \tag{4}$$

the relation we used in Refs. 1 and 2.

Suppose that ω_0 is not very much larger than ω_p, then the phase velocity of the plasma wave is much less than c. The resultant plasma wave can quickly trap electrons and saturates. In the course of interaction the nonlinear effects may change the phase velocity of the plasma wave. In the case of ω_p/ω_0 not small, the interaction of light waves and plasma is strong and the light waves suffer strong feed-back from the plasma. (The light

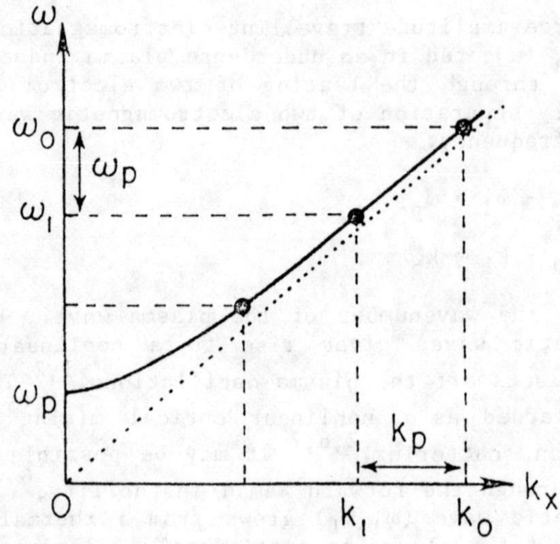

Fig. 3: Dispersion relation for the EM waves. Two laser beams beat.

waves may be called "plastic" or "soft" in this case.) In our case of $\omega_p/\omega_0 << 1$, the interaction of light waves and plasma is less intense and the characteristics of light waves are largely preserved. As we shall see later, the ratio of the energy density of the electrostatic plasma wave to the electromagnetic wave is $(\omega_p/\omega_0)^2$, i.e. the light wave dominated. Therefore, the plasma wave remains reinforced or "regulated" by the beating two laser beams

$$\omega_0 = k_0 c/\sqrt{1-\omega_p^2/\omega_0^2} \sim k_0 c$$

$$\omega_1 = k_1 c/\sqrt{1-\omega_p^2/\omega_1^2} \sim k_1 c \quad .$$

(The light waves may be called "hard" in our case.) Since the phase velocity v_p is very close to c for $\omega_p/\omega_0 << 1$, the electron trapping and, therefore, saturation, occur only when the electrostatic wave grows up to an amplitude so large that it becomes relativistic. The other important consequence of $\omega_p/\omega_0 << 1$ is that particles will be in phase with the plasma wave for a long time and achieve a large amount of acceleration, because, again, the phase velocity $v_p \sim c$ and the particles would not exceed v_p easily.

Let us consider the energy gain of an electron trapped in the electrostatic wave with phase velocity $v_p = \omega_p/k_p$. We go to the rest frame of the photon-induced longitudinal wave (plasma wave). Since the wave has the approximate phase velocity Eq. (4), $\beta = v_p/c$ and $\gamma = \omega_0/\omega_p$. Note that this frame is also the rest frame for the photons in the plasma: in this frame the photons have no momentum and the photon wavenumber is zero. The Lorentz transformations of the momentum four-vectors for the photons and the plasmons (the plasma wave) are

$$\begin{pmatrix} \gamma & i\beta\gamma \\ -i\beta\gamma & \gamma \end{pmatrix} \begin{pmatrix} k_0 \\ i\omega_0/c \end{pmatrix} = \begin{pmatrix} 0 \\ i\omega_p/c \end{pmatrix}, \qquad (5)$$

$$\begin{pmatrix} \gamma & i\beta\gamma \\ -i\beta\gamma & \gamma \end{pmatrix} \begin{pmatrix} k_p \\ i\omega_p/c \end{pmatrix} = \begin{pmatrix} k_p/\gamma \\ 0 \end{pmatrix}, \qquad (6)$$

where the right-hand side refers to the rest frame quantities with respect to the plasma wave ($k_p^{wave} = k_p/\gamma$), k_0 is the photon wavenumber in the laboratory frame and the well-known dispersion relation for the photon in a plasma $\omega_0 = (\omega_p^2 + k_0^2 c^2)^{1/2}$ was used. Equation (5) is reminiscent[8] of the relation between meson and the massless (vacuum) photon: Eq. (5) indicates that the photon in the plasma (dressed photon) has the rest mass ω_p/c, because the electromagnetic interaction shielded by plasma electrons can reach only the collisionless skin depth c/ω_p in the plasma. This is just as the nuclear force reaches the inverse of the meson mass and Yukawa predicted the meson energy as $\omega = \sqrt{c^2/a^2+k^2c^2}$, where a is the nuclear radius. Compare:

Meson	↔	Vacuum Photon
(interaction length a)		(interaction length ∞)
$\omega = \sqrt{\frac{c^2}{a^2} + k^2c^2}$		$\omega = kc$
Photon in Plasmas	↔	Vacuum Photon
(interaction length c/ω_p)		(∞)
$\omega = \sqrt{\omega_p^2 + k^2c^2}$		$\omega = kc$.

At the same time, the Lorentz transformation gives the longitudinal electric field associated with the plasmon as invariant ($E_L^{wave} = E_L$).

The electrostatic wave amplitude can be evaluated by a few different (independent) ways yielding the same result. An argument resorting to the wave breaking limit was employed in Ref. 2. Here let us discuss in terms of the available electron density. When most of electrons are bunched as a result of the beat electromagnetic waves, we may estimate the electrostatic field by assuming most of the electrons give rise to this field:

$$\nabla \cdot E = -4\pi en \quad \text{or} \quad k_pE_L = -4\pi en \quad , \tag{7}$$

where n is the electron density. The maximum electrostatic field is, therefore,

$$E_L = m\omega_pc/e \quad . \tag{8}$$

We may derive the maximum electrostatic field Eq. (8) by another method. As we shall discuss in Sec. III, the electrostatic wave saturates only when the trapping width of the wave becomes wide enough to begin trapping the tail of electrons. This condition may be written as

$$v_p - v_e \lesssim v_{tr} = \left(\frac{eE_L}{m\omega_p} v_p\right)^{1/2} \quad , \tag{9}$$

where v_e is the electron thermal velocity. If we neglect v_e in comparison with v_p and approximate v_p by c, Eq. (9) now yields Eq. (8).

The condition (8) is valid even if the trapped electrons are highly accelerated as long as the bulk of electrons remain nonrelativistic. However, when the bulk of electrons obtain kinetic energy, say the perpendicular energy, then the formula needs corrections.[9] According to Ref. 9, the obtainable electrostatic field increased to a value $E_L^{rel} \simeq \gamma_p^{3/2}\, mc\omega_p/c$, where γ_p is the effective relativistic factor. When strong heating of electrons occurs, mismatching of conditions (1) and (2) arises and we need more detailed study on the attainable electric field in this case.

The electric potential due to the plasma wave evaluated in the laboratory frame is

$$e\phi = e\int_0^{\lambda} E_L dx = e\,\frac{m\omega_p c}{e}\left(\frac{c}{\omega_p}\right) = mc^2 \quad . \tag{10}$$

Going to the wave frame, we obtain the potential in the wave frame

$$e\phi^{wave} = \gamma e\phi = \gamma mc^2 \quad . \tag{11}$$

This energy in the wave frame correponds to the laboratory energy by the Lorentz transformation

$$\begin{pmatrix} \gamma & -i\beta\gamma \\ i\beta\gamma & \gamma \end{pmatrix} \begin{pmatrix} \gamma\beta mc \\ i\gamma mc^2 \end{pmatrix} = \begin{pmatrix} 2\gamma^2\beta mc \\ imc\gamma^2(1+\beta^2) \end{pmatrix} , \tag{12}$$

where the right-hand side refers to the laboratory frame quantities. Thus we obtain the maximum energy electrons can achieve by the plasma wave trapping as

$$W^{max} = \gamma^{max}/mc^2 = 2\gamma^2 mc^2 = 2\left(\frac{\omega_0}{\omega_p}\right)^2 mc^2 \quad . \tag{13}$$

The time to reach energies of Eq. (13) may be given by

$$t_a \simeq W^{max}/ceE_L = 2\left(\frac{\omega_0}{\omega_p}\right)^2/\omega_p \tag{14}$$

and the length of acceleration to reach the Eq. (13) energy as

$$\ell_a \simeq 2\omega_0^2\, c/\omega_p^3 \quad . \tag{15}$$

For the glass laser of 1μ wavelength shone on a plasma of density $10^{18}(10^{17})\mathrm{cm}^{-3}$, it would require under the present mechanism a power of $10^{18}(10^{18})$ W/cm^2 to accelerate electrons to energies W^{max} of $10^9(10^{10})$eV over the distance of 1(30)cm with the longitudinal field E_L of $10^9(3 \times 10^8)$V/cm. For the CO_2 laser of 10μ wavelength, these numbers scale accordingly.

To demonstrate the present mechanism for electron acceleration, we have carried out computer simulations employing $1\frac{2}{2}$ - D (one spatial and three velocity and field dimensions) fully self-consistent relativistic electromagnetic code.[10] Two parallel electromagnetic waves (ω_0, k_0) and (ω_1, k_1) are imposed on an initially uniform thermal electron plasma. The direction of the photon propagation as well as the allowed spatial variation is taken as the x-direction. The system length is $L_x = 1024\Delta$, the speed of light $c = 10\omega_p\Delta$, the photon wavenumber $k_0 = 2\pi \times 68/1024\Delta$, the number of electrons 10240, and the particle size 1Δ with a Gaussian shape, and the ions are fixed and uniform, where Δ is the grid spacing. The thermal velocity $v_e = 1\omega_p\Delta$. The photon frequencies are taken as $\omega_0 = 4.29\omega_p$ and $\omega_1 = 3.29\omega_p$, while the amplitudes are $v_i \equiv eE_i/m\omega_i = c(i = 0 \text{ or } 1)$.

Figure 4 shows the phase space of electrons accelerated by the beat plasma wave $k_p \simeq \omega_p/c$. High energy electrons are seen in every ridge of each length of the resonantly excited electron plasma wave. The horizontally stretched arms in Fig. 4(a) are separated by length $\lambda = 2\pi/k_p$. The maximum electron energy was $85mc^2$ in this case, higher than the value given by Eq. (13). One reason for this discrepancy may be that we now have two intense electromagnetic waves so that magnetic acceleration associated with $v_0 \times B_1$, and $v_1 \times B_0$ also begins to play a role. The distribution function $f(p_\parallel)$ or $f(\gamma_\parallel)$ is shown in Fig. 4(b), exhibiting strong main body heating as well as a high energy tail. Figure 5 shows the electrostatic field profile in space at $t = 30\ \omega_p^{-1}$ (an early time). The field amplitude already reached $E_x \simeq E_L = m\omega_p c/e$. One can also see its coherent field pattern. The observed wavelength is $2\pi/k_p = 2\pi/(k_0-k_1)$.

The second case is that of injection of a wavepacket of a single photon (ω_0, k_0), whose packet length $L_t = \lambda/2 = \pi c/\omega_p$, as discussed in Refs. 1 and 2. Using the same code with parameters $L_x = 512\Delta$, $c = 5v_e$, the photon wavenumber $k_0 = 2\pi/15\Delta$, the number of electrons 5120, $eE_0/m\omega_0 = eB_0/m\omega_0 = c$, $L_t = \pi c/\omega_p$, $p_0 = eE_0/\omega_0$ and $\omega_0 = (\omega_p^2 + k_0^2c^2)^{1/2}$, we start the system with electromagnetic pulse in the plasma with initial conditions

$$E_y = E_0 \sin k_0(x-x_0)$$

$$B_z = B_0 \sin k_0(x-x_0)$$

$$p_y = p_{thermal} + p_0 \cos k_0(x-x_0)$$

for the period of $x = [50\Delta,\ 81.4\Delta]$ and $x_0 = 50\Delta$. With the assignment, the wavepacket has a spectrum in k with a peak around $k = k_0$ and $\omega = (\omega_p^2 + k_0^2c^2)^{1/2}$, and propagates in the forward x-direction approximately retaining the original polarization.

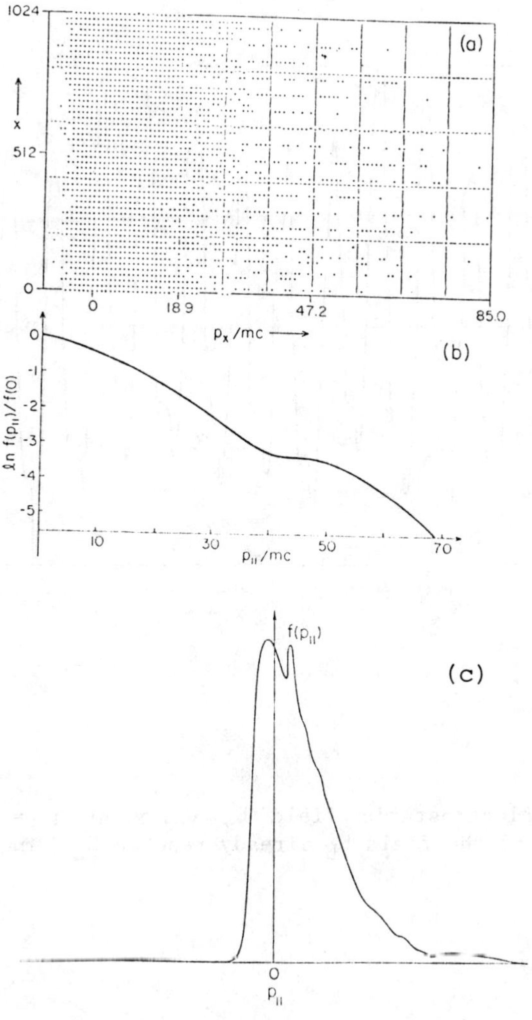

Fig. 4: Photon beat acceleration by two laser beams (ω_0, k_0) and (ω_1, k_1). (a) The electron phase space (x, p_x) at $t = 240\omega_p^{-1}$. The maximum $\gamma_\parallel$ for electrons is 85 in this case. (b) The logarithm of the electron distribution function at $t = 135\omega_p^{-1}$. (c) The electron distribution function at $t = 135\omega_p^{-1}$.

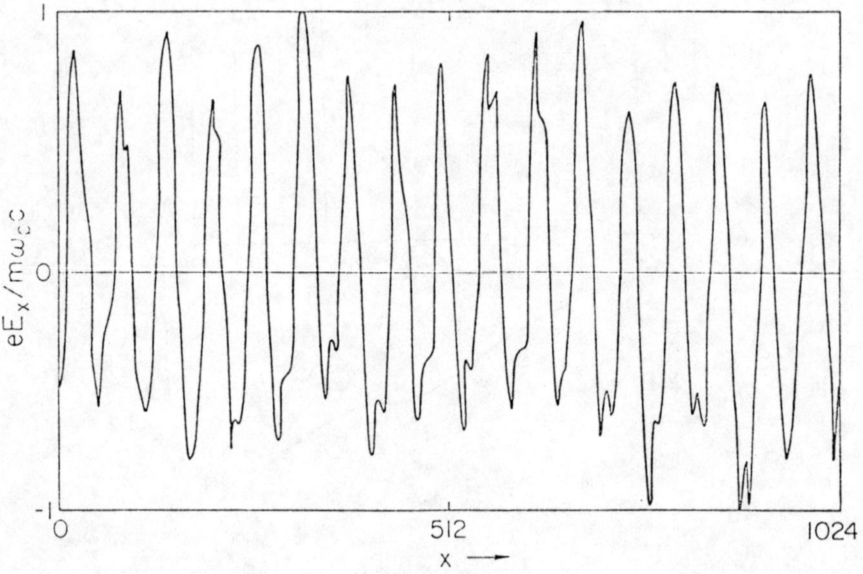

Fig. 5: The electrostatic field E_x vs. x at $t = 30\omega_p^{-1}$. The amplitude of the field E_L already reached $E_L \sim m\omega_p c/e$.

Figure 6 shows an early stage of the system development. The phase-space plot [p_y vs. x in Fig. 6(b)] indicates a strong modulation in the p_y distribution within the photon packet location. The kink structure extends beyond the packet ending. Figure 6(a) shows p_x vs. x. The intense longitudinal momentum oscillations are clearly appearing, beginning at the photon packet and extending to its initial starting point. This is the wake plasma wave set off by the photon packet seen in Fig. 6(b). As in Fig. 5, the long stretching arm-like phase-space pattern [Fig. 6(a)] appears with its momentum keeping increasing. The wake plasmon structure is also apparent in the longitudinal fields [Fig. 6(c)]. In this case the electrostatic field reaches values around $E_L \sim 0.6\ mc\omega_p/e$. In Fig. 7 we plot the maximum electron energy observed in our simulations as a function of $(\omega_0/\omega_p)^2$. The prediction Eq. (13) is written in a solid line to compare with the simulation values. Agreements are reasonable.

III. FORWARD RAMAN INSTABILITY AND EXPERIMENTS

We study the spectral distribution of photons in time. From the simulation run with two photons (ω_0, k_0) and (ω_1, k_1) we observe in Fig. 8 a clear cut energy cascade via multiple Raman forward scattering. The original waves with equal amplitude $E_i = mc\ \omega_i/e$ (i = o or 1) cascade toward smaller k as seen in Fig. 8(a) and 8(b). A small amount of energy is up-converted. The spectrum is sharply peaked at a particular discrete wavenumber $k_n = k_o - nk_p$ where n is an integer. The spectral intensity $S(k,\omega)$ for the electrostatic component shows overwhelming peak at $k = k_p$ and no significant energy in any frequency at the backscatter wavenumbers. This strongly suggests that all possible backscattering processes are suppressed or saturated at a very low level in our present problem. The electrostatic spectral density $S(k,\omega)$ shows peaks at $k = k_p$ as well as $k = nk_p$. All these observations confirm that the downward photon cascade is due to the multiple forward Raman scattering.

A similar downward photon cascade is observed in the case with a photon packet of one plasma wave (ω_0, k_0). Figure 9 shows the wavenumber spectrum of the electromagnetic pulse at successive times. The original smooth-shaped spectrum evolves into a multipeak structure with a roughly equal, but slightly increasing, separation in wavenumber as k approaches k_p. This again indicates that the photon (ω_0, k_0) decays into (ω_1, k_1), (ω_2, k_2)... by successive or multiple forward Raman instability.

The reason why the backscattering is suppressed but the forward scattering is very prominent is the following: When the backscattering plasma wave is excited, enhanced Landau damping or electron trapping by this plasma wave saturates it at a low level, thus limiting the backscattering to a small value. The wavenumber k_b of the plasma wave produced by the backscattering process in

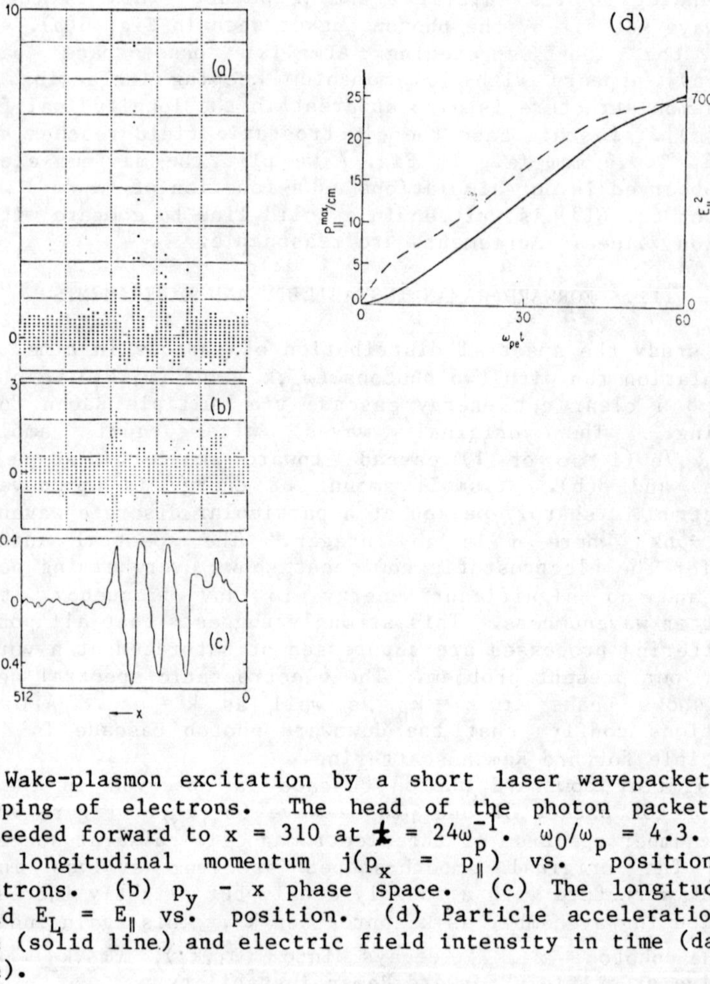

Fig. 6: Wake-plasmon excitation by a short laser wavepacket and trapping of electrons. The head of the photon packet has proceeded forward to $x = 310$ at $t = 24\omega_p^{-1}$. $\omega_0/\omega_p = 4.3$. (a) The longitudinal momentum j($p_x = p_\parallel$) vs. position of electrons. (b) $p_y - x$ phase space. (c) The longitudinal field $E_L = E_\parallel$ vs. position. (d) Particle acceleration in time (solid line) and electric field intensity in time (dashed line).

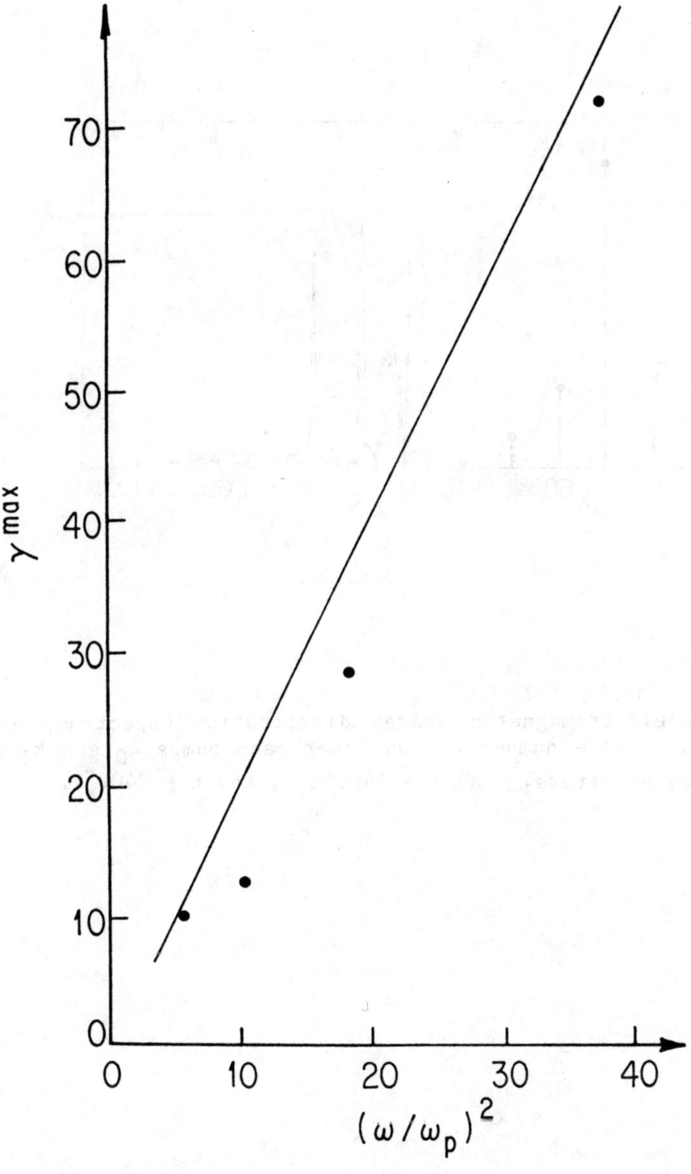

Fig. 7: Maximum electron energy vs. $(w_0/\omega_p)^2$ in the short wavepacket case. The dots are from simulations and the solid line is from Eq. (13).

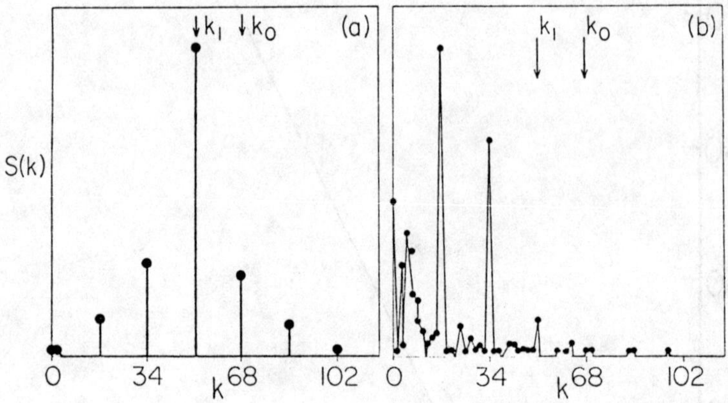

Fig. 8: The electromagnetic energy distribution (spectrum) as a function of mode numbers. Two laser beam pumps k_0 and k_1 are indicated by arrows. (a) $t = 142.5\omega_p^{-1}$, (b) $t = 240\omega_p^{-1}$.

the case of $\omega_p/\omega_0 \ll 1$ is $k_b = 2k_0$. The phase velocity of the backscattering plasma wave is

$$v_p = \frac{\omega_p}{2k_0} = \frac{c}{2}\frac{\omega_p}{\omega_0} . \tag{16}$$

Trapping of electrons by this wave begins happening when the trapping width v_{tr} becomes wide enough to reach the tail of the thermal electron distribution. An approximate trapping width may be written as

$$v_{tr} = \left(\frac{e\phi^b}{m}\right)^{1/2} = \left(\frac{eE_L^b}{m\omega_p} v_p\right)^{1/2} , \tag{17}$$

where the superscripts b refer to the backscattering electrostatic wave[11]. The formula is nonrelativistic, but is sufficient for the present purpose. Also recall the discussion given after Eq. (9). The condition that a large number of electrons are trapped is given[12] by

$$v_p - v_{tr} \leqslant 2v_e = 2(T_e/m)^{1/2} . \tag{18}$$

The maximum electrostatic wave amplitude is obtained for a cold plasma by setting $v_e = 0$:

$$\frac{eE_L^b}{m\omega_p} = \frac{c}{4}\frac{\omega_p}{\omega_0} . \tag{19}$$

In both cases of Fig. 8 and Fig. 9 we did not detect a large peak of electrostatic spectrum $S(k,\omega)$ at $k = 2k_0$. Thus the forward Raman instability or process appears to be the last parametric process to saturate in a hot underdense plasma. In fact it can be argued that it will saturate only when the original electromagnetic wave has completely cascaded by multiple forward Raman process to waves near $\omega \sim \omega_p$. In this case most of the electromagnetic energy may be extracted from the laser lights to electrostatic wave energy and eventually to kinetic energy. The idealized efficiency, therefore, may be given by $\eta = 1 - (\omega_p/\omega_0)^2$.

An experimental observation of the forward Raman instability and associated electron acceleration and heating has recently been done[6] in conjunction with the present concept and physical discussion. A CO_2 laser is shone on an underdense plasma producing electrons of energy up to 1.4 MeV. The laser power density is such that $eE_0/m\omega_0 c \sim 0.3$ and the frequencies are $\omega_p/\omega_0 \sim 0.46$. The plasma was created by the laser light shone on 130Å thick carbon foil producing the initial plasma temperature of ~20keV. In the experiment the laser emits only one beam so that the beat has to grow from the noise. It is, therefore, in general, possible to have other competing processes such as side scatter, backscatter, two plasmon decay simultaneously taking

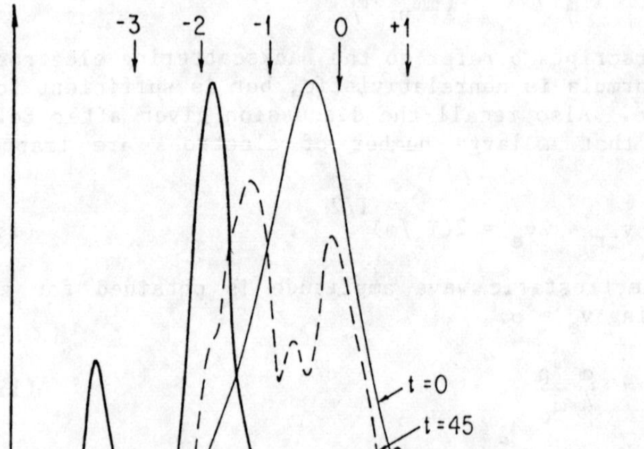

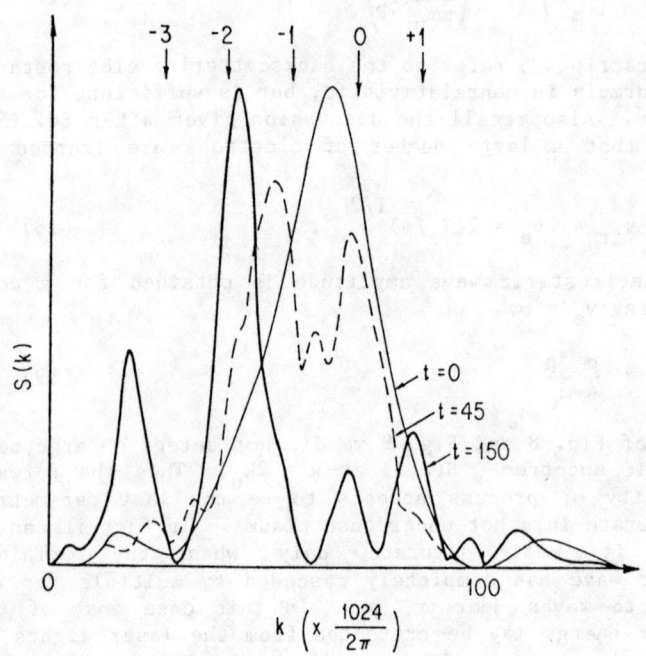

Fig. 9: Electromagnetic spectral intensity in wavenumber. The arrow with 0 indicates the rough position of the original laser beam wavenumber peak; n indicates $k' = k \pm nk_p$.

place. In spite of these competing processes, lower quivering velocity of the laser, and lower ω_0/ω_p, the experiment shows[6] high energy electrons in the forward direction.

Simulations are carried out in order to see the wave spectrum and to compare the distribution function of electrons with the experiment. Using a similar set-up as before, we set the plasma parameters same as the experiment:[6] $T_e \sim 20$keV, and uniform plasma, $\omega_p/\omega_0 \sim 0.46$, the propagating electromagnetic wave has $eE_0/m\omega_0c \sim 0.3$. The electron parallel distribution function $f(p_\parallel)$ along with the electrostatic wave spectra are displayed in Fig. 10. The temperature and the maximum electron energy observed in the simulation distributions are similar to the experimentally measured values. For example, simulations show the forward electron maximum energy of 1.3 MeV and temperature of 100keV in comparison with the experimental values of 1.4MeV and 90 ~ 100keV, respectively. In the backward direction, simulations show the electron maximum energy of 0.9MeV and temperature of 60keV compared to the experimental values of 0.8MeV and 40 ~ 50keV. The electrostatic wave spectrum, Fig. 10(b), shows that the backscattering mode k_b (which grows initially) is swamped by other modes with a smaller wavenumber, the most intense of which is the plasma wave associated with forward scattering k_p. In addition, there are some wavenumbers which are less than k_p. Thus the heated electron distributions obtained by the experiment[6] and the simulations agree well with most of the electron heating due to the forward Raman instability, but is not so much due to the backward process. In the present case the phase velocity of the backscattering plasma wave $\omega_p/k_b \sim 1.6v_e$. Thus this wave is heavily Landau damped to begin with and as it grows in amplitude, more and more electrons will be trapped by it and and the damping will grow. In view of Eq. (18), therefore, the experimental as well as simulation results are reasonably well understood.

IV. ULTRARELATIVISTIC CASES

As the laser beam becomes more intense, the accelerated electrons become more numerous and they are more energetic. When the laser beam is ultrarelativistic ($eE_0/m\omega_0c > 1$) and the laser wave packet is localized, such a wavepacket exerts a large ponderomotive force on the plasma and can create a local vacuum.[13] The intense electromagnetic wave pulse pushes the plasma forward and expands the region of plasma-plowed area (vacuum or very low density plasma). Since the expanding electromagnetic pulse acts like a piston which reflects incoming particles, we can evaluate how much momentum transfer takes place during the process. Equating the electromagnetic pressure (piston pressure) to the momentum exchange by electrons which are pushed by the piston, we obtain

$$P = \frac{E_0^2}{4\pi} = 2nv_g p \simeq 2nv_g \gamma mc \quad , \qquad (20)$$

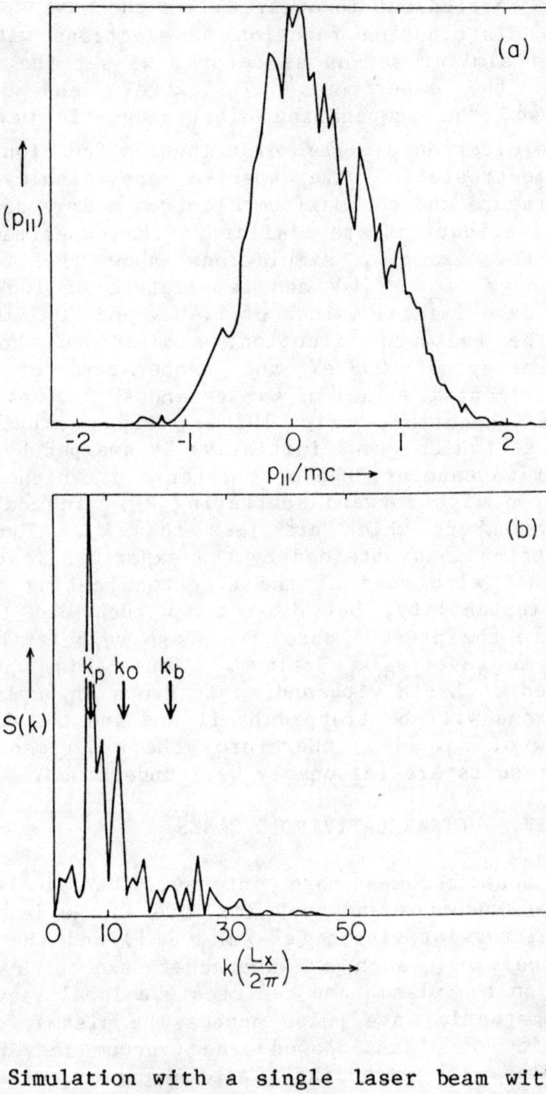

Fig. 10: Simulation with a single laser beam with $\omega_p/\omega_0 = 0.46$.
(a) Electron momentum distribution function at $t = 250\omega_p^{-1}$.
(b) The electrostatic mode spectrum at $t = 100\omega_p^{-1}$.

where v_g is the velocity of the electromagnetic wave front. From Eq. (20), the particle energy γ^{max} is calculated as

$$\gamma^{max} \simeq \frac{1}{2}\left(\frac{c}{v_g}\right)\left(\frac{\omega_0}{\omega_p}\right)^2\left(\frac{eE_0}{m\omega_0 c}\right)^2 \quad , \tag{21}$$

which resembles Eq. (13) except for a factor of $\frac{1}{2} \sim 2$ when $eE_0/m\omega_0 c = 1$ [which was the case for Eq. (13)].

Simulations have been performed to study this parametric dependence.[13] Again the same type of set-ups and code as in the short wavepacket case is used. Parameters we use are $L_x = 1024\Delta$, $v_e = 1\ \omega_{pe}\Delta$, the initial photon wavenumber $k_0 = 2\pi/10\Delta$, $c = 5\ \omega_{pe}\Delta$, 1024Δ electrons and ions each, and $eE_0/m\omega_0 c$ is varied from 1 to 20. The packet length is chosen to be $L_t \simeq \pi c/\omega_p$. The obtained energy scaling is displayed in Fig. 11, showing $\gamma^{max} \propto \nu^2 = (eE_0/m\omega_0 c)^2$. The coefficient of the proportionality is also reasonably fit with Eq. (21). More details are shown in Ref. 13. In a similar study Sullivan and Godfrey also investigated[9] the accelerated electron energy dependence on the laser field strength.

V. DISCUSSION

We have introduced the concept of laser electron accelerator using two parallel intense laser beams and a resultant beat plasma wave. We discussed various characteristics of this scheme and demonstrated it via computer simulations and to a certain degree via experiments. Although these investigations are preliminary, they are certainly encouraging. In the present paper we propose and demonstrate the way to accelerate particles to high energies without losing regular structure of the field. This is important since in a very high energy accelerator it is likely that any material cannot withstand the very strong accelerating field, the system has to regulate itself; in the present case the plasma and the laser beams have to regulate themselves. The regularity of the accelerating fields is kept by the reinforcing two intense laser beams. Since the plasma is underdense, the laser beams are able to impress their periodic structure on the plasma and the plasma oscillation. Their frequencies and wavenumbers are $\omega_0 = k_0 c/\sqrt{1-\omega_p^2/\omega_0^2} \sim k_0 c$ and $\omega_1 = k_1 c/\sqrt{1-\omega_p^2/\omega_1^2} \sim k_1 c$ and k_0, k_1 reinforcing the plasma frequency and wavenumber. The plasma particle-wave interaction cannot destroy the structure easily either until the forward Raman instability saturates, which does not happen at a low level. This is because the excited plasma wave has a phase velocity very close to c. Once the particle is accelerated and becomes relativistic, it will stay with the accelerating field for a long time in a coherent fashion. Since the density of electrons is high, the created field can be quite respectable of the order of 10^9 V/cm.

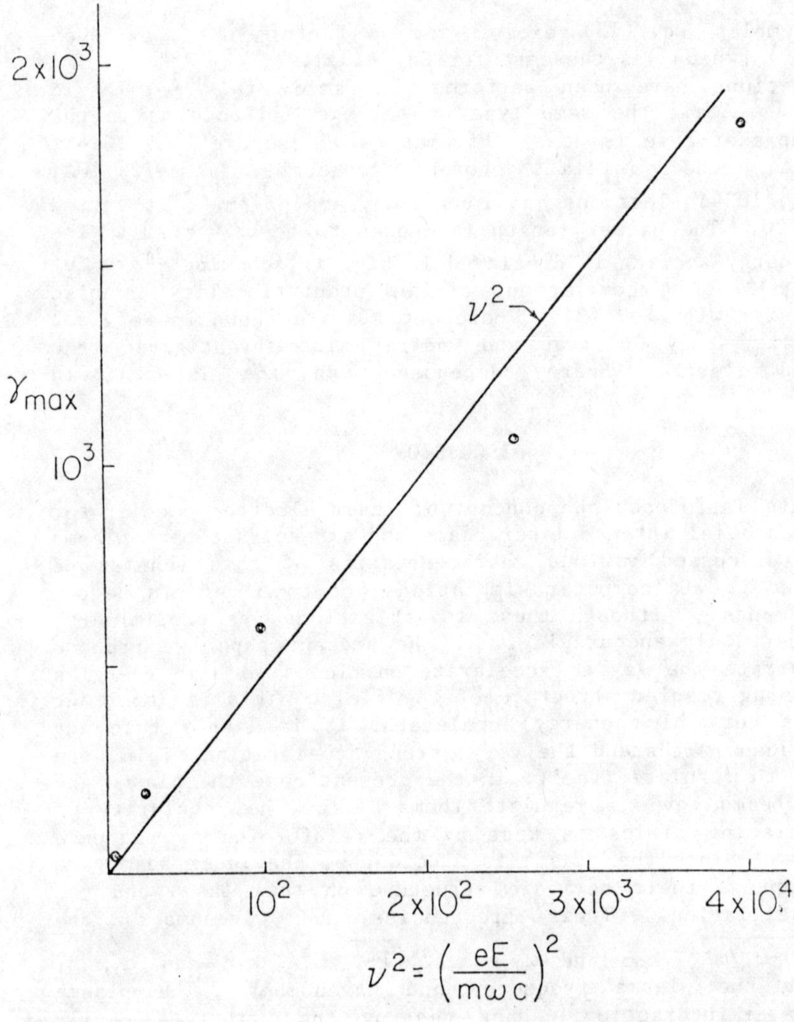

Fig. 11: The maximum electron energy vs. laser electric field intensity $(eE_0/m\omega_0 c)^2$ in the case of ultrarelativistic laser beam. The straight line is $\gamma^{max} = \frac{1}{2}\left(\frac{c}{v_g}\right)\left(\frac{\omega_0}{\omega_p}\right)^2 \nu^2$.

This scheme may be useful not only for accelerating electrons to high energies in large flux, but also for bunching electrons with a regular structure. A large regular electric field of 10^9V/cm range propagating with phase velocity very close to c is certainly attractive for accelerating ions, as well. One possible way to do this is to preaccelerate ions to a moderately to highly relativistic energies and then to inject onto this field. A schematic diagram is depicted in Fig. 12. In order to accelerate to multi-TeV energies it is likely to need many modulars of such an accelerator, since the laser focus and other problems may arise. In this case it may be a technical challenge to make ion clumps in phase with the positive electric field in all those modulars. Even if we assume that we can accelerate ions almost all the time, with the field of 10^9V/cm, it would take 10^5cm to reach an energy of 100TeV. If necessary, a plasma of gradually decreasing density, for example, might be useful, since the group velocity of the laser light increases; or by slowly changing other plasma parameters.

Scaled-down experiments may be possible by injecting preaccelerated electron beams. If the beam velocity v_b is slightly less than the phase velocity of the plasma wave, the plasma wave amplitude does not have to grow to grab the bulk electron distribution, but simply the beam electrons and then accelerate them. For this the necessary laser power is $P \simeq P_0/4\gamma_b^4$, where $P_0 = cE_0^2/4\pi$ and γ_b is the relativistic factor of the electron beam. The trapped electrons (originally injected beam electrons) would reach energies of $W \simeq W_0/4\gamma_b^4$, where W_0 is the maximum energy electrons gain if they are in the bulk [i.e. Eq. (13)].

There are a number of physics questions associated with the present concept. Obviously we have not addressed two or three dimensional phenomena. Among these, the self-magnetic field effects generated by the accelerated electrons and their currents have to be investigated. The filamentation instability and other types of higher dimensional instabilities may take place. The sidescattering instability may increase emittance of the beam. It is not clear at this moment whether these instabilities are helpful or not to the acceleration if they occur. Problems of laser focusing along with the coherency of laser and created plasma waves are important practical questions to be attacked. Another important set of questions is what is the optimal combination of the plasma temperature, density, laser frequency etc., for acceleration. The answer to this may very well depend on the purpose of the experiment. One should also investigate the optimal strategy to accelerate ions to a very high energy.

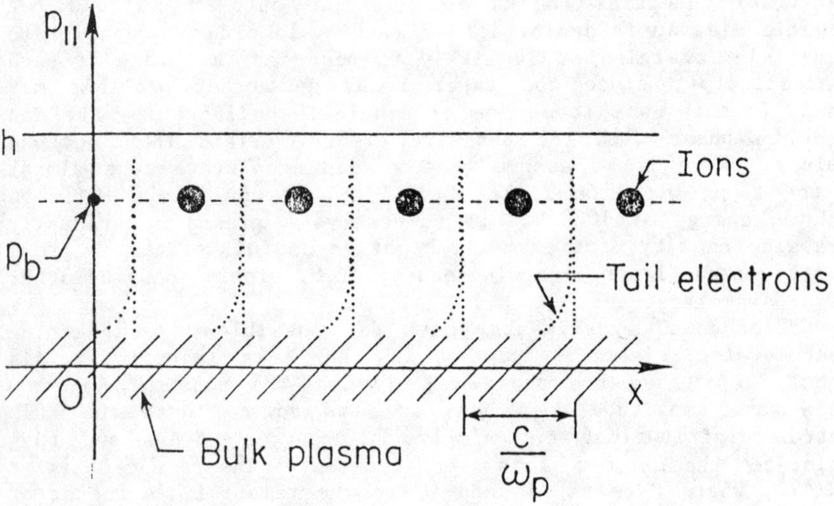

Fig. 12: Schematic phase space diagram with bulk electrons, accelerated tail electrons, and injected preaccelerated ions at the right phase. p_{ph} is the phase momentum of the plasma wave excited by the beating two laser beams. p_b is the injected ion beam momentum. The ion clump separation is c/ω_p.

ACKNOWLEDGMENTS

The authors would like to thank Drs. C. Joshi, D. Sullivan, A. Sessler, W. Willis, M. N. Rosenbluth, and N. Bloembergen for stimulating discussions, encouragements and interests. The work was supported by the U. S. Department of Energy, Contract No. DE-FG05-80-ET53088 and National Science Foundation Grants ATM81-10539 and PHY79-01319.

REFERENCES

1. T. Tajima and J. M. Dawson, IEEE Trans. Nucl. Science NS-26, 4188(1979).

2. T. Tajima and J. M. Dawson, Phys. Rev. Lett. 43, 267(1979).

3. T. Tajima and J. M. Dawson, IEEE Trans. Nucl. Science NS-28, 3416(1981).

4. B. Bernstein and I. Smith, IEEE Trans. Nucl. Science 3, 294(1973).

5. N. Kroll, A. Ron, and N. Rostoker, Phys. Rev. Lett. 13, 83(1964).

6. C. Joshi, T. Tajima, J. M. Dawson, H. A. Baldis, and N. A. Ebrahim, Phys. Rev. Lett. 47, 1285(1981).

7. Y. R. Shen and N. Bloembergen, Phys. Rev. 137, A1787(1965).

8. H. Yukawa, Proc. Phys. Math. Soc. Jpn. 17, 48(1935).

9. D. J. Sullivan and B. B. Godfrey, IEEE Trans. Nucl. Science NS-28, (1981).

10. A. T. Lin, J. M. Dawson, and H. Okuda, Phys. Fluids 21, 1995(1974).

11. Sometimes v_{tr} is defined as $\sqrt{2}$ times the value of Eq. (17).

12. J. M. Dawson and R. Shanny, Phys. Fluids 11, 1506(1968).

13. M. Ashour-Abdalla, J. N. Leboeuf, T. Tajima, J. M. Dawson, and C. F. Kennel, Phys. Rev. A23, 1906(1981).

PLASMA LASER ACCELERATOR: LONGITUDINAL DYNAMICS, THE PLASMA/LASER INTERACTION, AND A QUALITATIVE DESIGN

R. D. Ruth*
Lawrence Berkeley Laboratory
University of California
Berkeley, California 94720

A.W. Chao†
Stanford Linear Accelerator Center
Stanford University
Stanford, Califonia 94305

ABSTRACT

In this paper we present our studies on a plasma laser accelerator. First we look at the longitudinal dynamics and the trapping of particles in the potential well due to the longitudinal electric field in a plasma density wave. Next we study the plasma/laser interaction to obtain power requirements. Lastly, we qualitatively design a plasma/laser accelerator with parameters somewhat more modest than existing suggestions.

INTRODUCTION

There has been recent interest in particle acceleration with a combination of a medium (plasma) and a laser[1,2]. The laser is used to generate a plasma wave of a particular phase velocity, v_p. The longitudinal electric field of the plasma wave is then used for acceleration.

In this context many questions arise. First, if we are given a plasma wave, we need to understand the longitudinal dynamics of a test particle in the electric field generated by the wave. This leads to questions concerning particle trapping, small oscillations and "enclosed" phase space area. Secondly, we must understand how the plasma wave is generated by the E.M. wave of the laser. What is the growth rate of the wave, and what are the power requirements on the laser to yield a plasma wave of a given amplitude? Lastly, we need to identify the parameter range of interest for high energy accelerator (say e^-) applications.

In the first section we treat longitudinal dynamics (ignoring transverse effects), in the second section we study the plasma/laser

* This work was supported by the Director, Office of Energy Research, Office of High Energy and Nuclear Physics, High Energy Physics Division, U.S. Department of Energy, under Contract No. DE-AC03-76SF00098.

† This work was supported by the U.S. Department of Energy under Contract No. DE-AC03-76SF00515.

0094-243X/82/910094-18$3.00

interaction, and in the last section we attempt a qualitative accelerator design. In all sections the test particle is an electron.

1) LONGITUDINAL DYNAMICS

a) The Equations of Motion

Consider a plasma density wave (ions stationary) given by

$$n_e = n_0 + n_1 \cos(kz - \omega_p t) \quad . \tag{1}$$

(n_0 and n_1 are constants, $n_1 > 0$)

From this form the phase velocity is given by

$$v_p = \omega_p / k \quad . \tag{2}$$

It is also obvious that, in this model, since $n_e > 0$

$$n_1 \max = n_0 \ . \tag{3}$$

The electric field due to such a wave is found from Maxwell's Equations to be

$$\vec{E} = \frac{4\pi\, n_1 e}{k} \sin(kz - \omega_p t)\, \hat{z} \quad . \tag{4}$$

We can also write the potential energy of a test particle in such a wave as

$$e\,\phi = \frac{4\pi\, n_1 e^2}{k^2} \cos(kz - \omega_p t) \quad . \tag{5}$$

The motion of a particle in this potential is obtained from Hamilton's Equations with the Hamiltonian:

$$H = e\,\phi\,(z - v_p t) + c\sqrt{m^2c^2 + p_z^2} \tag{6}$$

where motion in the transverse degrees of freedom has been neglected. One of our tasks is to find constants of the motion which lead to curves in phase space that close and so, thereby, trap particles in the interior.

It is obvious that H is not a constant of the motion. However, it is possible to go to a new coordinate system in which the new Hamiltonian will be a constant of the motion. Consider the canonical transformation given by the generating function:

$$(z,\, p_z) \to (z',\, p_z') \tag{7}$$

$$F_2(z,\, p_z',\, t) = (z - v_p t)p_z' \quad ,$$

which yields

$$z' = z - v_p t$$

$$p_z' = p_z \tag{8}$$

$$H' = H - v_p\, p_z' \; .$$

This is simply a coordinate system which moves with the wave. The new Hamiltonian is

$$H' = e\,\phi(z') + c\sqrt{m^2c^2 + p_z'^2} - v_p p_z' \;, \tag{9}$$

and is a constant of the motion. Defining

$$\alpha = \frac{4\pi\; n_1 e^2}{m\; c^2 k^2} \tag{10}$$

we have

$$H' = \alpha\; mc^2 \cos kz' + c\sqrt{m^2c^2 + p_z'^2} - v_p p_z' \; . \tag{11}$$

Note that at n_1 max = n_0 we have

$$\alpha_{max} = (v_p/c)^2 \simeq 1 \; . \tag{12}$$

In Eq. (11) we have not expanded the p_z' terms into a power series to simplify the problem since we anticipate the application of this Hamiltonian for particles with non-relativistic to ultra-relativistic velocities.

Comparing Eq. (11) with the Hamiltonian for a pendulum or for a particle in an accelerator, we know that part of the phase space will be enclosed and correspond to trapped motion (oscillations) while another part will not be; these are separated by a separatrix. To find the value of H' on the separatrix simply note that the fixed points of H' lie at

$$k\,z' = n\,\pi$$

and

$$\frac{p_z'}{m\gamma} = v = v_p \; . \tag{13}$$

Evaluating H' at the unstable fixed point yields the value

$$H_s' = mc^2(\alpha + 1/\gamma_p), \quad \left[\gamma_p^2 \equiv 1/(1 - v_p^2/c^2)\right] \tag{14}$$

which is the value of H' on the separatrix.

So given the initial conditions of an individual particle we can calculate H'. If $H' < H_s'$, the particle is trapped; if $H' > H_s'$, it is not trapped.

In Fig. 1 we sketch several sets of separatrices for differing values of α. The peculiar shape of the "buckets" is due to relativistic effects. Ultra relativistic particles tend to be inside the separatrix since they travel with $v \sim v_p \sim c$, while non-relativistic particles tend to drop out of the bucket due to their slower velocities.

b) Emittances

It is interesting to calculate the phase space area enclosed in the trapped region. This gives us the maximum possible value of the trapped longitudinal emittance.

The area inside the separatrix is given by

$$\text{Area} = \varepsilon = \oint_{\text{separatrix}} P_z(z)dz = \int_0^{2\pi/k} dz(P_+(z) - P_-(z)) \tag{15}$$

where $P_\pm(z)$, the upper and lower branches of the curve in phase space corresponding to the separatrix, is easily seen to be

$$\frac{P_\pm(z)}{m\,\gamma_p v_p} = (2\,\gamma_p\,\alpha\,\sin^2(k\,z/2) + 1) \pm \frac{\gamma_p c}{v_p}\left[4\,\alpha^2\,\sin^4(kz/2) + \frac{4\,\alpha\,\sin^2(kz/2)}{\gamma_p}\right]^{\frac{1}{2}} . \tag{16}$$

The integral above can be simplified considerably with the result

$$\varepsilon = \frac{16\gamma_p^2\,mc\,\alpha}{k}\int_0^1 (1 + \frac{1}{\alpha\gamma_p} - x^2)^{\frac{1}{2}}\,dx . \tag{17}$$

If for cases of interest α is not too much smaller than unity and $\gamma_p \gg 1$ ($v_p \sim c$), then we find

$$\varepsilon \simeq \frac{4\pi\,\gamma_p^2\,\alpha\,mc}{k}(1 + \frac{1}{\alpha\gamma_p} + \ldots) . \tag{18}$$

c) Trapping

Electrons in the plasma will be trapped by the travelling plasma wave if the initial conditions of the electrons are inside of the separatrix. In a cold plasma, the initial distribution in phase space is that of a line distribution:

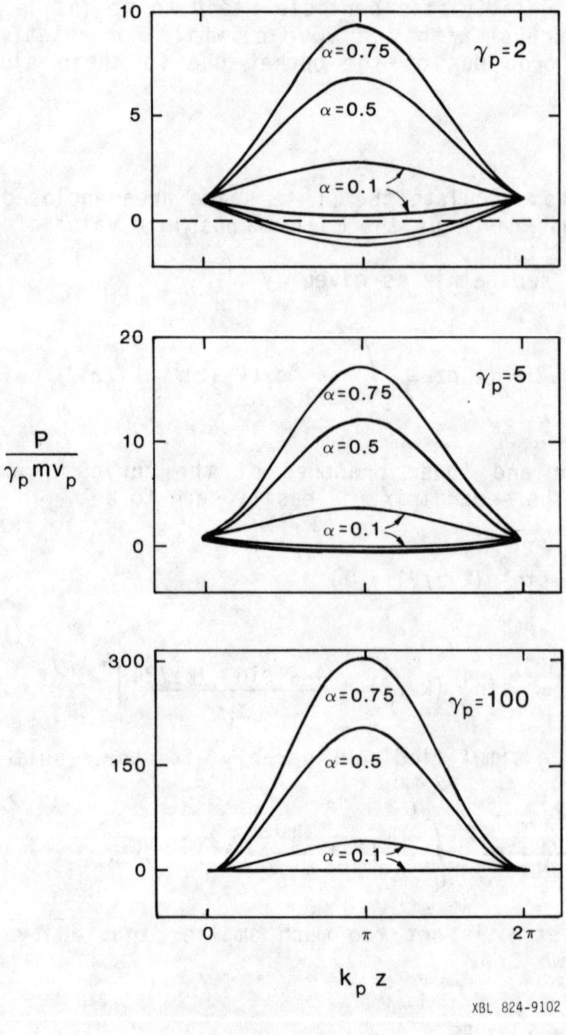

Fig. 1 Separatrices in phase space for different α and γ_p. Note that the wave breaking limit is $\alpha_{max} = 1 - 1/\gamma_p$, and also that the nested curves correspond to different peak electric field strengths ($eE_z = \alpha mc^2/k_p$).

$$f(p) = \frac{kN}{2\pi}\,\delta(p)$$
$$\int_0^{2\pi/k} dz \int_{-\infty}^{\infty} dp\; f(p) = N \qquad (19)$$

Whether cold electrons will be trapped by the plasma wave depends on whether the separatrix touches (and crosses) the line $p = 0$; the condition for this to happen is $p_-(z) \le 0$ at $kz = \pi$, where $p_-(z)$ is given by Eq. (16). In other words, no cold electrons will be trapped by the plasma wave if

$$p_-(\pi/k) \geqq 0 \qquad (20)$$

or equivalently

$$2\alpha + \frac{1}{\gamma_p} \le 1\ . \qquad (21)$$

If (21) is not satisfied, the trapping fraction of electrons by the plasma wave is determined by the fraction of the line $p = 0$ which is inside of the separatrix. Thus the trapping fraction is given by

$$T_f = \frac{\pi - kz_0}{\pi}\ , \quad \text{where } p_-(z_0) = 0 \qquad (22)$$

or after some algebra,

$$T_f = 1 - \frac{2}{\pi}\sin^{-1}\sqrt{\frac{1}{2\alpha}\left(1 - \frac{1}{\gamma_p}\right)} \qquad (23)$$

As α increases beyond the value in Eq. (21), the trapping fraction starts to increase from zero; as α reaches the maximum value $(v_p/c)^2 \approx 1$, the trapping fraction reaches $\approx 50\%$.

Note that the plasma density in Eq. (1) assumes that the trapped particles do not perturb the electrostatic potential well significantly. In this sense, the calculations are only self-consistent if the trapped fraction of particles is sufficiently small. Note also that the trapping fraction applies to the longitudinal dimension only. When transverse motions are included, the actual trapping could be lower.

d) Small Amplitude Oscillations

If we inject a test particle into the plasma wave near the velocity $v \simeq v_p$ and near the position $z' = \pi/k$, it will execute small oscillations at some frequency, ω_s, with no average acceleration. It is easy to show that if we let

$$x = z' - \pi/k$$
$$p_x = p_z' - m\gamma_p v_p \tag{24}$$

where x, p_x are small, then we obtain the new Hamiltonian

$$H'' \simeq \text{const} + \frac{p_x^2}{2\, m\gamma_p^3} + \frac{\alpha\, mc^2k^2x^2}{2} \tag{25}$$

which yields a small oscillation frequency of

$$\omega_s = \frac{\sqrt{\alpha} \quad \omega_p}{\beta_p \ \gamma_p^{3/2}} \tag{26}$$

(Notice the $\gamma_p^{3/2}$ due to the relativistic longitudinal mass).*

If we anticipate using this as an accelerator, then we, of course, should not inject particles at the stable fixed point. Rather we should inject in such a way that at the end of acceleration, the injected beam sits close to the top of the separatrix. This necessarily involves injection away from the stable fixed point. This point will be discussed further in the section on phase slippage.

2) THE PLASMA/LASER INTERACTION

a) The "ponderomotive force" of the EM wave

Consider a laser wave packet striking a plasma or, for simplicity, let us follow Ref. 1 and use 2 frequencies beating. Let the 2 frequencies be such that

$$\omega_1 - \omega_2 = \omega_p \tag{27}$$

and $\omega_1, \omega_2 >> \omega_p$. $(\omega_1, \omega_2 \approx \omega)$

Then a plasma wave will be generated with a wave number given by

$$k_p = k_1 - k_2 \tag{28}$$

and using the dispersion relation for an EM wave in a plasma ($\omega^2 = k^2c^2 + \omega_p^2$) we find

$$v_p \equiv \frac{\omega_p}{k_p} \simeq c\left[1 - \frac{1}{2}\left(\frac{\omega_p}{\omega}\right)^2\right] \tag{29}$$

---

* Pointed out by Phil Morton

So the phase velocity of the plasma wave matches the group velocity of the EM wave. Because of this matching the laser can drive the plasma on resonance at its natural frequency.

To calculate the force exerted by the laser, consider a superposition of 2 plane waves such that

$$\vec{E} = \frac{\vec{E}_0}{\sqrt{2}}\left[\sin\,(k_1 z - \omega_1 t) + \sin\,(k_2 z - \omega_2 t)\right] \tag{30}$$

Using the transverse gauge ($\vec{\nabla} \cdot \vec{A} = 0$) we find the vector potential

$$\vec{A} = -\frac{cE_0}{\sqrt{2}}\left(\frac{1}{\omega_1}\cos\,(k_1 z - \omega_1 t) + \frac{1}{\omega_2}\cos\,(k_2 z - \omega_2 t)\right) \tag{31}$$

with B given as usual by $\vec{B} = \vec{\nabla} \times \vec{A}$. The intensity of the laser is simply given by

$$I = \frac{E_0^2}{8\pi}\, c \tag{32}$$

The force on a particle due to this field can be calculated with the Hamiltonian

$$H = \frac{(\vec{p} - e/c\,\vec{A})^2}{2m} \tag{33}$$

Note that this is non-relativistic; this approximation is good so long as $eE_0/\omega \ll mc$ and not too bad if $eE_0/\omega \simeq mc$.

To find the average force for times $\gg \frac{1}{\omega}$, we simply average over a time $2\pi/\omega$ to obtain

$$\overline{H} = \frac{\vec{p}^{\,2}}{2m} + \frac{e^2}{c^2}\overline{\left(\frac{\vec{A}^2}{2m}\right)} \tag{34}$$

or using Eq. (31) we find

$$\overline{H} = \frac{\vec{p}^{\,2}}{2m} + \frac{e^2\,E_0^2}{4\,m\omega^2}\cos\,(k_p z - \omega_p t) \tag{35}$$

where we have retained the interference term since $\omega \gg \omega_p$, and we have dropped constant terms. The second term in this Hamiltonian is just the "ponderomotive" potential for this problem. The force in z direction is thus given by

$$F_z^L = \frac{k_p e^2 E_0^2}{4 m\omega^2} \sin (k_p z - \omega_p t) \quad . \tag{36}$$

b) The Plasma Response

The simplest approach in this section is to consider the response of a cold plasma with stationary ions. In the one dimensional linear* regime it is easy to see that the perturbed electron density obeys the equation

$$\frac{\partial^2 n_1}{\partial t^2}(z,t) + n_0 \frac{\partial}{\partial z}\left[\frac{F_z^L}{m} + \frac{e E_1}{m}\right] = 0 \tag{37}$$

where E_1 is the field due to n_1, and n_0 is the background electron density. Evaluating E_1 and using Eq. (36) we find

$$\frac{\partial^2 n_1}{\partial t^2} + \frac{n_0 k_p^2 e^2 E_0^2}{4 m^2 \omega^2} \cos (k_p z - \omega_p t) + \omega_p^2 n_1 = 0 \quad . \tag{38}$$

Now search for a solution of the form

$$n_1(z,t) = f(k_p z - \omega_p t) \sin(k_p z - \omega_p t) \quad . \tag{39}$$

The plasma wave is thus represented as a simple harmonic oscillator driven on resonance by the laser. Requiring $n_1 = 0$ ahead of the laser pulse, we obtain the solution

$$f = \frac{-k_p^2 e^2 E_0^2 n_0}{8m^2 \omega^2 \omega_p^2} (k_p z - \omega_p t) \tag{40}$$

which yields a growth rate for a given z of

$$\frac{\dot{f}}{n_0} = \frac{k_p^2 e^2 E_0}{8m^2 \omega^2 \omega_p} \quad . \tag{41}$$

The time, τ, to reach a longitudinal electric field of

$$E_z = \alpha\, m\, c\, \omega_p / \beta_p \tag{42}$$

---

* Note that these restrictions must be kept in mind for actual calculations.

is then given by

$$= \frac{8\ m^2 \omega^2 c^2 \alpha}{e^2 E_o^2 \omega_p} \quad . \tag{43}$$

Let us summarize the picture by looking at a particular point, z, along the accelerator. We see the laser beat wave and the plasma wave going by. The plasma wave is out of phase from the beating electric field by 180° and is growing at a rate given in Eq. (41). After the laser pulse has passed, the plasma wave remains at the amplitude in Eq. (42) provided the pulse length and coherence length of the laser is given by Eq. (43). Note that since the plasma is cold, the group velocity of the plasma wave is zero, and there is no damping.

The process of beating is effective in reducing the instantaneous laser power. Compared with the case of a shock excitation of a plasma wave, the beating excitation requires less instantaneous laser power because the plasma wave is excited more gently but over a longer time. However, this process requires that during the time, τ, the plasma frequency remains in resonance with the beat frequency, $\omega_1 - \omega_2$. This may put a severe limitation on frequency errors within the system.* Fortunately there are indications that the nonlinearity of the plasma wave will not introduce changes in frequency.[3]

To write Eq. (43) in terms of power/energy requirements note that

$$\frac{E_o^2 c}{8\pi} = W / \frac{\pi}{2} \sigma^2 \ , \tag{44}$$

where W is the laser power and σ is the gaussian spot size radius. Then we find for a laser pulse length, , the energy requirement

$$E_{Laser} = W\tau = \frac{\alpha\ m^2 c^2 \omega^2}{2\ e^2 \omega_p} \sigma^2 \ , \tag{45}$$

and we simply divide by τ to obtain the power. Since it is the intensity, $\sim E_o^2$, which enters into the growth rate, the spot size is a critical factor. In section 3b we discuss spot size requirements and obtain the final formula for laser power.

3) DISCUSSION: A PLASMA/LASER ACCELERATOR

In this section we use the results of the previous sections to put forward a qualitative design of a high energy plasma/laser accelerator. First, however, we need to discuss a few other aspects of acceleration.

---

*As pointed out by Lloyd Smith

a) Phase Slippage

Since the emphasis in this paper is on high energy electron acceleration, we avoid the problem of changes in particle velocity. However, since plasma wave will travel slower than, for example, an injected high energy electron bunch, we must consider phase slippage. We certainly don't want the plasma wave to trap the electron bunch and decelerate it!

Consider a plasma wave with a velocity

$$v_p \simeq \left(1 - \frac{1}{2\gamma_p^2}\right) c \tag{46}$$

where γ_p is much less than the γ of the accelerated particle. Then the velocity difference between wave and particle is just

$$\frac{\Delta v}{c} = \frac{1}{2\gamma_p^2} = \frac{1}{2}\left(\frac{\omega_p}{\omega}\right)^2 \tag{47}$$

which yields a phase slippage of

$$\delta = \left(\frac{\omega_p}{\omega}\right)^2 \frac{Lk_p}{2} \tag{48}$$

where L is the length of the accelerator or acceleration stage. Solving for the plasma frequency we find

$$\omega_p = \left(\frac{2\delta c\ \omega^2}{L}\right)^{1/3} \tag{49}$$

The plasma frequency is at our disposal; however, we don't want the phase to slip too much during acceleration. Is there an optimal phase shift δ which maximizes the acceleration for a given length L? The acceleration gradient is given by

$$e\,E_z = \alpha\, m\, c\, \omega_p \qquad (\beta_p \simeq 1); \tag{50}$$

however, this must be modified by averaging over the phase slip. The best we can do is therefore

$$\begin{aligned} e\,E_z &= \alpha\, m\, c\, \omega_p \frac{1}{\delta}\int_{-\delta/2}^{\delta/2} \cos\Theta\, d\Theta \\ &= \alpha\, m\, c\, \omega_p \frac{2\sin(\delta/2)}{\delta} \end{aligned} \tag{51}$$

Then using Eq. (49) we find

$$e E_z \propto \frac{\sin \delta/2}{\delta^{2/3}} \quad . \tag{52}$$

Due to Eq. (52) there is an optimum phase shift given approximately by

$$\delta \simeq \frac{5\pi}{8} \tag{53}$$

which, in turn, determines the plasma frequency for a given length L.

$$\omega_p = \left(\frac{5\pi}{4} \frac{c\,\omega^2}{L}\right)^{1/3} \tag{54}$$

and the effective accelerating electric field:

$$e E_z = \alpha\, m\, c\, \omega_p \ (.85) \tag{55}$$

It is also possible to optimize the final particle energy while holding ω_p fixed. The final energy is given by the product eE_z in Eq. (51) and L from Eq. (48). The optimum choice is then found to be $\delta = \pi$ and the accelerator length is chosen according to Eq. (48). The point here is that the phase slippage consideration, although necessary, has a rather broad optimization range.

b) Scaling Laws

From the previous section we know the plasma frequency. What is the final energy for an acceleration stage of length L? This is simply given by

$$E_{max} = e\, E_z\, L = \alpha\, m\, c\, \omega_p\, L\ (.85)\ . \tag{56}$$

Solving Eq. (54) for the length L and using Eq. (56) we find

$$\gamma_m = E_{max}/mc^2 = \alpha\, \frac{5\pi}{4} \left(\frac{\omega}{\omega_p}\right)^2 (.85)\ . \tag{57}$$

In order for Eq. (56) and Eq. (57) to be valid, the intensity of the laser must be about constant over the length of an acceleration stage. With this in mind we set the Rayleigh length,

$$R = L/2 \tag{58}$$

which yields a spot size

$$\sigma^2 = R\lambda/\pi$$

$$= L\lambda/2\pi \tag{59}$$

$$= \frac{5\pi\ c^2\omega}{4\ \omega_p^3}$$

where, again, Eq. (54) has been used to eliminate L. So using Eq. (45) we find the laser energy to be

$$W\tau = \frac{5\pi\alpha}{8} \frac{m^2c^5}{e^2\omega_p}\left(\frac{\omega}{\omega_p}\right)^3 \tag{60}$$

We have ignored the depletion of the laser pulse in the model used to calculate the laser/plasma interaction. To see if this is consistent let's calculate the total energy deposited in the plasma. This is simply given by the energy density times the volume, or

$$P.\ E. = \frac{E_z^2}{8\pi} \frac{\pi\sigma^2}{2} L\ . \tag{61}$$

Substituting from Eq. 's (50), (54) and (59) we find

$$P.\ E. = \frac{\alpha^2 25\pi^2}{256} \frac{m^2c^5}{e^2\omega_p}\left(\frac{\omega}{\omega_p}\right)^3 \tag{62}$$

Therefore, using Eq. (60), we find the fraction of energy deposited by the laser

$$\frac{P.E.}{W\tau} \simeq \frac{\alpha}{2}\ . \tag{63}$$

Thus for a somewhat small α, the depletion effects are moderate.

Lastly, we would like to calculate the number of particles, N, which can be accelerated. Conservation of energy yields

$$N\ E_{max} = \varepsilon\ P.E.$$

where E_{max}, P.E. are evaluated for one stage of the accelerator and ε is an efficiency factor.

Then using Eq.'s (57) and (62), we find

$$N = \frac{5\ \varepsilon\ \alpha}{(.85)\ 128} \frac{\lambda_p}{r_e}\left(\frac{\omega}{\omega_p}\right),\quad \left(r_e = \frac{e^2}{mc^2}\right)\ . \tag{65}$$

It is interesting to note that as we decrease the length of an acceleration stage, ω_p increases. This means higher acceleration gradients but lower energy requirements on the laser pulse! (This also decreases the number of particles which can be accelerated). The decrease in the laser energy is due to two effects. The first is due to the resonant effect driving the plasma. From Eq. (43) the growth time

$$\tau \propto \frac{1}{E_o^2 \omega_p} \tag{66}$$

For larger ω_p, a smaller laser intensity yields the same growth time. The second effect is due to the spot size;

$$\sigma^2 \propto \frac{1}{\omega_p^3} \quad . \tag{67}$$

For a larger ω_p, we need much less power to yield the same intensity because the spot size decreases rapidly.

c) The Basic Design

In this section we will use the results of the previous sections to put forward a qualitative design for a high energy plasma/laser accelerator. In our view the plasma should be used as an acceleration element only and not as the source of the beam; thus, an injected beam is necessary. In addition the preceding sections suggest that a higher laser frequency is better for high energy applications. With this in mind we fix the laser to be

$$\begin{gathered} \text{Nd: Glass Laser} \\ \lambda = 1.06\ \mu\text{m} \\ \omega = 1.78 \times 10^{15}\ /\text{sec} \end{gathered} \tag{68}$$

The relations derived in the preceding sections are very restrictive; if we now specify the length and number of acceleration sections, everything else is determined. So consider a possible e^- accelerator:

$$\begin{gathered} \text{1 km in length, 100 acceleration stages.} \\ \text{injection energy} > 10\ \text{GeV} \end{gathered} \tag{69}$$

Note that we assume that optimal phasing can be achieved at each stage.

The next step is to restrict α. In our case we would like to avoid trapping since this would contaminate the accelerated phase space. For a cold plasma this is accomplished if

$$\alpha \leq \frac{1}{2}\left(1 - \frac{1}{\gamma_p}\right) \tag{70}$$

or since we anticipate that $\gamma_p >> 1$ we adopt

$$\alpha = \frac{1}{2} \quad . \tag{71}$$

The optimal plasma frequency can be determined by Eq. (54):

$$\omega_p = 7.20 \times 10^{12}/\text{sec}, \ \lambda_p = .26 \text{ mm} \quad . \tag{72}$$

This is turn sets the density of our plasma to be

$$n_0 = 1.6 \times 10^{16}/\text{cm}^3 \quad . \tag{73}$$

To drive the plasma wave above we must use the beating between 2 frequencies as suggested in Ref. 1. It is this resonance effect which forces the wave to grow throughout the entire coherence length, τ. It is interesting to note at this point that a SLAC size bunch is only a bit too long to fit into one half of a plasma wavelength.

The accelerating field in each case can now be calculated from Eq. (55):

$$e\,E_z = 5.1 \text{ GeV/m} \tag{74}$$

which yields a final energy for a 1 km accelerator of

$$E_F = 5.1 \text{ TeV} \quad . \tag{75}$$

Now we come to the question of spot size. Applying Eq. (59) we find

$$\sigma = .13 \text{ cm} \quad . \tag{76}$$

Note that this spot size is much larger than the optical wavelength due to the long Rayleigh length necessary for the accelerator, and is also 5 times larger than the plasma wavelength. This is important for reducing transverse effects.

Before addressing pulse length and power lets look at the total laser energy necessary to drive the plasma wave. From Eq. (60) we find

$$W\tau = 1.7 \times 10^4 \text{ joules} \times 100 \text{ stages} \tag{77}$$

This energy is quite large. Therefore, it is important to let τ be as large as possible. On the other hand there will be frequency errors in the system. If we assume somewhat arbitrarily and optimistically that the frequency errors are held to 0.1%, then we can select

$$\begin{aligned} \tau &= 1000/\omega_p \\ \tau &= .14 \text{ ns} \end{aligned} \tag{78}$$

and we find

$$W = 1.2 \times 10^{14} \text{ watts} \times 100 \text{ stages} \tag{79}$$

Of course there is a trade off between W and τ. If the frequency errors are larger, the instantaneous power necessary will be larger.

Lastly, let's calculate the number of particles which can be accelerated in each case. Using Eq. (65) we find

$$N = \varepsilon \times 5.3 \times 10^{11} \tag{80}$$

The efficiency ε is hard to estimate; however, since the beam size must be somewhat less than the plasma wave size ε would probably be less than 10%.

Given the above designs the accelerator works as follows: The laser pulse is injected into the plasma with the electron bunch at its tail at the proper phase. In practice one might have to inject a long bunch and only extract the accelerated part. The bunch is simply accelerated through one stage and reinjected into the next stage.

This exercise in designing a high energy plasma wave accelerator has been an attempt to discover a "reasonable" accelerator without stretching present technology too far. The numbers seem to fall into place somewhat naturally. On the other hand there are many questions to be answered before such a scheme could be realistically contemplated.[4]

The most important of these is the excitation of plasma waves by EM waves. Of course, much work has been done on this subject;[5] however, the emphasis in past work has been mostly on the heating of already hot plasmas. We would like the wave to persist rather than "thermalize", so past results need to be reapplied to the case at hand. To make the connection with present capabilities, in the Appendix we qualitatively design an experiment using existing facilities.

ACKNOWLEDGEMENTS

We would like to thank Frank Cole, Phil Morton, A. Ruggiero and Lloyd Smith for helpful discussions during and after the workshop. We would especially like to thank Andy Sessler for his guidance in the media group and for helpful suggestions and also Paul Channel and the staff at Los Alamos for their hospitality during the workshop.

REFERENCES

1. T. Tajima and J.M. Dawson, Phys. Rev. Lett., 43, No. 4, 267 (1979).

2. D.J. Sullivan and B.B. Godfrey, IEEE Trans. Nucl. Sci. 28, 3395 (1981).

3. E. Lee, private communication, June 1982.

4. See the "Report of the Working Group on Media Accelerators", Andrew M. Sessler, this conference.

5. See for example, Physics of Laser Driven Plasmas, Heinrich Hora, John Wiley and Son Inc., 1981, and B.I. Cohen, A.V. Kaufman, and K.M. Watson, Phys. Rev. Lett. 29, 581(1972).

APPENDIX

It is instructive to design an experiment using an existing laser. We select again a Nd: glass laser with a power level

$$W = 1.2 \times 10^{13} \text{ watts} \quad \text{(Rochester)}$$

$$\omega = 1.78 \times 10^{15}/\text{sec}$$

$$\lambda = 1.06 \ \mu\text{m}$$

We choose a plasma density somewhat higher than that suggested in the text:

$$n_o = 3.7 \times 10^{16} \text{ cm}^{-3}$$

$$\omega_p = 1.10 \times 10^{13}/\text{sec}$$

$$\lambda_p = 0.17 \text{ mm} \quad .$$

However, we will not address the question of how to obtain two laser line such that $\omega_1 - \omega_2 = \omega_p$.

The accelerator length is chosen to be

$$L = 2.8 \text{ m}$$

and the spot size, from Eq. (59), is

$$\sigma = .68 \text{ mm} ,$$

which is about 4 times the plasma wavelength. So in this case we would expect to see some transverse effects.

If we somewhat arbitrarily assume the frequency errors in the system are about 1%, then coherence between the plasma wave and the beating laser light can be maintained for

$$\tau = 100/\omega_p$$

$$\tau = 9.1 \text{ ps}$$

The energy is this length of the laser pulse is

$$W\tau = 110 \text{ J};$$

then from Eq. (60) we find

$$\alpha = 0.017 .$$

This relatively small value of α means that the wave will be quite linear. The effective acceleration gradient is obtained from Eq. (55)

$$eE_z = 0.27 \text{ GeV/m},$$

which yields a final energy gain due to acceleration of

$$eE_zL = 0.76 \text{ GeV} .$$

THE LASER FOCUS ACCELERATOR*

by H. Hora, D.A. Jones, E.L. Kane** and B. Luther-Davies***.
Department of Theoretical Physics, The University of New South Wales,
Kensington - Sydney, Australia.

ABSTRACT

The well-known phenomenon of generating very energetic ions (MeV and more) from laser produced plasmas is analysed for the use of alternative accelerators. In laser-plasmas an ion acceleration is possible to MeV energies by the electrostatic double layer of the surface of a plasma of keV temperature, however the number of ions in the Debye sheath is small and cannot account for the observed 10^3 times more accelerated ions than in the Debye sheath. Another mechanism is the nonlinear force acceleration. This resulted in a transfer of 50% of the laser energy at uniform irradiation of a spherical pellet into 60 keV ions. The nonlinear force and the relativistic properties of the optical constants cause two types of self-focussing where the high energy densities in the focus can lead to MeV ion emission of large quantities in full agreement with experiments. In order to extend these models by a sophisticated numerical treatment by 1 to 2 orders of magnitude, rather safe predictions were possible of how to produce 10^{10} eV heavy high Z ions with 100 MeV/nucleon. Bursts of 10^3 Ampere and 10^{-11} sec duration will provide special new technique in nuclear and high energy physics. Combination with the nonlinear force driven double layer method should result in less dense TeV ions with 10^{10} to 10^{11} eV/nucleon.

INTRODUCTION

There is a need to look for alternative accelerators if the next generation of electron beams beyond 300 GeV is considered. As Willis[1] pointed out, one has either to look for classical accelerators of the dimension of hundreds of kilometers or one would have to look for other accelerator concepts, of which the use of lasers is of interest. Most of the schemes there were extensions of the accelerator work using microwaves[2], combining gratings (Smith-Purcell effect)[3] with standing wave[4], surface wave[5], inverse Cerenkov effect[6], or by the inversion of the wiggler-field-electron laser[7,8].

Including laser interaction with plasmas, a very potential method for an electron accelerator is the coupling of two-laser frequencies whose difference fits the plasma frequency[9]. Coupling of electrons and ions is involved in the concept of Willis for ion acceleration[10], or the cooling of ions is achieved[11], or use is made of a sequence of foci[12].

Differing from most of the above mentioned very sophisticated combinations of phenomena, an immediate acceleration of electrons and ions to more than 10 MeV has been observed in laser produced plasmas of high density, mostly due to self-focussing processes involved automatically in this interaction. Though the processes in this interaction are very complicated, an analysis has led to an extension of the experimental results by one to two orders of magnitude to the generation of GeV heavy ions. Links exist to the above mentioned acceleration concepts, especially from the theories of the generation of high laser intensities in a focus or to the nonlinear forces (ponderomotive forces) driving the electrons directly by the laser field without thermalization or gasdynamics.

In the following, a short review will be given of the basic mechanism involved in laser-plasma interaction, starting with the gasdynamic and electrostatic surface-type interactions before the nonlinear forces are explained. Then the analytical models of self-focussing will be described which have enabled some general conclusions to be made on the expected high ion energies. The occurrence of a counteracting mechanism then required the establishment of a very general two-dimensional computer code for the propagation of a laser beam in a plasma and for the resulting time dependence of accelera-

* Work supported in part by ARGC Grant 75/15538, NERDDC Grant 79/9433, and U.S. Dept. of Energy Grant DE-A508-81 DP-40139 (cooperation with Prof. G.H. Miley, Univ. Illinois).

** from Applied Science Inc., Alexandria, Va.

*** from Australian National Univ., Canberra.

tion of plasma and the achieved ion energies.

This model is limited to macroscopic magneto-hydrodynamics. Limitations are then given by the Debye length, while thermodynamic equilibrium is of minor importance as the dynamics is determined by the nonlinear electrodynamic forces. The code covers the relativistic effects of optics, e.g. with respect to the quiver motion of the electrons in the very high intensity laser field. The macroscopic net motion of the plasma is, however, limited to subrelativistic motion only. The numerical code is close to the limit of computer capacities. This does not mean that we have well planned strategies to improve the numerics with respect to relativistic dynamics or a semi-two-fluid inclusion of space charges, but we are aware that the interpenetration of particle jets is the clear limitation of the technique. Any chance to arrive at a kinetic (distribution function type) or single particle (simulation-type) description for the necessary two spatial dimensions and beam interaction dynamics might be considered as impossible at the moment. The inverse situation is well-known from the Tajima-Dawson frequency mixing accelerator: in this case[9], the simulation type computation exhausted the computer capacity, though the dynamics were presented in one spatial dimension only and no answer could be given from the codes as to how the radial (lateral) properties of the beam-plasma interaction evolved[13]. For our case of the laser focus accelerator, however, an extension of the experimental values by two orders of magnitude are necessary only, which may explain in a rather safe way some of the conclusions we may be able to draw in our case.

2. THERMOKINETIC LASER INTERACTION AND THERMAL ELECTROSTATIC DOUBLE LAYERS

If plane electromagnetic waves (laser) are incident perpendicularly on a solid or the then created plane plasma, the interaction can be described by classical plasma optics and the purely thermokinetically determined gasdynamics, if the laser intensities are below a certain threshold I^* above which nonlinear and anomalous processes will start and will then dominate over the thermokinetic effects. This nonlinear threshold has been determined approximately[14] by the laser wave length λ

$$I^* \cong 10^{14}/\lambda^2 \text{ W/cm}^2 \ [\lambda \text{ in } \mu\text{m}] \qquad [1]$$

where a very weak dependence on the plasma temperature and a modification by absorption has been neglected[15]. Below this threshold, anomalies and nonlinearities may happen if the laser pulse duration is too short for achieving thermal equilibrium of the energy deposited in the plasma, or if a deviation of the assumed plane geometry permits a self-focussing of the laser beam in the plasma. In Chapter 4 we shall see that this can happen with laser beams whose power exceeds 1 MW only. This was observed indeed in the early stages of laser-plasma experiments: below one megawatt pulses, an emission of ions[16] of a few eV energy occurred (corresponding to a few ten thousand degrees of plasma temperature), while laser pulses a little above one megawatt power produced ions of 10 keV energy[17]. A similar jump of electron emission from the classical space charge limited current densities of 100 mA/cm^2 below 1 MW laser pulses to 100 A/cm^2 above 1 MW was observed[18]. Our explanation[15] is that the self-focussing in plasmas appearing at 1 MW and above, results in laser intensities $I > I^*$ within the laser filament and produces nonlinear particle acceleration.

Nowadays, a self-focussing-free plane wave or spherical wave interaction with plane targets or spherical pellets respectively can be realized experimentally quite well. The purely thermokinetic behaviour has been confirmed then for $I < I^*$. In this way, thin layers of dense material have been accelerated up to velocities of 2×10^7 cm/sec (Mach 600, the fastest laboratory velocity of compact material ever achieved[19]). The first numerical analysis of the thermokinetic behaviour of laser-plasma interaction was published by Mulser[20] and Rehm[21]. A strong ablation of hot (plane) plasma occurs while a compression of plasma below the plasma corona of light interaction is driven by the net dynamics. Even at higher density than the critical density of the corona, a plane-front compression by a factor four was found numerically. The compressed material is finally a thick block (essentially different from the usual shock wave theory). For more details see section 10.1 of Reference[15]. We shall not consider these cases for laser focus acceleration, because the thermokinetic particle energies are rather below 10 keV. It should be mentioned that one concept of laser-pellet compression for fusion restricts

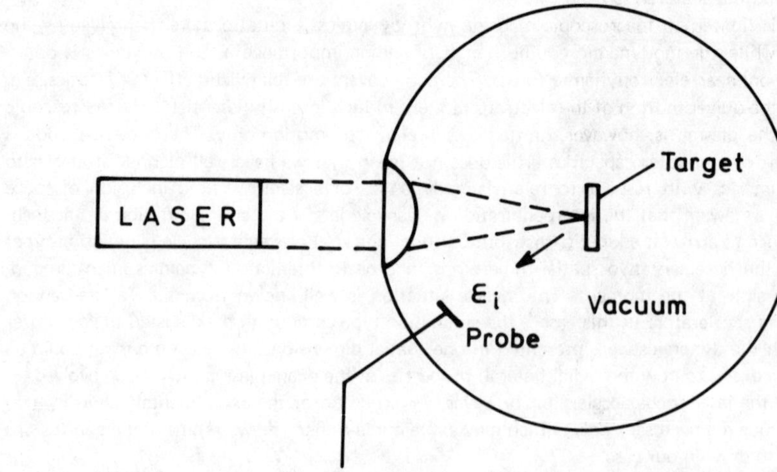

Fig. 1 Typical experiment of laser irradiation of a solid target in a vacuum. A probe is connected with an oscilloscope whose trace in time results in a first peak due to photoelectrons from the probe emitted by UV radiation of the target. After this time-zero-peak, no signal is seen until the first ions arrive. Time of flight results in ion spectra which at high intensities show different peaks for ions of different charge Z, see Fig. 1.9 of Ref.[15] following A.W. Ehler, J. Appl. Phys. **46**, 2464 (1975).

the laser intensities below I* (10^{14} W/cm^2 for Nd glass or 2×10^{15} W/cm^2 for UV laser pulses) to avoid any anomalies or nonlinearities[22]. After a lot of confusing observations at higher laser intensities, this concept is the lowest common denominator for a fusion strategy. This concept is prohibitive for CO_2 lasers. An alternative to the thermokinetic compression is to use the nonlinearities for $I >> I^*$ to directly compress the plasma[15], as optical energy can be converted directly (isentropically without heat loss) into mechanical motion of thick blocks of compressed plasma[23]. This can be used for CO_2 laser compression or for high intensity neodymium glass laser compression[24].

Being aware of the limited possibilities of electron or ion acceleration from plasmas at laser irradiation at moderate intensities $I < I^*$ for thermokinetic conditions, nevertheless an electrostatic mechanism will be present at the plasma periphery to accelerate ions to high energies, however relatively small numbers of ions only. What had to be neglected completely in the above mentioned macroscopic thermokinetic magnetohydrodynamics, was the electrostatic double layer mechanism at the periphery of the laser produced plasma when expanding into the vacuum.

When Linlor[17] measured the surprisingly high ion energies of several keV from targets irradiated by a few MW power, speculations were made as to whether these ion energies were due to electrostatic effects. It is obvious that the expanding plasma produced at the target surface, has a temperature T and ions and electrons have nearly the same energy ε_i and ε_e equal to the thermal energy ε_{th}. Due to the different masses m_i and m_e of the electrons, their velocities v_e and v_i are very different.

$$\frac{m_i v_i^2}{2} = \frac{m_e v_e^2}{2} = \varepsilon_{th} \qquad [2]$$

At the plasma periphery the faster electrons will leave the plasma and generate a negative space charge area far out while positive charges will remain in the plasma surface (Fig. 2). Fast electrons from the plasma interior will be reflected by the generated fields within the positive ions to provide a usual space charge neutral internal plasma which will be assumed as homogeneous. These considerations were formulated in the earlier stages of laser-plasma interaction theory[25] and used as a transparent way to derive the Debye length[26]. Related to a cross section of 1 cm^2 of Fig. 2, the generated maximum

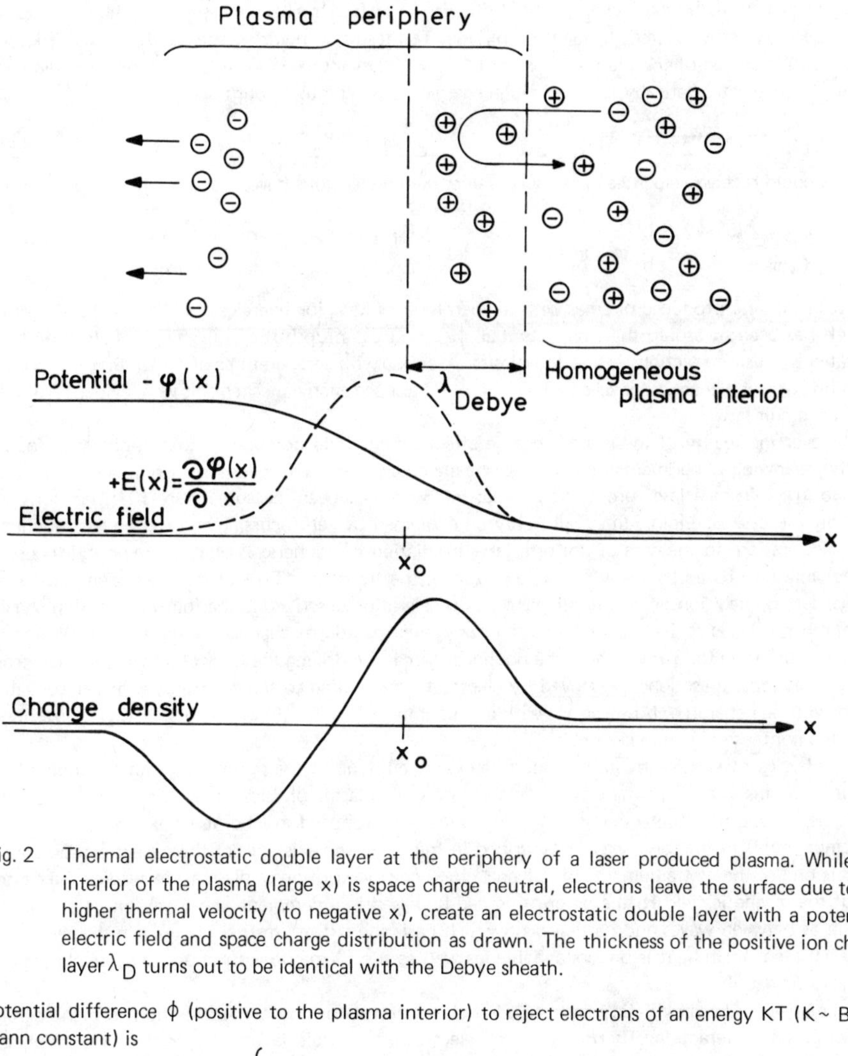

Fig. 2 Thermal electrostatic double layer at the periphery of a laser produced plasma. While the interior of the plasma (large x) is space charge neutral, electrons leave the surface due to the higher thermal velocity (to negative x), create an electrostatic double layer with a potential, electric field and space charge distribution as drawn. The thickness of the positive ion charge layer λ_D turns out to be identical with the Debye sheath.

potential difference ϕ (positive to the plasma interior) to reject electrons of an energy KT (K~ Boltzmann constant) is

$$KT = e\phi = e\int |\mathbf{E}|\, dx \cong e|\mathbf{E}|\,\lambda_D \qquad [3]$$

where the electric field **E** due to ϕ was used. The positive charges within a thickness λ_D of the surface with an ion density n_i, ion charge Z and an electron density n_e in the plasma interior (assumed as $Zn_i = n_e$), will produce the electric field given by

$$4\pi\lambda_D\, n_e e = \oint \mathbf{E}^2 \cdot d^2\mathbf{a} = |\mathbf{E}| \qquad [4]$$

where for the last step we have **E** = 0 at the plasma interior, **E** is perpendicular to the surface normal at the cylinder surface parallel to x and **E** only counts at the outside 1 cm^2 area at x_0. Elimination of $|\mathbf{E}|$ from Eqs. (3) and (4) results in

$$\lambda_D = (KT/4\pi e\, n_e)^{\frac{1}{2}} \qquad [5]$$

which is the Debye length.

The existence of the space charge area at the plasma surface has been measured by the deflection of an ion beam which was fired through the plasma. The resulting fields[27] were in the range of kV/cm and a positive charge of the plasma interior of 10 kV was measured[28]. The ions could be calculated to be speeded up electrostatically[29] to very high energies. Taking into account

$$\varepsilon_{th} = \frac{M_e V_e^2}{2} = \frac{M_i V_i^2}{2} \quad ; \quad v_e^2 = v_i^2 \, m_i/m_e \, , \qquad [6]$$

the ions could receive velocities up to such values as the electrons have, reaching their maximum ion energies

$$\varepsilon_{i,max} = \varepsilon_{th} \, m_i/m_e = \frac{3}{2} \, kT \left(\frac{m_i}{m_e} \right) \qquad [7]$$

Indeed, from laser produced plasmas with temperatures of keV, ion energies of MeV have been measured[28]. It has been measured that the potential ϕ produces an upshift of the energy of alpha particles generated by fusion reactions inside the plasma. Plasmas with temperatures of some 100 eV produced an upshift of alpha energies by several 100 keV[30]. The space charge mechanisms were evident from the Child-Langmuir law[31].

Whatever the reality of these space charge effects, they could not explain quantitatively Linlor's[17] and the following experiments: the Debye sheath could explain fast ions of a number of less than 10^{10} to 10^{11} from a laser produced plasma, but what has been measured were 10^{15} fast ions and more. In the case of Linlor, this could only be explained by self-focussing and subsequent nonlinear force acceleration. In the case of uniform laser irradiation of a spherical pellet where no self-focussing was possible (as proved by X-ray pin hole pictures), the transfer of 50% of the laser energy into non-thermal fast 60 keV ions[32] is a nonlinear force acceleration described in the following section. Again, this number of ions is 10^4 times more than any electrostatic mechanism would permit. We should mention that in the following, where the nonlinear forces are driving the thick blocks of dense electrons between the ions, these ions are moved by electrostatic coupling to the electrons, and surface Debye sheaths will be generated, however not with a temperature T in Eq. (5) but with an energy corresponding to the nonthermal nonlinear forces.

The physics of the electrostatic double layers as built up at the periphery of an expanding laser produced plasma is an important phenomenon in cosmic plasma physics. It was not before 1976 that these phenomena were detected[33] and systematically reproduced in laboratory experiments (Torven and Lindberg[33]) in the same way as described in Fig. 2[26]. The existence of these electrostatic double layers is one of the new arguments of Alfven[34] for a new description of plasmas: instead of the central role of the magnetic field **H**, the description can be based in a complementary way on j (as a similar dualism as between waves and particles) on which basis the extraterrestrial plasma and the generation of the striated north-light is understandable, as well as the filamentary structure of solar plasma and inter galactic nebulae.

Summarising: thermokinetic processes are not important for particle acceleration to high energies at laser-plasma interaction. Thermally driven electrostatic double layers can produce very high ion energies but only for a much smaller number of ions than observed in laser-plasmas if $I \gg I^*$.

NONLINEAR FORCES FOR ACCELERATION TO HIGH ENERGIES

At laser interaction with plasmas at sufficiently high laser intensities I, several anomalous and nonlinear phenomena will occur. Significant processes are the resonance absorption and parametric instabilities which will be discussed below. As the laser plasma interaction theory was the tool to derive the general formulation of the nonlinear electrodynamic forces in plasma, these steps should be performed first. These forces - frequently called ponderomotive forces in an imprecise way - are of essential importance in many of the laser acceleration schemes mentioned before and which are not discussed in this contribution of laser-plasma interaction.

In a plasma, the complex dielectric constant or the complex refractive index $\tilde{n}$ is given from the dispersion relation of the Maxwellian equation as for a monochromatic electromagnetic wave of frequency ω

$$\tilde{n}^2 = 1 - \frac{\omega_p^{\,2}}{\omega^2 + \nu^2} (1 + i\,\nu/\omega) \qquad [8]$$

where the plasma frequency ω_p is given by

$$\omega_p^{\,2} = \frac{4\pi\, e^2 n_e}{m} \qquad [9]$$

using the mass m, the charge e and the density n_e of the electrons, and where ν is the Coulomb collision frequency of the electrons in the plasma[15].

$$\nu = \frac{Z n_e \pi e^4}{3^{3/2}\, m^{1/2}\, (KT)^{3/2}} \qquad [9a]$$

The use of this classical direct current collision frequency for the high frequency case is justified only by the quantum mechanical inverse bremsstrahlung theory if the stimulated emission is included. Without stimulated emission, a $T^{-1/2}$ - slope is the quantum electrodynamical result similar to the $T^{-1/2}$-slope of the collision described as de Broglie diffraction[36].

Nonlinearities of the optical constant (8) appear[15], when $I > I^*$. In this case one has to substitute

$$KT = KT_{th} + \varepsilon_{osc} \qquad [9b]$$

where ε_{osc} is the quiver energy of the electron in the laser field

$$\varepsilon_{osc} = m_o c^2 \,[\, (1 + A(I)\, I/I_r)^{1/2} - 1 \,] \qquad [10]$$

$$= \begin{cases} \dfrac{e^2}{2m}\dfrac{|E|^2}{\omega^2}, & \text{if } I \lesssim I_r \\[2ex] \dfrac{ec}{\omega}|E|, & \text{if } I \gtrsim I_r \end{cases}$$

where the relativistic threshold intensity I_r is that where $\varepsilon_{osc} = m_o c^2$ (m_o ~ rest mass of the electron)

$$I_r = \frac{3 m_o^{\,2} \omega^2 c^3}{8\pi e^2} = 3.6 \times 10^{18}/\lambda^2 \text{ W/cm}^2 \qquad [10a]$$

where the laser wave length λ is given in micrometers. A(I) is a correction factor between 1 (at $I \lesssim I_r$) and 1.06 (for $I >> I_r$) if linear polarized light is used[37] or is always unity for circularly polarized light. Any relativistic modification of Eq. (8) is not only given by T (Eqs. 9b to 10a) but also by the relativistic change of the electron mass m in Eqs. (8) to (10a)[15]

$$m = m_o [1 + 3A(I)\; I/I_r]^{1/2} \qquad [10b]$$

Using the high frequency dispersion relation Eq. (8) in the ponderomotive force density derived by Kelvin (Thomson) 1845[35] for static fields in materials for the time averaged high frequency fields,

$$\mathbf{f}_{NL} = \left(\frac{1-\tilde{n}^2}{8\pi}\right) \{\, \mathbf{E}\cdot\underline{\nabla}\,\mathbf{E} - \mathbf{E} \times (\underline{\nabla} \times \mathbf{E}) \,\}$$

and by using a vector identity for the bracket containing the fields, we arrive at

$$\mathbf{f}_{NL} = \frac{(\tilde{n}^2 - 1)}{16\pi}\, \underline{\nabla}\, |E|^2 \qquad [11]$$

This expression was used in many of the contributions of these proceedings as the force acting on the electrons by the laser field. As described before[15], this force has been called the field gradient force, electrostrictive force (of laser interaction), and is identical with the Lorentz force

$$\mathbf{f}_{NL} = \frac{1}{c}\,\mathbf{j} \times \mathbf{H} \qquad [12]$$

and called the j-cross-H force. The confusion with the expression ponderomotive force can be seen from the fact that Russian authors call all forces ponderomotive forces which all others call Lorentz-force[37]. The term nonlinear force of $\mathbf{f}_{NL}$ is now common in the laser-plasma literature and textbooks [15, 38]. The special derivation of Eq. (11) for high frequency fields in plasmas was given by Weibel, and Gapunov and Miller[40] and is sometimes called the "Miller force". Our first application in laser plasmas[41] arrived at the same formula (11) of 1845 as used before, but taking into account the WKB approximation $\mathbf{E} = \mathbf{E}_v/\tilde{n}^{1/2}$ with the vacuum amplitude $\mathbf{E}_v$

$$\mathbf{f}_{NL} = \frac{(\tilde{n}^2-1)}{16\pi}\,E_v^{\,2}\underline{\nabla}\frac{1}{\tilde{n}} = \frac{1}{8\pi}\,\underline{\nabla}\,(E^2+H^2) \qquad [13]$$

immediately showed that an explosion of the plasma was due to the spatial variation of the refractive index.

The difficulties with the nonlinear force at obliquely incident radiation on a plasma without collisions were due to the fact that no forces parallel to the surface should appear as no recoil is given. This criterion of shear-free motion led to the most general expression of the nonlinear force[42].

$$\mathbf{f}_{NL} = \frac{1}{c}\,\mathbf{j}\times\mathbf{H} + \frac{1}{4\pi}\mathbf{E}\,\underline{\nabla}\cdot\mathbf{E} + \frac{1}{4\pi}\,\underline{\nabla}\cdot(\tilde{n}^2-1)\,\mathbf{EE} \qquad [14]$$

which is algebraically identical with

$$\mathbf{f}_{NL} = \nabla\cdot\left(\underline{\underline{T}} + \frac{(\tilde{n}^2-1)}{4\pi}\,\mathbf{EE}\right) - \frac{\partial}{\partial t}\,\frac{\mathbf{E}\times\mathbf{H}}{4\pi c} \qquad [15]$$

using the Maxwellian stress tensor

$$\underline{\underline{T}} = (\mathbf{EE} + \mathbf{HH} + (E^2+H^2)\,\underline{\underline{I}}\,/2\pi)/4\pi \qquad [16]$$

Eq. (14) is an extension of Schluter's two fluid equation[43] by two nonlinear terms (two of the three terms given by differentiation of the last expression in Eq. (14)). The direct derivation (Appendix 3 of Ref.[15]) from the Euler equations of the electron and ion fluids indicated that a special interpretation of the dielectric displacement $\mathbf{D} = \tilde{n}^2\mathbf{E}$ is required. Eq. (15) is identical with a formulation of Landau and Lifshitz[44] which however was not valid for a plasma with dispersion. Our treatment proved the validity of Eqs. (14) and (15) for dispersive and dissipative plasmas[42] where ponderomotive and (collision determined) non-ponderomotive terms could be distinguished[45]. The completeness was proved from the criterion of shear-free motion.

The predominance of the nonlinear force for intensities above an intensity I* where the gas kinetic force due to the pressure $p = (n_i + n_e)\,KT$

$$\mathbf{f} = -\underline{\nabla}p = \mathbf{f}_{NL}(I = I^*) \qquad [17]$$

was discussed before[15,42] and led to Eq. (1). Eqs. (14) and (15) reduce to Kelvin's formula of 1845, Eq. (11) for the very special case of a collisionless plasma and a plane electromagnetic wave perpendicularly incident on a plane inhomogeneous plasma.

The force density in a plasma for $I > I^*$

$$\mathbf{f} = -\underline{\nabla}p + \mathbf{f}_{NL} \qquad [18]$$

is mainly determined by the spatial behaviour of $\tilde{n}$. As described in ref.15 for the special case of perpendicular incidence, the refractive index n in a plasma is always less than unity, therefore $\mathbf{E}^2$ or $\mathbf{E}^2 + \mathbf{H}^2$ swells by a factor 10 or 100 over its vacuum value (experimentally: 700, see Wong and Stenzel[15]) and the gradients drive the plasma corona to the vacuum (ablation), while the other side moves as a thick block of plasma to the interior of the plasma in a kind of a dielectric explosion. This swelling $S = 1/\tilde{n}$ defines the increase of the usual radiation pressure for driving the plasma. Between the ablative and compressive blocks of plasmas, a density minimum (caviton) with a steepening of the density profile will occur, as first derived numerically by Shearer, Kidder and Zink[46] and reproduced later numerically by several treatments[15], even by single particle simulation[47]. The observation of the density minimum was the best proof of the action of the nonlinear force (see Chapter 10 of Ref.[15]).

Under stationary subrelativistic $I < I_r$ conditions, the equation of motion can be integrated (Chapter 9, Ref.[15]) showing that the electrons receive an energy of translation ε_e which is given by the difference of the quiver energy at the maximum swelling and the vacuum (index v) quiver energy $\varepsilon_{osc,v}$

$$\varepsilon_e = \frac{1}{2}(\varepsilon_{osc,max} - \varepsilon_{osc,v}) = \frac{\varepsilon_{osc,v}}{2}\left(\frac{1}{|\tilde{n}|} - 1\right) \qquad [19]$$

where[15] (in cgs units) from Eq. (10)

$$\varepsilon_{osc,v} = \frac{I_v}{cn_{ec}} = \frac{e^2}{2m}\frac{E_v^2}{\omega^2} \qquad [20]$$

using the cut-off density which is $n_e = n_{ec}$ where $\omega_p = \omega$ in Eq. (9). The block of electrons is being moved over a single Debye length, but is then electrostatically coupled to the ions whose inertia determines the motion. The ion energy after this acceleration is then

$$\varepsilon_i = Z\,\varepsilon_e = \frac{Z\,\varepsilon_{osc,v}}{2}\left(\frac{1}{|\tilde{n}|} - 1\right) \qquad [21]$$

For relativistic intensities it was shown[48], that Eq. (21) is valid using a value $\varepsilon_{osc,v}$ as if I is subrelativistic, as long as ε_i is not relativistic.

The consideration was simplified to plane wave interaction at perpendicular incidence. Apart from a simple block motion which we have proved in numerical detail to be used for highly efficient laser compression of plasma,[23] the generation of density ripples will occur by the partial, growing, standing light wave causing high reflectivity and parametric decay instabilities. These were globally described and categorized by F.F. Chen as caused by the nonlinear forces[15, 49]. These instabilities produce the backscattering of higher or half - or mixed harmonics to the wave of frequency ω as observed. The amount of energy transferred in this way is relatively small[50] and decreases at higher intensities[51]. Another complication occurs at oblique incident. Any standing wave structure produces a striated motion[52] and dynamic dissipation. For p-polarization, a local maximum of the E-component in the evanescent wave field perpendicular to the surface is built up which causes Denisov's resonance absorption as derived from the magnetic fields[53]. The consequent derivation from the E-fields for a collisionless plasma was achieved by White and Chen[53] and for the case with collisions[54] a jump of a pole from $-\infty$ to $+\infty$ was derived. The properties for a most general density profile led to the Denisov length of the critical layer[54]. The dynamics could be explained by quiver drift[54] and a consequent description by the nonlinear forces arrived at a conical ion emission[55]. The phenomenon of resonance absorption decreases for higher laser intensities reaching relativistic quiver motion[56]. The complexity of all these phenomena should not be forgotten in the following discussions.

SELF-FOCUSING OF LASER BEAMS IN PLASMAS

While ideal plane (or spherical) wave fronts might be necessary for the application of lasers for nuclear fusion and any self-focusing would have to be avoided, many other applications, such as treatment of materials, make use of self-focusing. Especially at very high intensities the generation of MeV ions or the expected electron-positron pair production and similar effects are of interest.

There is an essential and basic problem we must mention at the beginning. This is the difficulty of an exact description of the laser beam. We shall see that several contradictions will arise if a solution for the complex nonlinear problems is not based on the fully exact description.

Historically, the first quantitative theory of self-focusing of laser radiation, resulting in a threshold for the laser power, was successful for dielectric materials (nonionized solids, liquids, and gases)[57].

The essential mechanism was the nonlinearity of the dielectric constant. For self-focusing in plasmas, there was no similar nonlinearity of the dielectric constant which could be used. The first successful way of deriving the self-focusing threshold was an application of nonlinear forces[58]. To arrive at a first qualitative threshold for the laser power it has to be mentioned that the first ideas on self-focusing of lasers in plasma was published by Askaryan[58]. He considered the energy momentum flux density of the laser beam $(\mathbf{E}^2 + \mathbf{H}^2)/8\pi$ by which the whole plasma has been expelled and where the pressure is then balanced by the plasma pressure profile acting against the center of the laser beam. The balance should be given in this way in the form of an equation

$$\frac{E^2 + H^2}{8\pi} = n_e K T_e \left(1 + \frac{1}{Z}\right) \qquad [22]$$

Askaryan was able to compare the necessary optical intensities to balance or compensate the gasdynamic pressure.

In order to derive the threshold for self-focusing of a laser beam in a plasma, three physical mechanisms (Fig. 3) have to be combined. Assuming that the laser beam has a Gaussian intensity profile along the y-axis while propagating in x-direction, the generated nonlinear force $\mathbf{f}_{NL}$ in the y-direction has to be compensated by the thermokinetic force $\mathbf{f}_{th}$[59]

$$f_{th} = f_{NL} \; ; \quad \nabla \cdot \left(\mathbf{T} - \frac{n^2 - 1}{4} \mathbf{EE}\right) = \nabla n_e KT(1 + Z) \qquad [23]$$

where use is made of Eq. (15). The second physical mechanism is the total reflection of the laser beam components starting under an angle α_o from the center of the beam and being bent into a parallel direction to the axis due to the density gradient of the plasma. The third condition is the diffraction

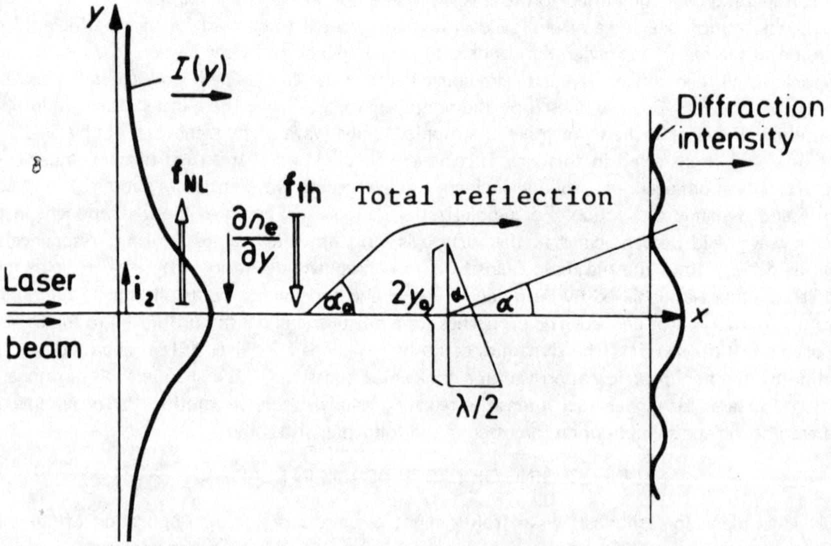

Fig. 3 Scheme of a laser beam in plasma of a lateral intensity decrease I(y), producing nonlinear forces $\mathbf{f}_{NL}$ in the plasma, rarifying the axial regions until being compensated by the thermokinetic force $\mathbf{f}_{th}$ of the gasdynamic pressure. The density gradiant causes a total reflection of partial beams. The diffraction condition for permitting partial beams of an angle of propagation less than total reflection for achieving the first diffraction minimum is the final condition for deriving the self-focusing threshold[59].

requirement that the main part (e.g., as defined by the first diffraction minimum) of the beam has to have an angle of propagation α, which is less than the angle of total reflection. These three conditions are sufficient to calculate the threshold.

A Gaussian density profile including the refractive index $\tilde{n}$ is described by the formula

$$\overline{E}_y^2 = \frac{E_v^2}{2|\tilde{n}|} \exp\left(-\frac{y^2}{y_o}\right); \quad \overline{H}_z^2 = |\tilde{n}|^2 \, \overline{E}_y^2 \qquad [24]$$

y_o can be interpreted as the radius of the laser beam. This is only an approximation of the exact Maxwellian formulation. In analogy to the suggestions of Askaryan[58] the nonlinear force in the direction of y is

$$f_{NL} = -\frac{1}{8\pi} \nabla(\overline{E}_y^2 + \overline{H}_z^2) \qquad [25]$$

It is important to note that this formulation is approximately valid only for conditions of strong swelling, that is, for refractive index differing from unity, which is the case for plasma exceeding the cutoff density in the region around the laser beam. Only by use of a Maxwellian exact beam with the necessary longitudinal components of **E** and **H**, can the radial force be calculated to arrive finally at the same formula[25] which was naively (and incorrectly, (see Eq. (15)) assumed[15]. The calculations[59] were repeated by several authors, on the basis of different assumptions, with always the same result achieved.

Using Eq. (24) in Eq. (25), the maximum nonlinear force in the y-direction is

$$\overline{f_{NL}} = i_y \frac{1 + \tilde{n}^2}{16\pi\tilde{n}} \frac{E_v^2}{y_o} \sqrt{2} \exp\left(-\frac{1}{2}\right) \qquad [26]$$

If this has to be compensated by a thermokinetic force under the assumption of a spatially constant plasma temperature (the general treatment for varying temperature was studied extensively by Sodha and co-workers[60]), the formulation is found

$$f_{th} = -i_y kT_{th}\left(1 + \frac{1}{Z}\right)\frac{dn_e}{dy} \qquad [27]$$

Equating this force and the nonlinear force of Eq. (12.5) provides an expression for the electron density gradient of the plasma at the laser beam.

$$\frac{\partial n_e}{\partial y} = \frac{\sqrt{2/\exp(1)}}{16\pi KT_{th}} (1 + n^2) \frac{E_v^2}{y_o |n| (1 + 1/Z)} \qquad [28]$$

The second physical condition of total reflection is given by the refractive index in the center of the beam, n, and its value at y_o, for which with Eq. (28) is formulated[59]

$$\sin\left(\frac{\pi}{2} - \alpha_o\right) = \frac{|n|}{|n_{y_o}|} \qquad [29]$$

Using the following Taylor expansion, for the case of a negligibly small collision frequency

$$n_{y_o} = n + \frac{\partial n}{\partial y} y_o; \quad n^2 = 1 - \omega_p^2/\omega^2 \qquad [30]$$

gives from Eq. (24)

$$\sin\alpha_o = \left(\frac{2}{n}\frac{\partial n}{\partial n_e}\frac{\partial n_e}{\partial y} y_o\right)^{\frac{1}{2}} \qquad [31]$$

If - as a third physical condition - a particular wave with an angle α for the first minimum of diffraction has to be reflected totally, the condition is found

$$\sin\alpha = \frac{\pi c}{2\omega y_o} \leq \sin\alpha_o \qquad [32]$$

Expressing the right-hand side by Eq. (31) and using there Eq. (28) and the relation for the electrical laser field amplitude $E_{vo} = c_1 P^{1/2}/y_o$ (where P is the average laser power and c_1 is a constant of

1.63×10^{-5} cgs) one arrives at[59]

$$P \geq \frac{(\pi c)^2 n^3 m_e}{e^2\,[2/\exp(+1)]^{\frac{1}{2}}\, c_1^2\, (1 + n^2)} \qquad [33]$$

It is remarkable that this threshold for self-focusing is a laser power and not an intensity. This is surprising, but not strange, as the threshold for the self-focusing of a laser beam in a dielectric nonionized medium is also a power and not an intensity[57], although both processes are basically different.

For an evaluation of Eq. (33), one can use the value of n given by Eq. (8), valid for temperatures above 10 eV. Expressing the plasma temperature T in eV and the laser power P in watts, one arrives at

$$2P \geq \begin{cases} 1.46 \times 10^6\, T^{-5/4} & \text{for } \omega_p \lesssim \omega \\ 1.15 \times 10^4\, T & \text{for } \omega_p < \omega \end{cases} \qquad [34]$$

This was the first quantitative theory of the nonlinear force self-focusing or, as it was initially called[59], the ponderomotive self-focusing. This result was rederived by several authors[61] and fully reproduced.

A further surprising result[59] is the fact that the power threshold for self-focusing in plasma is very low, in the range of megawatts or less. This is in agreement with measurements first published by Korobkin and Alcock[62]. The measurements of Richardson et al. [63] especially demonstrated in detail that the beam center shows a depletion of plasma. Another success of the theory is the agreement of the measured beam diameters[62] of a few microns for a laser power of 3 MW. If one can assume that the stationary conditions for self-focusing are reached when all plasma is moved out of the center of the laser beam, the electromagnetic energy density is then equal to $n_e(1 + 1/Z)$ KT. This is the case for densities close to the cutoff density, where the laser intensity is equal to the threshold intensity I^*, as given by Eq. (1) for neodymium glass laser radiation. It is evident that the beam has then to shrink down to such a diameter to reach the necessary 10^{14} W/cm^2 from a laser power of 3 MW. The resulting beam diameter is then a few micrometers, in full agreement with the measurements.

Another type of self-focusing happens[64] if the relativistic effects are considered. The relativistic change of the electron mass, due to oscillation energies close to or above $m_o c^2$ causes a modification of the optical constants, as shown in Eqs. (8) to (10b). For the optical constants, Eq. (8) has to be used while for the absolute value of the refractive index, the following equation is used.

$$|\tilde{n}| = \left[\left(1 - \frac{\omega_p^2}{\omega^2 + \nu^2}\right) \left(\frac{\nu}{\omega} \frac{\omega_p^2}{\omega^2 + \nu^2}\right) \right]^{1/4} \qquad [35]$$

With this relativistic intensity dependence, the effective wavelength of propagating laser radiation in a plasma is then given by

$$\lambda = \frac{\lambda_o}{|\tilde{n}(I)|} \qquad [36]$$

where λ_o is the vacuum wavelength. In Fig. 4, a Gaussian-like intensity profile of a laser beam moving through a homogeneous plasma is considered. The relativistic refractive index results in the condition

$$|n(I_{max})| > |n(I_{max}/2)| \qquad [37]$$

showing that the effective wavelength[36] is shorter for the higher laser intensity in the center of the beam than at the lower intensity of the half maximum intensity value. As shown in Fig. 4, an initially plane wave front is then bent into a concave front, which tends to shrink down to a diffraction limited beam diameter of about one wavelength. From the geometry of Fig. 4 this shrinking can be approximated by an arc resulting in a self-focusing length ℓ_{SF}

$$\ell_{SF} = \left[d_o \left(\rho_o + \frac{d_o}{4} \right) \right]^{\frac{1}{2}} \qquad [38]$$

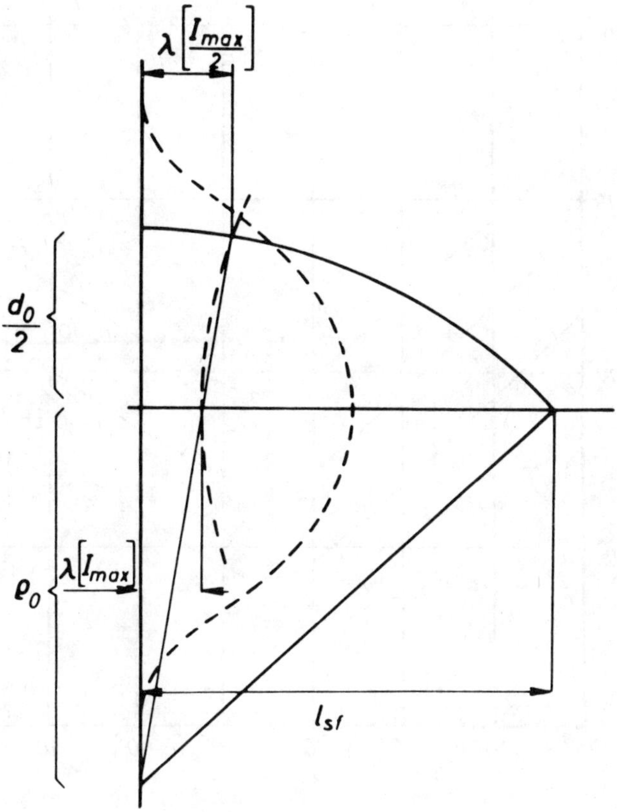

Fig. 4 Evaluation of the relativistic self focusing length from the initial beam diameter d_o and from the effective wavelengths. The relativistic effects cause a shorter wavelength at the maximum laser intensity I_{max} than at the half maximum intensity[64].

d_o is the initial beam diameter, and the radius of the arc with ρ_o is given by the effective wavelengths of the various intensities. From the geometry of Fig. 4, the following relation is derived:

$$\frac{|n(I_{max}/2)|^{-1}}{(d_o/2 + \rho_o)} = \frac{|n(I_{max})|^{-1}}{\rho_o} \qquad [39]$$

In combination with Eq. (38), this results in the ratio of the self-focusing length related to the beam diameter[64]

$$\frac{\ell_{SF}}{d_o} = 0.5 \left(\frac{|n(I_{max})| + |n(I_{max}/2)|}{|n(I_{max})| - |n(I_{max}/2)|} \right)^{\frac{1}{2}} \qquad [40]$$

Using the exact absolute value of the refractive index n, as given by Eq. (36), with the intensity dependent relativistic values of the plasma frequency and the collision frequency, a numerical evaluation of Eq. (40) is given in Fig. 5 for neodymium glass laser radiation for plasma densities of 10, 1, and 0.1% of the nonrelativistic cutoff density value. It is remarkable that the self-focusing length is as low as seven times the beam diameter for 10% of the cutoff density if the laser intensity is 3×10^{18} W/cm^2.

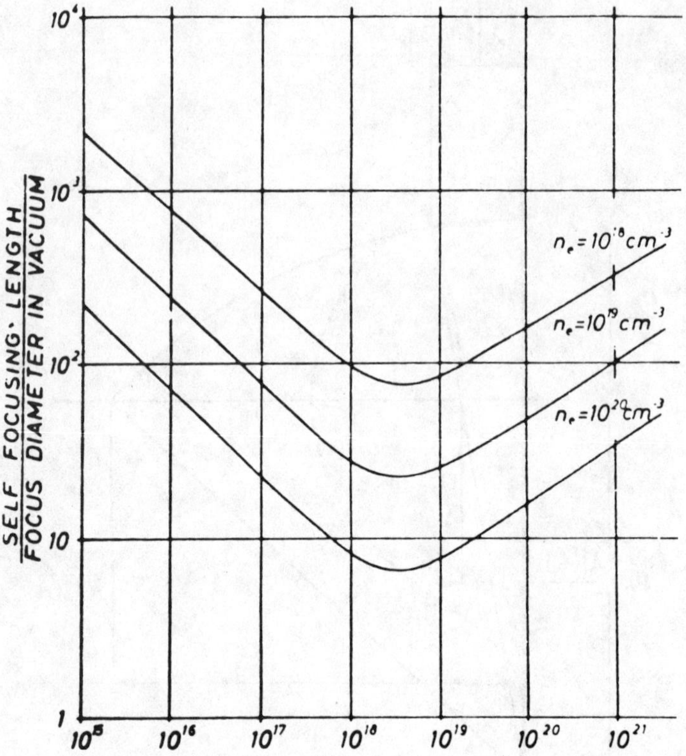

Fig. 5 Calculated self-focusing lengths divided by the laser beam diameter for neodymium glass laser radiation for various plasma densities depending on the laser intensity[64].

This intensity is the relativistic threshold corresponding to an electron oscillation energy of m_oc^2. It is further interesting to note that the process of the relativistic self-focusing also occurs for laser intensities that are much less than the relativistic threshold, even 1000 times less. This phenomenon of the occurrence of relativistic effects at intensities much lower than the relativistic threshold was not new, as could be seen from relativistic instabilities in plasmas.

The relativistic self-focusing has its maximum effect at the relativistic threshold. Its effect is lower for higher intensities. This can be easily understood from the fact that, at these higher intensities, there is an intensity dependent increase of the cutoff density. Eq. (9b), so to speak, the plasma becomes transparent for propagating waves at densities, where the non-relativistic conditions would require evanescent waves.

The extension of Eq. (40) to higher densities, shown in Fig. 5, is possible numerically. Simultaneously, the dependence on the plasma temperature and on the degree of ionizations is included. It is very surprising that self-focusing lengths of the same value as the initial beam diameter result, see Fig. 6[48]. For lower intensities, a numerical cutoff of the plots is observed, where nevertheless the action of relativistic self-focusing is still working for intensities of less than 1% of the relativistic threshold. This is a remarkable result. While the well-known difficulties of designing an optical lens systems for focusing a laser beam in vacuum limit the minimum beam diameters to about 10 wavelengths, the plasma at the cutoff density realizes the very fast shrinking of a laser beam down to one

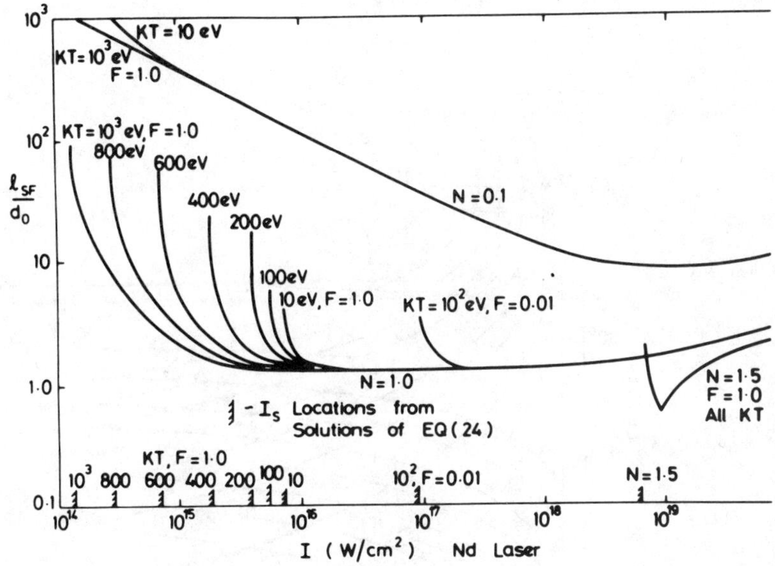

Fig. 6 Ratio of the self-focusing length l_{SF} over the initial laser beam diameter d_o for laser intensities near the relativistic threshold of 3×10^{18} W/cm^2 for neodymium glass laser radiation for varying plasma temperatures. The plasma density is equal to the nonrelativistic cutoff value ($N = n_e/n_{ec} = 1$) and 10% of this value ($N = 0.1$), respectively. The factor F is given by an effective collision frequency $\nu_{eff} = \nu/F^{2/3}$ to understand an eventual increase by anomalous effects[48].

wavelength diameter automatically by relativistic plasma effects.

As soon as the conditions for relativistic self-focusing are reached, the effects of very high laser intensities in the focused filaments open the door to very interesting high-intensity effects. These effects are indeed not appropriate for laser fusion, and one has to know how to avoid the relativistic self-focusing in the case of laser fusion. For the physics of higher energies rather than for laser fusion, however, the fast shrinking of the laser beams to a diameter of a wavelength is very desirable.

One question is how the oscillation energy of the electrons can be increased in the relativistically self-focused filament, if the fact is taken into account that the oscillation energy of the electrons increases by a square root law in the laser intensity only at superrelativistic intensities. This change in the exponent of I is a very important reason to seek to understand the laws of blackbody radiation and to discuss a derivation of the fine structure constant from basic physical laws[65]. This lower power increase of the electron energy, however, results in some disadvantages in reaching the highest possible electron oscillation energies. Another disadvantage is the fact that, if the focusing is performed in plasma densities closer to the cutoff density, the larger the effective wavelength the larger is the effective beam diameter at self-focusing. The relation of short self-focusing length is then influenced negatively by the necessary higher laser power. All these factors together have been evaluated[48] and the result is achieved in Fig. 7. The focusing of a neodymium glass laser beam in vacuum is assumed to be down to a diameter of d_o = 30 wavelengths, which seems to be realistic. The resulting maximum oscillation energy of the electrons is then given for the various intensities I_v of the laser beams in such a vacuum focus of 30 wavelengths diameter, where the plasma density has been varied with $N = n_e/n_{ec}$ between 0.1 and 0.99. The resulting oscillation energies and self-focusing lengths are given in the diagram of Fig. 7.

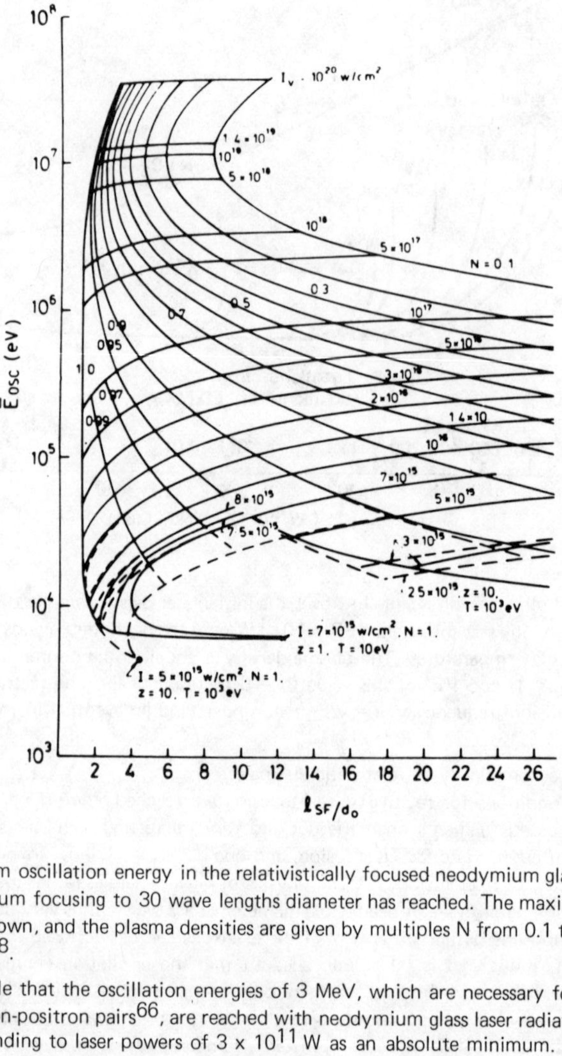

Fig. 7 Maximum oscillation energy in the relativistically focused neodymium glass laser beam, where the vacuum focusing to 30 wave lengths diameter has reached. The maximum laser intensities I_v are shown, and the plasma densities are given by multiples N from 0.1 to 0.99 of the cut off density[48].

It is remarkable that the oscillation energies of 3 MeV, which are necessary for a quantitative production of electron-positron pairs[66], are reached with neodymium glass laser radiation of only 5×10^{17} W/cm^2, corresponding to laser powers of 3×10^{11} W as an absolute minimum. The laser beam then has to be sufficiently smooth that only one filament is being produced. The laser powers with short rise times, however, are within the state of the art[67].

A further result[68] is that the ion energy after nonlinear force acceleration from the extremely high intensity of laser beams with a diameter of one wavelength does not follow the relation of Z times the relativistic electron oscillation energy. The fact is that the ions have Z times the oscillation energy of the electrons as if these were following the subrelativistic law as long as the ion energies are subrelativistic. Using the relativistic threshold intensity I_r, Eq. (10a), the ion energies ε_i of translation after being accelerated by the nonlinear forces from the relativistic self-focused filament are from Eq. (21)

$$\varepsilon_i = \frac{Z_i mc^2 (I/I_r)}{4} \approx 3 \times 10^{-6} ZP \qquad [41]$$

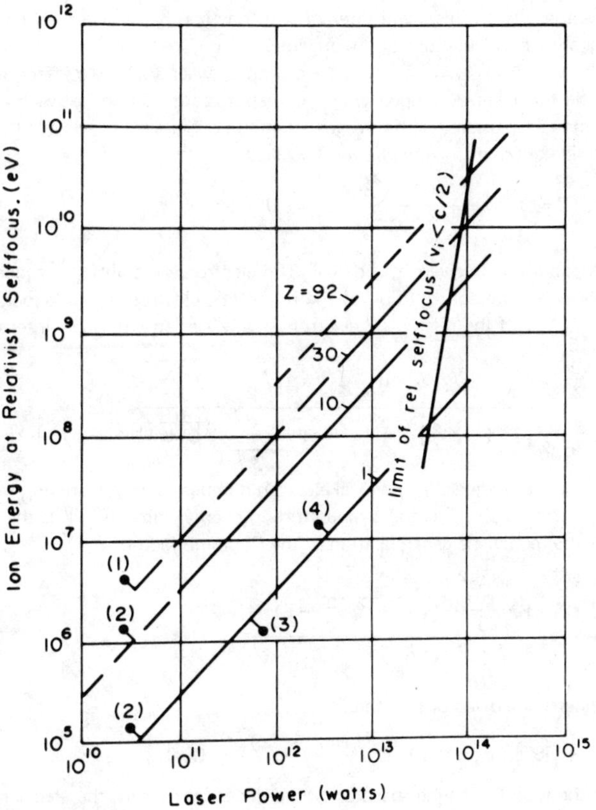

Fig. 8 Energy of Ions from a target with relativistic self-focusing according to Eq. (41)[68] depending on the laser power. There is no dependence on the wavelength. The dependence on the ion charge number is restricted by the expected degree of ionization. The measured ion energies correspond to (1)[70], (2)[69], (3)[72], and (4)[71].

Depending on the laser power, the ion energy is independent of the wavelength. The result is given in Fig. 8. Historically, the measurement of the MeV ion energies[69] by Ehler, in full agreement with the later ones, Fig. 8, was very transparent, although several authors could not believe in the MeV ion energies. Luther-Davies et al.[70] measured the MeV ions and found agreement with the theory of self-focusing.

It is remarkable that the measured protons from a laser produced plasma of 15 MeV[71] would correspond to 450 MeV ions, which are 30 times ionized, if the same laser beam had been applied to a high-Z target. Ionizations of 40 and more are known from laser produced plasmas.

5. NUMERICAL CALCULATIONS

The nonlinear force self-focusing has a delay time due to the motion of the plasma from the beam center, while the relativistic self-focusing occurs almost instantaneously, in a time comparable with the period of the optical radiation. The calculations of the preceding section assumed ideal conditions, i.e. the nonlinear force self-focusing was calculated for nonrelativistic intensities, and the relativistic self-focusing was calculated under the assumption of an initially homogeneous plasma. For a high intensity laser pulse these two effects will interact with one another and destroy some of the assumptions made in the analytical calculations of the previous section.

In order to make a rigorous study of the competition between these two effects a two dimensional time dependent code has been developed which models the interaction between a high power laser beam and a fully ionized plasma[72]. The program combines Maxwell's laws for the electromagnetic field and the two fluid equations for the plasma. The electric field satisfies

$$\nabla^2 \underset{\sim}{E} - \nabla(\nabla\cdot\underset{\sim}{E}) - \frac{1}{c^2}\frac{\partial^2(\tilde{n}^2\underset{\sim}{E})}{\partial t^2} = 0 \qquad [42]$$

where the refractive index n is given by Eq. (8). The electron-ion collision frequency ν includes a generalisation to allow for the contribution of the relativistic electron quiver velocity to the temperature, given by Eq. (9b), and the relativistic electron mass variation with laser intensity, given by Eq. (10b). The result is[15]

$$\nu = \frac{\pi^{3/2} Z_I n_e e^4 \ln \Lambda_R}{2m_o^{1/2}[1+\Omega\psi^2]^{1/4}\left[2\left\{T+\frac{m_o c^2}{3}\sqrt{1+\Omega\psi^2}-1\right\}\right]^{3/2}} \qquad [43]$$

where Z_I is the average ion number, n_e the local electron number density, e and m_o the electron charge and rest mass, $\Omega = (e/m_o \omega c)^2$, T is the temperature in energy units and ψ is the amplitude of the electric field. Λ_R is the relativistic generalization of the Coulomb factor

$$\Lambda_R = \frac{3\left\{T+\frac{m_o c^2}{3}(\sqrt{1+\Omega\psi^2}-1)\right\}}{2 Z_I e^3 (\pi n_e)^{1/2}} \qquad [44]$$

By writing the electric field vector **E** in the form

$$\underset{\sim}{E} = \tfrac{1}{2}\left\{\hat{\underset{\sim}{k}}\,\psi(r,z,t,)\,e^{i(\omega t-k_o z)} + C.C.\right\} \qquad [45]$$

and separating out the rapidly varying space and time dependent terms, the electrodynamic equation for ψ can be written (in cylindrical coordinates) as

$$\frac{\partial\psi}{\partial z} + \frac{i}{2k_o r}\frac{\partial}{\partial r}\left(r\frac{\partial\psi}{\partial r}\right) + \frac{ik_o}{2}\left(\frac{\alpha(t)}{k_o^2}-1\right) + \frac{i\beta(t)}{2k_o}\frac{\partial\psi}{\partial t} = 0 \qquad [46]$$

where $\alpha(t) = \left(\frac{\omega}{c}\right)^2\left[1-\frac{\zeta(t)}{n_c}\left(\frac{1+i\nu/\omega}{1+(\nu/\omega)^2}\right)\right]$ [47]

$\beta(t) = \frac{2\alpha(t)}{i\omega}$ [48]

and $\zeta(t) = n_e(r,z,t)/\gamma^R$ [49]

For computational purposes the time derivative is removed using the following independent variable transformation

$$z' = z \quad \text{and} \quad t' = t - \frac{i\beta(t)\,z'}{2k_o} \qquad [50]$$

The electrodynamic wave equation then becomes

$$\frac{\partial\psi}{\partial z} = -\frac{i}{2k_o R(z,t)^2}\frac{1}{\eta}\frac{\partial}{\partial\eta}\left(\eta\frac{\partial\psi}{\partial\eta}\right) - \frac{ik_o}{2}\left(\frac{\alpha(t')}{k_o^2}-1\right)\psi \qquad [51]$$

where $\eta = r/R$ and R(z,t) is the maximum beam radius, which is calculated from a power balance using the Poynting vector and stepwise energy loss through absorption.

The plasma dynamics is described using the continuity and momentum conservation equations for the electron and ion fluids. The electron momentum equation contains the relativistic factor γ^R and the nonlinear force

$$\frac{m_o \partial(\gamma^R \underset{\sim}{v}_e)}{\partial t} = \frac{\underset{\sim}{f}_{NL}}{n_e} - e\underset{\sim}{E} - \frac{\nabla P_e}{n_e} + \frac{\underset{\sim}{P}_{eI}}{n_e} \qquad [52]$$

where

$$\underset{\sim}{f}_{NL} = -\frac{n_e e^2}{4 m_o \gamma^R \omega^2} \nabla(\psi^2) \qquad [53]$$

and is a relativistic generalization of the nonlinear force. $\underset{\sim}{P}_{eI}$ is the electron-ion collisional momentum transfer and P_e is the electron pressure term.

The ion momentum equation is similar to equation (52), but without the relativistic and nonlinear force effects. By eliminating the electric field between the two equations, using the ion continuity equation to eliminate some of the variables, and setting $n_e = Z_I n_I$, an equation for $N = n_e/n_o$ can be written

$$\frac{1}{\eta}\frac{\partial}{\partial \eta}\left(\eta \frac{\partial \ln N}{\partial \eta}\right) - \Gamma \frac{\partial^2 \ln N}{\partial \tau^2} - \sqrt{\Gamma}\frac{\partial}{\partial \tau}\left(\frac{V_r}{C_I}\frac{\partial \ln N}{\partial \eta}\right) = -G(\eta,z,\tau) \qquad [54]$$

Here the z derivatives have been neglected as well as several small nonlinear terms. $\tau = t/t_p$, where t_p is the total pulse time, $\Gamma = (R/t_p C_I)^2$, and $C_I = [(Z_I T_e + T_I)/M]^{1/2}$ is the ion-acoustic wave velocity. Time and radially dependent derivatives are specifically considered in the plasma motion solution while the z-dependency enters through the nonlinear electrodynamic equation. G is the electrodynamic source function, given by

$$G(\eta,z,\tau) = \frac{m_o c^2 Z_I}{2(Z_I T_e + T_I)} \frac{1}{\eta}\frac{\partial}{\partial \eta}\left(\eta \frac{\partial \gamma^R(\eta,z,\tau)}{\partial \eta}\right) \qquad [55]$$

A solution of equation (54) is constructed by assuming that an average value of the radial ionic velocity may be assigned. $< V_R >$ is calculated from number density conservation. The resulting equation is solved iteratively. The first order solution neglects the mixed derivatives and is solved exactly by a Hankel transform over the radial variable η. Once a first order solution for N has been found it is used to calculate the coupled term in τ and η. A second order solution is then obtained by placing the mixed derivative term on the right hand side of equation (54) and regarding it as a new source term.

Solution of the coupled set of equations, i.e. the ion acoustic equation (54) and the electrodynamic equation (51), then proceeds in a stepwise self-consistent manner. The electron number density is calculated from the analytic solution of equation (54) and this solution is then used to evaluate the terms α and β in the electrodynamic equation. This equation is then solved by standard finite difference techniques to obtain a new value for the electric field. This new value is then used in equation (55) to obtain a new source function for the ion-acoustic equation, which is then solved to find the change in density at the next time step. The solution proceeds in this stepwise manner with the time dependence provided by the ion acoustic equation and the z dependence by the electrodynamic equation. More detailed information about the coupled set of equations, and the numerical method used to solve equation (51) can be found in Ref. 72.

We now describe the results of the program when used to model the interaction of a 3×10^{11} Watt Nd glass laser beam of nearly Gaussian radial intensity profile with 30 μm diameter and 30 psec duration incident on an initially homogeneous hydrogen plasma at 80% cut off density.

Fig. 9 shows the density profile as a function of axial and radial distances within the plasma after 9.2 psec. The formation of a cylindrical density depression enclosing a residual plasma along the beam centre line is clearly visible and easily understood in terms of the radial Gaussian laser profile. As the radial nonlinear force is proportional to the gradient of the square of the field intensity, we expect the

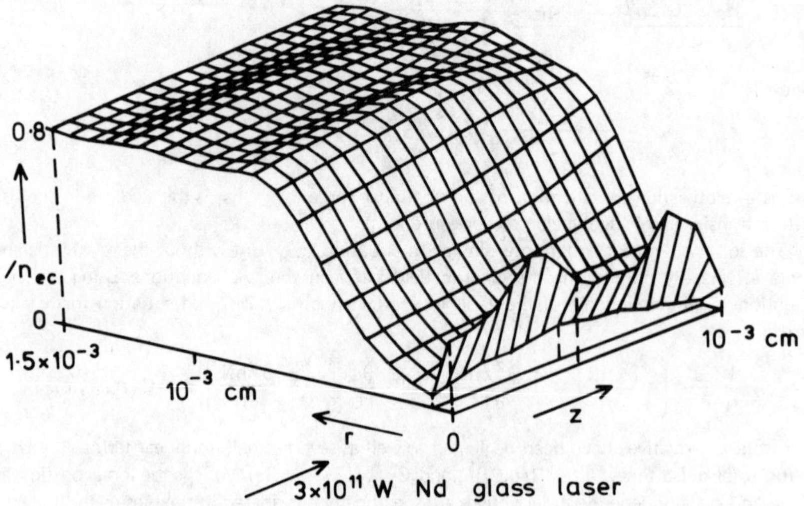

Fig. 9 Density profile in an initially homogeneous hydrogen plasma of 80% critical density after 9.2 psec interaction with a Nd laser pulse of peak power 3×10^{11} Watts and 30 psec duration. A cylindrical density depression has been formed with a small amount of plasma remaining along the beam centre line.

force to be maximum at some point between the centre line and beam edge. Fig. 10 shows the state of the plasma after 18.4 psec. The residual centre line plasma has been removed by fast axial motion.

The threshold for relativistic self-focusing is a function of the plasma temperature, density, and ion charge number (see Fig. 6). For the 100 eV hydrogen plasma at 80% cut off density studied in this section the threshold for Nd glass lasers is near 10^{15} W/cm^2 [73]. Our intensity is near 10^{16} W/cm^2, and this leads to a self-focusing length to beam width ratio of about twenty, which in this case means a self-focusing length of about 400 µm. From Eq. (34) the threshold for pondermotive self-focusing for a 100 eV plasma is about 2×10^6 Watts. Self-focusing lengths for the considerably higher beam powers used here have been calculated by Siegrist[74]. For a power of 3×10^{11} Watts and an input beam of 15 µm half-intensity radial width he finds a steady state self-focusing distance greater than 1000 µm. The maximum axial distance in Figs. 9 and 10 is only 10 µm, so the cylindrical density depression we see here is simply the beginning of the self-focussed beam.

We now describe the results of the interaction of a 5 psec pulse from a Nd laser of peak power 10^{13} Watts with a 38 times ionized tin target. We assume the target has been prepulsed so a uniform, homogeneous plasma has been produced. The plasma density is equal to the critical value and the electron and ion temperatures are both 100 eV. Because of the higher power and critical density we expect relativistic self-focusing to be much more pronounced in this case.

The leading edge of the pulse has almost a step function shape as this facilitates the formation of fast relativistic self-focusing before appreciable plasma motion occurs. For the first two time steps the power is quite low at 10^{10} Watts, or an intensity of 3×10^{15} W/cm^2. This is just within the threshold for relativistic self-focusing calculated by Kane and Hora[73], and the self-focusing length is approximately 35 µm. This self-focusing behaviour can be seen in Fig. 11 where we show the filament radius (defined as that value of the radius which contains 90% of the beam power) as a function of axial distance for the different time steps. For the first two time steps the filament radius decreases uniformly from 40 µm at the vacuum-plasma interface to about 2.5 µm at 23 µm axial depth.

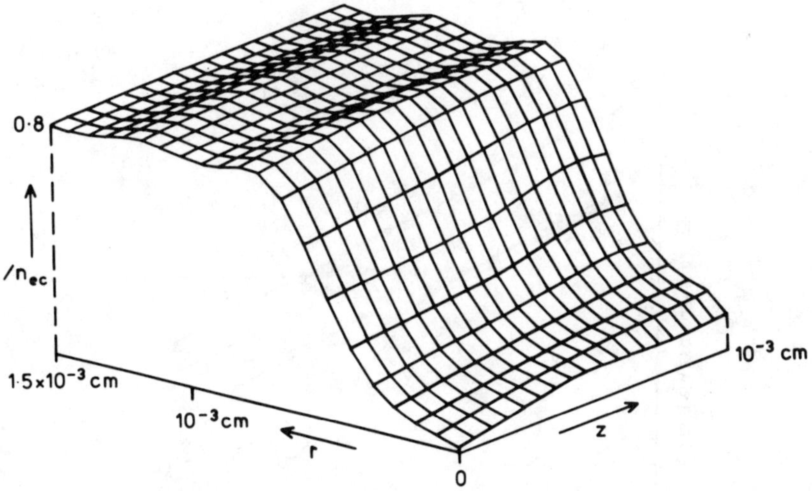

Fig. 10 Same conditions as for Fig. 9, but at a time of 18.4 psec.

There is a sharp decrease in filament radius between the second and third time steps due to the considerably higher beam power of 10^{13} Watts, or a vacuum center-beam intensity of 3×10^{18} W/cm^2. Kane and Hora have shown that the ratio of self-focusing length to half-intensity beam diameter at the critical density is nearly independent of temperature, ion charge number and parametric decay instabilities and has a value of approximately 1.0 in the range of 10^{17} to 10^{19} W/cm^2 (see Fig. 6). The focussed beam diameter for these conditions is expected to be of the order of the wavelength of the laser light. Hence in this case we expect to see self-focusing down to a diameter of the order of 1 μm in a distance of about 30 μm. The data in Fig. 11 show that the average value of the filament radius at the later time steps does converge to about 1 μm at 30 μm depth, but there is quite a scatter in the value of the filament radius at this depth. This is due to the radial nonlinear force induced plasma motion destroying the uniformity of the medium.

The gradient in electric field strength along the axis of the laser beam produced by the relativistic self-focusing also results in a strong nonlinear force which accelerates ions to high energies against the laser beam. Approximate calculations of the maximum ion energies obtainable by this effect have been made by Hora, Kane and Hughes[68], who show that the maximum energy depends only on laser power and wavelength, and on the ion charge number. For the 38 times ionized tin plasma irradiated by a 10^{13} Watt Nd pulse studied in the previous section the maximum ion energy should be several GeV. Fig. 12 shows the center line ion energy as a function of depth into the plasma for different times for the run described in the previous section. The energies were calculated using a finite difference form of the ion momentum equation. The observed maximum ion energy of 5 GeV is in good agreement with the more approximate estimates shown in Fig. 8.

The large degree of plasma motion in such a short time (of the order of picoseconds) is entirely consistent with the magnitude of the nonlinear force resulting from the relativistic increase in intensity by self-focusing. From plots of the filament radius in Fig. 11 we estimate the intensity to be approximately 10^{21} W/cm^2, and hence the electric field strength in cgs units is 3×10^9. The nonlinear force density results in an acceleration of

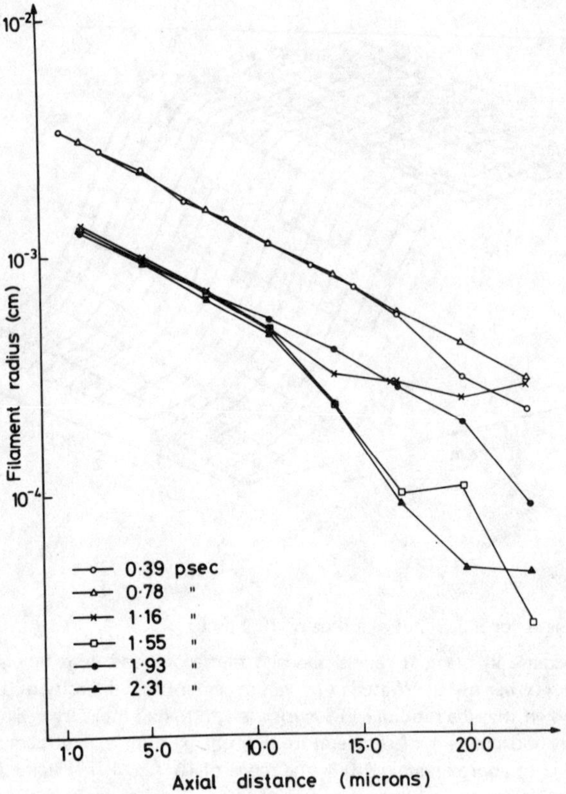

Fig. 11 A 10^{13} Watt Nd glass laser pulse from vacuum incident on Sn^{38+} plasma of 10^{21} cm^{-3} density. Filament radius as a function of depth into the plasma for various times.

$$\frac{dv}{dt} = \frac{1}{M_I n_I} \frac{d}{dz} \frac{1}{8\pi} (E^2 + H^2) \qquad [56]$$

Using $\Delta (E^2 + H^2)/8\pi = 2.38 \times 10^8$, $\Delta z = 2\mu m$, the ion mass of tin, and $n_I \cong 3.0 \times 10^{19}$ cm^{-3}, an acceleration of 2.7×10^{21} cm/sec^2 will result. This produces a change of the velocity within 0.5 psec of 1.1×10^{10} cm/sec or an energy of tin ions of 7.8 GeV, in rather good agreement with Fig. 12. The number of ions gaining this energy can be estimated from the plots to be 2.6×10^7, representing 5.8×10^{-2} Joules, and indicating that about 0.4% of the laser energy has been transferred to the GeV ions.

For a CO_2 laser under similar conditions we expect a maximum ion energy of 200 MeV for a proportionally longer path length. We are at present modifying the code to model CO_2 laser-plasma interaction dynamics so we can check this estimate. The present code though has still not exhausted all of parameter space for Nd glass laser-plasma interaction dynamics, further studies of the effect of the laser rise time on the competition between relativistic and nonlinear force self-focusing would be interesting, and more runs for different plasma species, charge states and densities should be made.

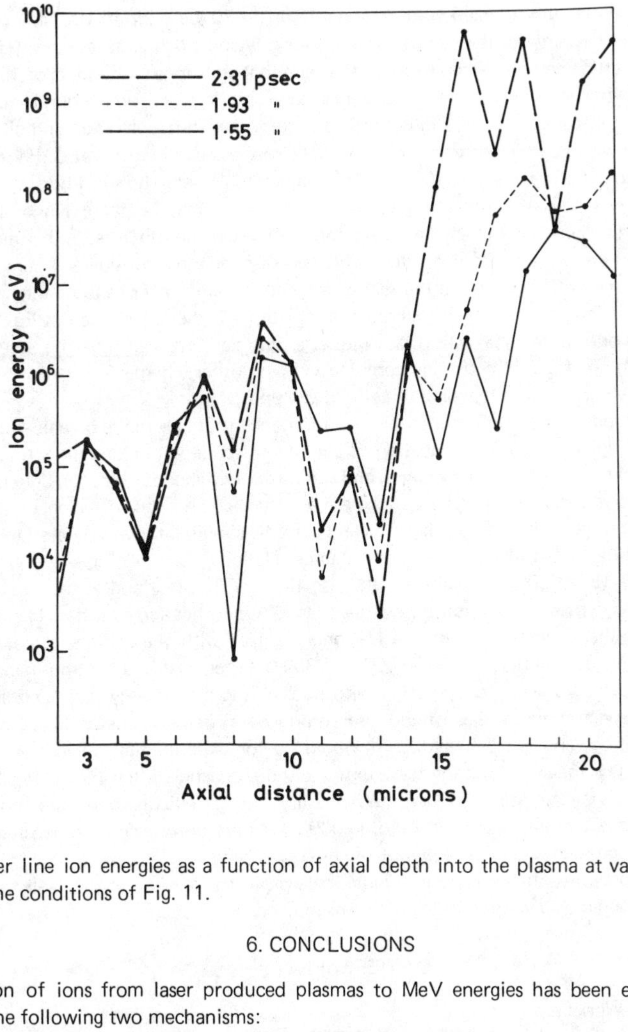

Fig. 12 Center line ion energies as a function of axial depth into the plasma at various times (psec) for the conditions of Fig. 11.

6. CONCLUSIONS

Acceleration of ions from laser produced plasmas to MeV energies has been established experimentally by the following two mechanisms:

a) Electrostatic acceleration within the double layer at the periphery of the expanding plasma. Though an experiment like that of Pearlman and Dahlbacka[28] would not be able to distinguish between a purely electrostatic acceleration, or a conical ion emission due to resonance absorption[55] or the relativistic self-focusing, nonlinear-force-swelling process, the electrostatic process cannot be excluded, as was made evident from the measured energy upshift of alphas[30]. It has to be noted that only a small number of fast ions can be produced, by many orders of magnitudes less than the usually observed fast ions.

b) After nonlinear-force or relativistic self-focusing occurs in the plasma, the laser beam can produce an acceleration of electrons and ions by the nonlinear forces. In agreement with a very general model, Fig. 8, Eq. (41), the measured 6 to 15 MeV ions from laser produced plasmas can be explained. These forces can explain the large numbers of generated fast ions.

In order to extrapolate by a factor of 10 to 100 over the experimental results for future accelerators, a detailed numerical study has been performed[72] where the counteracting mechanisms are elaborated from a two dimensional beam-plasma dynamics. With single laser pulses of 10^{13} to 10^{14} Watts, ion energies for nearly fully ionized gold or uranium up to 20 GeV (up to 100 MeV per nucleon) can be expected if the rise time of the laser pulse is sufficiently short (Nd-glass laser: $<$ 5 psec; CO_2 laser: $<$ 80 psec). An efficiency of 1% can be expected such that GeV ions with bursts of 10^{12} particles per shot of 10^{-10} sec duration are generated. Using an envelope of 1 cm^2 area around the focus, currents near k Amp are generated. At these high particle current densities, spallation and all kinds of heavy nuclear reactions with subsequent dense high energy processes can be generated. Photons of 50 MeV energy can be expected in bursts of 10^8 per cm^2 and per 10^{-10} sec. The short bursts have a density of 10^{18} photons/cm^2 sec. The plasma-nonlinear-force mechanism may not be extended beyond 50 GeV ion energies as then the Debye mechanisms are too strong. On the other hand, the (nonthermal) nonlinear force driven electrostatic double layer - which corresponds to a considerably large volume and is, therefore, not so much limited as the thermal electrostatic double layer - should then accelerate a reasonable number of ions to energies which are (m_i/m_e) times higher than given in Eq. (41), i.e. up to TeV ion energies and more. The only disadvantage is that electrons can never be accelerated beyond GeV energies from one focus without any combination with other schemes[1, 13].

The proposal which should immediately be followed up is:

a) A continuation of the numerical studies should concentrate on the re-organization and improvement of the existing code and an exhaustion of the present code with many more runs to understand the influence of the initial profiles, swelling, and optimized initial conditions for CO_2 laser interaction (1.5 man-years), the inclusion of relativistic plasma dynamics and the inclusion of the electrostatic fields as well as the generated magnetic fields and currents should be relatively easy (by 4 man-years). Experiments would be ideal at the Los Alamos National Lab. Antares laser with one beam of 0.8×10^{14} Watt, 150...250 psec pulse length (rise time $<$ 80 psec) and at the Australian National University 0.5 TW, 5 psec pulse duration Nd glass laser. A modification of other Nd glass lasers would need liaison with the Lawrence Livermore National Lab. or with the National Laser Users Facility, Univ. of Rochester. This activity of $100,000 to $300,000 per year is a minimum and very modest.

b) Simultaneously, studies in cooperation with nuclear and high energy physicists should be made to elaborate the specific advantages of the laser produced extremely intense heavy nuclei bursts of $< 10^{-10}$ sec duration, and what advantage it would be for experimental nuclear technique to have excitation within less than 10^{-11} seconds by extremely high particle intensities. After learning of any special advantages, the estimations of cost are necessary, though existing hardware at moderate prices is available (e.g. one Antares beam for $11,000,000). An initiation of contract work and workshops for this task could be funded by $100k p.afor the next two years.

There is the possibility, however, that a large scale project is feasible and so realistically promising that a strong increase of the volume of the activities may occur.

REFERENCES

1. W. Willis, this Workshop;
   Ugo Amaldi, Colliding Linacs, CERN Report EP/73-136, Nov. 6, 1979.
2. K. Shimoda, Appl. Opt. **1**, 33 (1962); H. Motz, Contemp. Phys. **20**, 547 (1979); J.D. Lawson, Free-electron Lasers, Nature **277**, 262 (1979).
3. S.J. Smith, and E.M. Purcell, Phys. Rev. **92**, 1069 (1953).
4. R.B. Palmer, J. Appl. Phys. **43**, 3014 (1972).
5. A. Lohmann, IBM Techn. Note TN5, San Jose, Oct. 3, 1962.
6. R.H. Pantell, this Workshop.
7. P. Srangle, and C.-M. Tang, IEEE Transact. Nucl. Sci. **NS-28**, 3346 (1981); M.V. Fedorov, and J.K. McIver, Optica Acta **26**, 1121 (1979).
8. J.D. Lawson, IEEE Transact. Nucl. Sci. **NS-26**, 4217 (1979).
9. T. Tajima, and J.M. Dawson, Phys. Rev. Lett. **43**, 267 (1979); C. Joshi, T. Tajima, J.M. Dawson, H.A. Baldis, and N.A. Ebrahim, Phys. Rev. Lett. **47**, 1285 (1981); C. Joshi, C.E. Clayton, and F.F. Chen, UCLA Rept. PPG-595 (1981).

10. W.J. Willis, in **Laser Interaction and Related Plasma Phenomena**, H. Schwarz and H. Hora eds. (Plenum, New York, 1977) Vol. **4B**, p. 991; W. Willis, CERN Report Nucl. Phys. Div. 75-9 (1975).
11. P.J. Channell, Laser Cooling of Heavy Ion Beams, J. Appl. Phys. **52**, 3791 (1981).
12. R. Rossmanith, Nucl. Instr. Meth. **154**, 29 (1978).
13. T. Tajima, presentation at this Workshop.
14. H. Hora, Opto Electronics **2**, 201 (1970).
15. H. Hora, **Physics of Laser Driven Plasmas** (Wiley, New York, 1981).
16. R.E. Honig, Appl. Phys. Lett. **3**, 8 (1963).
17. W.I. Linlor, Appl. Phys. Lett. **3**, 210 (1963).
18. G. Siller, K. Buchl, and H. Hora, **Laser Interaction and Related Plasma Phenomena**, H. Schwarz and H. Hora eds. (Plenum, New York, 1972) Vol. 2, p. 253.
19. B.H. Ripin, R. Decoste, S.P. Oberschain, S.E. Bodner, E.A. McLean, F.C. Young, R.R. Whitlock, C.M. Armstrong, J. Grun, J.A. Stamper, S.H. Gold, D.J. Nagel, R.H. Lemberg, and J.M. McMahon, Phys. Fluids, **23**, 1012 (1980).
20. P. Mulser, Z. Naturforsch **25A**, 282 (1970).
21. R.G. Rehm, Phys. Fluids **13**, 282 (1970).
22. Yu. V. Afanasyev, N.G. Basov, P.P. Volosebni, E.G. Gamali, O.N. Krokhin, S.P. Kurdeomov, E.T. Levanov, A.A. Samarski, A.N. Tikhonov, Plasma Phys. and Controlled Thermonucl. Res. 1974 (IAEA Vienna, 1975) Vol. 2, p. 559.
23. H. Hora, R. Castillo, R.C. Clark, E.L. Kane, V.F. Lawrence, R.D.C. Miller, M.F. Nicholson-Florence, M.M. Novak, P.S. Ray, J.R. Shepanski, and A.I. Tsivinsky, Plasma Physics and Controlled Thermonuclear Fusion Res. (IAEA, Vienna 1979), Vol. 3, p. 237.
24. C. Yamanaka, (invited paper) European Conference on Plasma Physics and Fusion, Moscow, Sept. 1981.
25. H. Hora, Colloquium, Max-Planck-Inst. Plasmaphysik, 1964.
26. H. Hora, **Laser Plasmas and Nuclear Energy** (Plenum, New York, 1975), p. 24.
27. C.W. Mendel, and J.N. Olsen, Phys. Rev. Letters **34**, 859 (1975).
28. J.S. Pearlman, and G.H. Dahlbacka, Applied Phys. Lett. **31**, 414 (1977).
29. J.E. Crow, P.L. Auer, and J.E. Allen, J. Plasma Phys. **14**, 65 (1975).
30. Y. Gazit, J. Delettrez, T.C. Bristow, A. Entenberg, and J. Soures, Phys. Rev. Letters **43**, 1943 (1979).
31. T.P. Donaldson, and P. Ladrach, Phys. Lett. **70A**, 419 (1979).
32. F.J. Mayer, R.K. Osborn, D.W. Daniels, and J.F. McGrath, Phys. Rev. Lett. **40**, 30 (1972); D.C. Slater, Appl. Phys. Lett. **31**, 196 (1977).
33. B.H. Quon, and A.Y. Wong, Phys. Rev. Letters **37**, 1393 (1976); P. Coakley, N. Hershkowitz, R. Hubbard, and G. Joyce, Phys. Rev. Letters **40**, 230 (1978); C. -G. Falthammer, S.-I. Akasofu, and H. Alfven, Nature **275**, 185 (1978); S. Torven, and L. Lindberg, TKITA-EPP-80-02 Report Royal Inst. Technol. Stockholm (1980).
34. Hannes Alfven, **Cosmic Plasma** (Reidel, Dordrecht, 1981).
35. Lord Kelvin, Cambr. and Dublin Math. J. Nov. 1845, see Electrostatics and Magnetism p. 32, London 1872.
36. H. Hora, Opt. Comm. **41**, 382 (1982).
37. C. Max, and F. Perkins, Phys. Rev. Lett. **27**, 1342 (1971).
38. Lorentz force. See e.g. A.V. Belorussov, A.I. Karchevshi, E.P. Potanin, and A.L. Ustinov, Sov. Phys. - Tech.Phys. **25**, 562 (1980), first column, 30th line.
39. F.F. Chen, Comments on Mod. Phys. **E1**, 81 (1972); **Plasma Physics** (Plenum, New York, 1975); H. Motz, **Physics of Laser Fusion** (Academic Press, London, 1979).
40. S. Weibel, J. Electr. Contr. **5**, 435 (1958); V.A. Gapunov, and M.A. Miller, Sov. Phys. JETP**7**, 168 (1958).
41. H. Hora, D.Pfirsch, and A. Schluter Z. Naturforsch. **22A**, 278 (1967).
42. H. Hora, Phys. Fluids **12**, 182 (1969).
43. A. Schluter, Z. Naturforsch **5A**, 72 (1950).

44. L.D. Landau, and E.M. Lifshitz, **Electrodynamics of Continuous Media** (Pergamon, Oxford 1966), p. 242.
45. R.D.C. Miller, and H. Hora, Plasma Physics **21**, 183 (1979); G.W. Kentwell, and H. Hora, Plasma Physics **22**, 1051 (1980).
46. J.W. Shearer, R.E. Kidder, and J.W. Zink, Bull. Amer. Phys. Soc. **15**, 1483 (1970).
47. K.G. Eastabrook, E.J. Valeo, and W.L. Kruer, Phys. Fluids **18**, 1151 (1975); A.L. Peratt, Phys. Rev. **A20**, 2555 (1979); I.D. Larsen, and J. Harte, Lawrence Livermore Lab. UCLR-79757-1 (1977).
48. E.L. Kane and H. Hora, **Laser Interaction and Related Plasma Phenomena**, H. Schwarz and H. Hora eds. (Plenum, New York, 1977) Vol. 4B, p. 913.
49. F.F. Chen, **Laser Interaction and Related Plasma Phenomena**, H. Schwarz and H. Hora eds. (Plenum, New York, 1974) Vol. 3A, p. 291; A.T. Liu, and J.M. Dawson, Phys. Fluids **18**, 201 (1971).
50. H.H. Chen, C. Grebogi, C.S. Liu, and V.K. Tripathi, **Plasma Physics and Controlled Nuclear Fusion Res.** 1978 (IAEA, Vienna, 1979) Vol. 3, p. 181.
51. J.L. Bobin, W. Woo, and J.S. De Groot, J. Physique **38**, 769 (1977).
52. H. Hora, Phys. Fluids **17**, 939 (1974).
53. N.G. Denisov, Sov. Phys. -JETP**4**, 544 (1957); R.B. White, and F.F. Chen, Plasma Phys. **16**, 565 (1974).
54. H. Hora, **Nonlinear Plasma Dynamics** (Springer, Heidelberg, 1979).
55. G.W. Kentwell, Phys. Fluids (submitted).
56. R. Dragila, Phys. Fluids **24**, 3212 (1981).
57. R.Y. Chiao, E. Garmire, and C.H.Townes, Phys. Rev. Lett. **13**, 479 (1964).
58. G.A. Askaryan, Sov. Phys. JETP, **15**, 1088 (1962); A.G. Litvak, Sov. Phys. JETP **30**, 364 (1970).
59. H. Hora, Z. Phys **226,** 156 (1969).
60. M.S. Sodha, A.K. Ghatak, and V.K. Tripathi, **Progress in Optics**, E. Wolf ed., (Acad. Press, New York, 1976) Vol. 13, p. 171; M.S. Sodha, L.A. Patel, and R.P. Sharma, J. Appl. Phys. **49**, 3707 (1978).
61. A.J. Palmer, Phys. Fluids **14** (1971), 2714; J.W. Shearer and J.L. Eddleman, Phys. Fluids **16** (1974), 1753; E. Valeo, Phys. Fluids **17** (1974), 1391; B. Bhat and V.K. Tripathi, J. Appl. Phys. **46** (1975), 1141; W.M. Mannheimer, Phys. Fluids **17** (1974), 1413; V. del Pizzo, B. Luther-Davies, and M.R. Siegrist, Appl. Phys. **14** (1977), 381; Yu, My, and K.H. Spatschek, Z. Naturforsch, **39A** (1974), 1736; R. Dragila and J. Krepelka, J. Physique **39** (1978), 617; J. Phys. **D11** (1975), 217.
62. V.V. Korobkin and A.J. Alcock, Phys. Rev. Lett. **21** (1968), 1433.
63. M.C. Richardson and A.J. Alcock, Appl. Phys. Lett. **18** (1971), 357.
64. H. Hora, J. Opt. Soc. Amer. **65**, 882 (1975).
65. H. Hora, and M.M. Novak, **Laser Interaction and Related Plasma Phenomena,** H.J. Schwarz et al., eds. (Plenum, New York) Vol. 4B (1977) p. 999.
66. J.W. Shearer, J. Garrison, J. Wong and J.E. Swain, **Laser Interaction and Related Plasma Phenomena**, H.J. Schwarz et al., eds. (Plenum, New York), Vol. 3B (1974), p. 803; H. Hora, Opto-Electron **5** (1973), 491.
67. B. Luther-Davies et al., European Conference on Laser Interaction with Matter, Schliersee, Germany, Jan. 1982.
68. H. Hora, E.L. Kane, and J.L. Hughes, J. Appl. Phys. **49**, 923 (1978). H. Hora, J.L. Hughes, J.C. Kelly, and B. Luther-Davies, U.S.A. Patent No. 510,700.
69. A.W. Ehler, J. Appl. Phys. **46** (1975), 2464.
70. B. Luther-Davies, and J.C. Hughes, Opt. Comm. **18** , 351 (1976).
71. R.A. Haas, J.F. Holzrichter, H.G. Ahlstrom, E. Storm and K.R. Manes, Opt. Comm. **18** (1976), 105.
72. R.A. Haas, J.F. Holzrichter, H.G. Ahlstrom, and K.R. Manes, Opt. Comm. **18**, 105 (1976); K.R. Manes, H.G. Ahlstrom, R.A. Haas, and J.F. Holzrichter, J. Opt. Soc. Amer. **67**, 717 (1977).
73. E.L. Kane, Ph.D. Thesis, Univ. New South Wales (1979); E.L. Kane and H. Hora, Aust. J. Phys.

34, 385 (1981).

74. E.L. Kane and H. Hora, **Laser Interaction and Related Plasma Phenomena** H.J. Schwarz and H. Hora eds. (Plenum, 1979), Vol. 4B, 913.
75. M.R. Siegrist, J. Appl. Phys. **48**, 1378 (1977).

REPORT OF THE WORKING GROUP ON FAR FIELD ACCELERATORS*

C. Pellegrini
National Synchrotron Light Source, Brookhaven National Laboratory
Upton, NY 11973

February 1982

ABSTRACT

This report describes the work of the Group on Far Field Accelerators. The work was focused on two accelerator schemes, the Inverse Free Electron Laser and the Two Wave Device. The possibilities and limitations of these two accelerators are discussed, as well as some of the requirements on the laser necessary to reach very high energies. A conceptual design of a single pass Inverse Free Electron Laser is presented.

I. INTRODUCTION

The very high electric field obtainable in a laser beam makes the idea of a laser accelerator attractive. This laser accelerator is even more attractive if the acceleration process does not require the presence of a plasma or other media or of a metallic structure at a wavelength distance from the beam, thus avoiding problems of electrical breakdown and beam intensity limitations typical of media or near field accelerators.

A far field accelerator, utilizing a laser beam in vacuum and with any material structure being at a distance from the laser and particle beam large compared to the radiation wavelength, offer some important advantages with respect to other laser accelerator concepts:

1) The acceleration region is far removed from all material boundaries, thus avoiding problems of material breakdown and plasma formation.
2) The accelerated beam is not constrained to move near a material boundary, thus avoiding problems of emittance limitations, and of intensity limitations due to the electromagnetic interaction of the beam with the material boundary.

The acceleration of particles utilizing a radiation field, laser field, and some combination of static magnetic fields has been proposed by many authors.[1-4] The work of the group on far field accelerators has been dedicated to the study of the Inverse

*Work supported by the U.S. Department of Energy under Contract No. DE-AC02-76CH00016.

Members of the Working Group were R. Freedman, L. N. Hand, H. Henke, C. Levey, H. Motz, R. H. Pantell, C. Pellegrini, A. Renieri, R. Rossmanith, A. Ruggiero, H. A. Schwettman, T. Smith, S. Solimeno, P. Sprangle, C.M. Tang, L.C. Teng.

Free Electron[5,6] (IFEL) and of the Two Wave Device[7] (TWD). It is worth noting that an experimental proof of the IFEL has been obtained recently at Los Alamos, Mathematical Sciences Northwest and TRW.[8] For these two systems the group has worked out a set of scaling laws to establish the maximum accelerating gradient obtainable. A conceptual design of an inverse free electron laser has also been done and has been very useful in understanding the limits on the maximum beam energy and other accelerator parameters.

There are two types of limits on the maximum beam energy, one related to the laser accelerating field and one to synchrotron radiation losses of the electron beam. To obtain high electron energies one needs a number of accelerating regions where the laser beam is focused to a small size. One acceleration region is shown in Fig. 1 for an IFEL; a laser beam is brought to a focus in an ondulator where the accceleration takes place. For a TWD the ondulator is to be replaced by a microwave structure.

A system with many acceleration regions can be obtained either by multiple focusing of one laser beam, as in Fig. 2, or by using a number of laser beams, each focused only once, as in Fig. 3. In the first case one needs to solve the problem of transmitting and focusing a high power laser beam, with a power of the order of 10^{13}-10^{14} W, over distances of the order of 10^3 m, periodically focusing the beam to a spot size of the order of 10^{-3} m. In the second case the main problem is to keep the phase coherence of the amplified laser beams so that the particles remain in step with the accelerating field.

The multiple focusing or multiple beam design are common also to other laser accelerator schemes and deserve a careful study. In this report we have chosen to simplify the accelerator design by considering only one acceleration region. This, of course, limits the maximum beam energy.

Synchrotron radiation losses are unavoidable in an inverse free electron laser accelerator, and are produced when the beam moves sideway in the ondulator. These losses can be minimized by a proper choice of ondulator parameter, but they remain the ultimate limitation to the maximum beam energy.

In what follows we will first discuss the basic physics of the Inverse Free Electron Laser and of the Two Wave Device. We will then obtain some scaling laws and discuss a conceptual design of a far field accelerator.

II. FAR FIELD ACCELERATORS

Consider one relativistic electron moving along the z-axis and subject, over a certain distance, to a transverse force F_x producing a transverse oscillation (Fig. 4). Assume

$$F_x = A\, e^{i(\omega_p t - K_p z)} \tag{1}$$

and

$$z = \beta_z ct \tag{2}$$

The transverse velocity is

$$\beta_x = -i \frac{A}{m\gamma c} \frac{1}{\omega_p - K_p \beta c} e^{i(w_p t - K_p z)} \tag{3}$$

where m is the electron mass, γ its energy in rest energy unit. Note that the energy change is proportional to $\beta_x F$ and has zero average value.

Add to the system a laser beam, propagating along z, which we represent as a plane electromagnetic wave

$$E_x = E_o e^{-i\omega_L(t-z/c) + i\phi_o} \tag{4}$$

We have now an energy exchange between the electron and the wave. The change in electron energy is

$$\frac{d\gamma}{dt} = \frac{eE_x \beta_x c}{m c^2} = -i \frac{eE_o Ae^{i[\omega_p - K_p \beta_z c - \omega_L + w_L \beta_z]t + i\phi_o}}{m^2 c^2 \gamma(\omega_p - K_p \beta_z c)} \tag{5}$$

The average value of $d\gamma/dt$ is non-zero if

$$\omega_L(1-\beta_z) = \omega_p - K_p \beta_z c \tag{6}$$

and this condition can also be obtained from energy and momentum conservation. If (6) is satisfied we can rewrite (5) as

$$\frac{d\gamma}{dt} = \frac{e E_o A}{m^2 c^2 \gamma(\omega_p - K_p \beta_z c)} \sin \phi_o \tag{7}$$

describing acceleration if $\sin \phi_o > 0$ or deceleration if $\sin \phi_o < 0$.

We will not discuss here the effect of oscillation around the phase ϕ_o, and the acceptance in energy of the accelerator. This discussion can be found in reference 4 for the IFEL case. We limit the discussion to the evaluation of the acceleration rate described by (7) for a given ϕ_o

To describe the electron oscillations and the energy change it is convenient to introduce the following quantities: electron oscillation wavelength

$$\lambda_o = \frac{2\pi c}{\omega_p - K_p \beta_z c} \tag{8}$$

the force parameter

$$K = \frac{A\lambda_o}{2\pi m c^2} \tag{9}$$

Using (8), (9), we can write the amplitude of electron transverse velocity, β_x, as

$$\bar{\beta}_x = \frac{K}{\gamma} \tag{10}$$

and the electron oscillation amplitude as

$$a = \frac{\lambda_o K}{2\pi\gamma} \tag{11}$$

With the help of (8), (9) and (10) we can rewrite the maximum accelerating gradient (7) as

$$mc^2 \frac{d\gamma}{dz} = \frac{e E_o K}{2\gamma} \sin \phi_o \tag{12}$$

III. THE IFEL AND TWD

In Section 2 the transverse force acting on the beam has not been specified. We consider now two particular ways to produce this force: one with a magnetic ondulator, the other with a slow TE electromagnetic wave propagating nearly parallel to the electron beam. The first case corresponds to an IFEL, the second to a TWD.

In the IFEL case[4] the force produced by an ondulator with a magnetic field B_w, is obtained by taking in (1).

$$A = e B_w \beta_z \tag{13}$$

$$\omega_p = 0, \quad K_p = \frac{2\pi}{\beta_z \gamma_o} \tag{14}$$

We then have from (9)

$$K = \frac{e B_w \lambda_o}{2\pi m c^2} \tag{15}$$

The synchronism condition (6) becomes

$$\omega_L = \frac{2\pi c}{\lambda_o(1-\beta_z)} \tag{16}$$

Since $\beta_z = (1-\beta_x^2)^{1/2} \approx 1 - (1+K^2)/2\gamma^2$, (16) can also be written in the more familiar form

$$\lambda_L = \frac{\lambda_o(1+K)}{2\gamma^2} \quad (17)$$

For a slow TE wave propagating at an angle θ to the z axis, with velocity c/n and an electric field amplitude E_p, we have[6]

$$A = e\,E_p\,(\cos\theta - \beta_z n) \quad (18)$$

$$K_p = \frac{n\,\omega_p\,\cos\theta}{c} \quad (19)$$

$$\omega_L = \omega_p\,\frac{1-n\,\beta_z\cos\theta}{1-\beta_z} \quad (20)$$

$$\lambda_o = \frac{2\pi\,n\,\cos\theta}{K_p(1-n\,\beta_z\cos\theta)} \quad (21)$$

$$K = \frac{e\,E_p\,\lambda_o\,(\cos\theta - \beta_z n)}{2\pi\,mc^2} \quad (22)$$

IV. SCALING LAWS

In the TWD case if one chooses

$$1 - n\,\beta_z\cos\theta \approx 0 \quad (23)$$

i.e. wave velocity nearly equal to particle velocity, the quantities λ_o, K, β_x become large and $d\gamma/dz$ increases. Pantell and Smith[6] have shown that one can get a solution of the electron equations of motion in the combined slow wave and laser field, such that

$$1-n\,\beta_z\,\cos\theta \propto \gamma^{-2/3} \quad (24)$$

From (24) and (18) to (22) one can obtain the scaling of the electron oscillation and of the acceleration rate with energy

$$\lambda_o \propto \gamma^{2/3} \quad (25)$$

$$K \propto \gamma^{2/3} \quad (26)$$

$$\beta_x \propto \gamma^{-1/3} \quad (27)$$

$$a \propto \gamma^{1/3} \tag{28}$$

$$\frac{d\gamma}{dz} \propto \gamma^{-1/3} \tag{29}$$

More exactly for the acceleration rate one has[6]

$$mc^2 \frac{d\gamma}{dz} = 2\text{x}10^{-2} \left(\frac{\lambda_L E_p}{\gamma}\right)^{1/3} \left(\frac{n^2-1}{n}\right)^{1/3} e E_o \frac{eV}{m} \tag{30}$$

with E_p and E_o in V/cm and λ_L in μm.

When γ changes n and/or θ have to be adjusted to satisfy the synchronism condition (20).

The scaling of the acceleration rate, (29), has a weak dependence on γ. However, as γ increases the oscillation amplitude increases, as shown by (28), and this may require a larger dimension of the laser beam which would reduce E_o.

In the IFEL case the acceleration rate is given by equation (12), with K defined in (15) and the synchronism condition by (16) or (17). Since both K and/or λ_o have to be a function of γ to satisfy (17), the dependence of the acceleration rate on γ is more complicated than γ^{-1}. As an example, let us assume that the ondulator magnetic field is kept constant and that one changes λ_o to satisfy (17) for a given λ_L. For K > 1 one obtains the scaling

$$\lambda_o \propto \gamma^{2/3} \tag{31}$$

and all other quantities scale as in (26) to (29).

Assuming again for simplicity K > 1, one obtains for $d\gamma/dz$ the expression

$$mc^2 \frac{d\gamma}{dz} = \frac{e E_o}{\gamma^{1/3}} \left(\frac{e B_w \gamma_L}{\pi mc^2}\right)^{1/3} \sin\phi_o, \; B_w = \text{constant} \tag{32}$$

As an example, if we assume for the TWD λ_L = 10 μm, E_p = 10^6 V/m, E_o = 10^{11} V/m, n = 1.5, $\gamma = 10^3$ we obtain from (30) an accelerating field

$$mc^2 \frac{d\gamma}{dz} = 400 \text{ MeV/m}$$

In the IFEL case if we use the same laser wavelength and electric field and we assume B_w = 0.12 T and λ_o = 50 cm we obtain at γ = 10^3

$$mc^2 \frac{d\gamma}{dz} = 600 \text{ MeV/m}$$

V. LASER FOCUSING AND RADIATION LOSSES

A simple far field accelerator can be designed considering a single acceleration region of length L and assuming a gaussian laser beam focalized to a waist w_o related to the radiation wavelength and the length L by[9]

$$2\pi w_o^2 = L\gamma \tag{33}$$

The change in beam area is then given by

$$w^2 = w_o^2 (1+ \frac{4z^2}{L^2}) \tag{34}$$

For a laser beam of total power W the electric field is given by

$$E_o = \left(\frac{2WZ_o}{\pi w^2} \right)^{1/2} \tag{35}$$

when $Z_o = 377\ \Omega$.

Since in the acceleration region, $|z| < L/2$, the waist variation is small, we simplify the calculation by assuming $w = w_o$ and

$$E_o = \left(\frac{4W\ Z_o}{\lambda_L L} \right)^{1/2} \tag{36}$$

Using (30), (36), we obtain for a single pass TWD of length L and assuming that the final energy, γ_f, is much larger than the initial energy, γ_o,

$$\gamma_f \approx 10^{-4}\ (WL)^{3/8}\ E_p^{\ 1/4}\ \lambda_L^{-1/8} \tag{37}$$

Note the weak dependence on the laser wavelength, λ_L. For a laser power $W = 10^{14}$ W, $\lambda_L = 10$ µm and $E_p = 10^6$ v/m we obtain from (37)

$\gamma_f = 2.4$ GeV for $L = 4$ m

$\gamma_f = 30$ GeV for $L = 4\text{x}10^3$ m

showing that the system is very effective for short acceleration length (an average accelerating field of 600 MeV/m for L = 4 m) but not for large L. This is a consequence of (33) which implies a large w_o, and hence a small electric field, as given by (36), for large L. To accelerate to high energies it is necessary to use multiple acceleration region as in Fig. 2 or 3. Synchrotron radiation losses are small for a TWD.[7] For the same value of λ_L,

W and E_p used before and for $\gamma = 10^6$ these losses are smaller than 1 MeV/m.

Synchrotron radiation losses can be an important factor in an IFEL. In this case the energy lost per unit ondulator length is

$$S = \frac{8}{3} \pi^2 \frac{r_e}{\lambda_o^2} K^2\gamma^2 mc^2 \tag{38}$$

The condition

$$mc^2 \frac{d\gamma}{dz} > S \tag{39}$$

can be written as

$$e E_o > \frac{8}{3} \pi^2 \frac{r_e}{\lambda_o^2} K \gamma^3 mc^2 \tag{40}$$

In the case discussed previously in which one keeps B_w = constant in the ondulator and $\lambda_o \propto \gamma^{2/3}$ we obtain

$$e E_o \propto \gamma^{7/3} \tag{41}$$

A scaling with λ_o = constant and $B_w \propto \gamma$ gives a less favorable result, $e E_o \propto \gamma^4$, equal to that of a circular accelerator. In the case $B_w = 0.1$ T = constant we obtain a synchrotron radiation loss, at $\gamma = 10^5$, $S = 4\times10^4$ eV/m. At the same energy the ondulator period is 12 m, and the electron oscillation amplitude is 2.4 mm.

VI. AN IFEL ACCELERATOR

In this section we want to summarize some of the previous results for an IFEL and make a conceptual design of a single pass IFEL accelerator. We first rewrite the formulae we want to use:

a) synchronism condition, eq. (17)

$$\lambda_L = \frac{\lambda_o(1+K^2)}{2\gamma^2} \tag{42}$$

b) acceleration rate, eq. (12)

$$mc^2 \frac{d\gamma}{dz} = \frac{e E_o K}{\gamma} \sin \phi_o \tag{43}$$

c) synchrotron radiation losses; using (38) and (39) we write this as

$$\frac{d\gamma}{dz} > \frac{8}{3} \pi^2 \frac{r_e}{\lambda_o^2} K^2 \gamma^2 \tag{44}$$

d) transverse beam oscillation amplitude, eq. (11)

$$a = \frac{\lambda_o K}{2\pi\gamma} \tag{45}$$

e) accelerator length, eq. (33), and corresponding laser electric field for given laser power, W, eq. (36)

$$L = \frac{2\pi w_o^2}{\lambda_L} \tag{46}$$

$$E_o = \left(\frac{4WZ_o}{\lambda_L L}\right)^{1/2} \tag{47}$$

We integrate (43), (42) assuming, for simplicity, K > 1 and in the two cases λ_o = constant or B_w = constant. This gives for the final electron energy, $mc^2\gamma_f$, the value

$$mc^2(\gamma_f - \gamma_o) = e\left(\frac{8W Z_o L}{\lambda_o}\right)^{1/2} \sin\phi_o, \quad \text{if } \lambda_o = \text{constant} \tag{48}$$

or

$$\gamma_f^{4/3} = \gamma_o^{4/3} + \frac{4}{3}\frac{e}{mc^2}\left(\frac{4W Z_o L}{\lambda_L}\right)^{1/2}\left(\frac{e B_w \lambda_L c}{\pi mc^2}\right)^{1/3} \sin\phi_o, \tag{49}$$

if B_w = constant

γ_o being the initial value. Since γ_f increases with z along the accelerator, λ_o and/or B_w have also to increase. Notice that there is no dependence on λ_L in (48) and a very weak dependence in (49).

The maximum beam current, I_B, is determined either by energy conservation

$$I_B < \frac{W}{mc^2\gamma_f} \tag{50}$$

or by the condition that the defocusing electric field produced by the beam iteself be less than the external focusing and accelerating field.[10] This last condition is in most cases weaker than (50).

The final longitudinal beam energy spread is determined by the depth of the trapping potential,[4]

$$\frac{\Delta\gamma}{\gamma} = \left(\frac{e\ E_o\ K\ \lambda_o}{\pi\ mc^2\ \gamma^2}\right)^{1/2} \left(\cos\phi_o - (\frac{\pi}{2} - \phi_o)\mathrm{som}\phi_o\right)^{1/2} \qquad (51)$$

Using eqs. (42) to (51) we can make a conceptual design of a 4 GeV IFEL accelerator, obtaining the parameters of Table I. The laser parameters used in this table correspond to an existing Nd laser.

VII. CONCLUSIONS

The TWD and IFEL are similar in the basic physics principles. The substitution of the static ondulator magnetic field of the IFEL with a slow microwave field in the TWD reduces synchrotron radiation losses at high electron energies.

Both the TWD and IFEL can produce GeV beams in a single pass with an accelerator length of the order of tens of meters. To produce higher energies it is essential to use a system with many acceleration regions and solve the related laser beam focusing problem.

Further studies are needed to design a multistage accelerators for very high energies, optimizing the variation of the refractive index for a TWD or of the ondulator parameters for an IFEL, to minimize the total accelerator length and the synchrotron radiation losses.

It is worth noting that in an accelerator based on the IFEL or TWD concept, the electron beam is bunched on the scale of the laser wavelength and can produce intense spontaneous radiation at the harmonics of the laser frequency.

TABLE I

	λ_o =10 cm (constant)	B_w =1 T (constant)
Laser Parameters		
Power, W		$2 \cdot 10^{13}$ W
Pulse duration, τ		1 ns
Spot size, w_o		0.25 cm
Wavelength, λ_L		1 μm
Electric field, E_o		$2.8 \cdot 10^{10}$ v/m
Interaction Length, L		39 m
Ondulator Parameters		
Period, λ_o	10 cm	3.8 → 23 cm
Magnetic field	0.31 → 3.8 T	1 T
Synchronous phase ϕ_o	$\pi/3$	$\pi/3$
Electron Beam Parameters		
Energy, $m_oc^2\gamma_f$	250 MeV → 4.2 GeV	250 MeV → 3.8 GeV
Current, I_B	<5 KA	<5 KA
Beam radius, r_B	0.2 cm	0.2 cm
Average accelerating field	101 MeV/m	90 MeV/m
Oscillation amplitude, a	0.007 cm	10^{-2} cm
Energy spread	10^{-4}	10^{-4}
Synchrotron radiation loss at γ_f	300 keV/m	20 keV/m

REFERENCES

1. A. A. Kolomenskii and A. N. Lebedev, Soviet Physics JETP 23, 733 (1966).
2. R. B. Palmer, J. Applied Physics 43, 3014 (1972).
3. H. Motz, Contemporary Physics 20, 547 (1979).
4. W. B. Colson and S. K. Ride, Applied Phys. 20, 61 (1979).
5. N. M. Kroll, P. L. Morton, and M. N. Rosenbluth, IEEE Journal of Quantum Electronics QE-17m 1436 (1981).
6. P. Sprangle and C. M. Tang, IEEE Trans. Nucl. Sci. NS-28, 3, 3340 (1981).
7. R. H. Pantell and T. I. Smith, Laser driven electron acceleration by means of two-wave interaction, Stanford University Preprint (1981).
8. H. Boehmer *et al.*, Phys. Rev. Letters 48, 141 (1982) and private communications.
9. See for instance A. Yariv, Quantum Electronics, Wiley, New York 1975.
10. V. K. Neil, Lawrence Radiation Laboratory Report UC1D-17985, (1978).

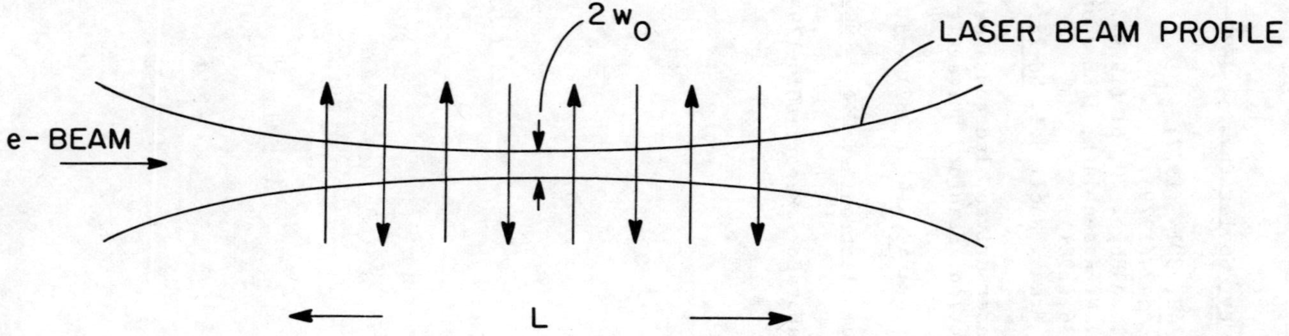

Fig. 1 Schematic representation of an acceleration region. A laser beam and an electron beam traverse an ondulator magnetic field of length L. The laser beam profile is assumed gaussian with a waist radius w_0.

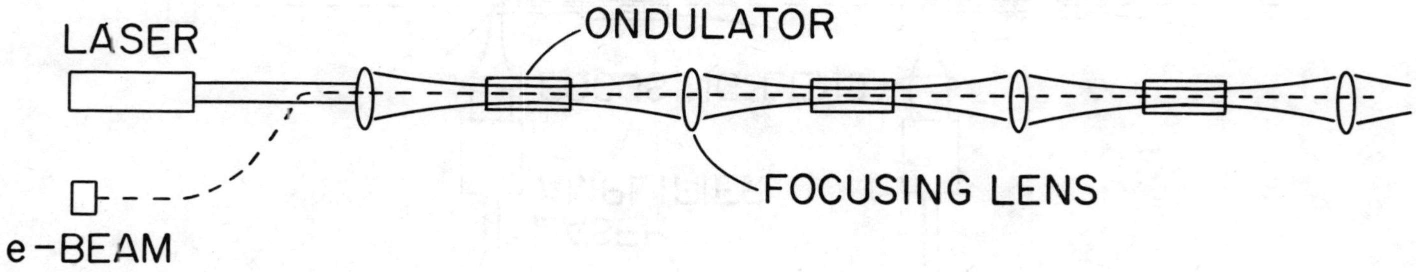

Fig. 2 Conceptual design of an accelerator with multiple acceleration regions and laser beam refocusing.

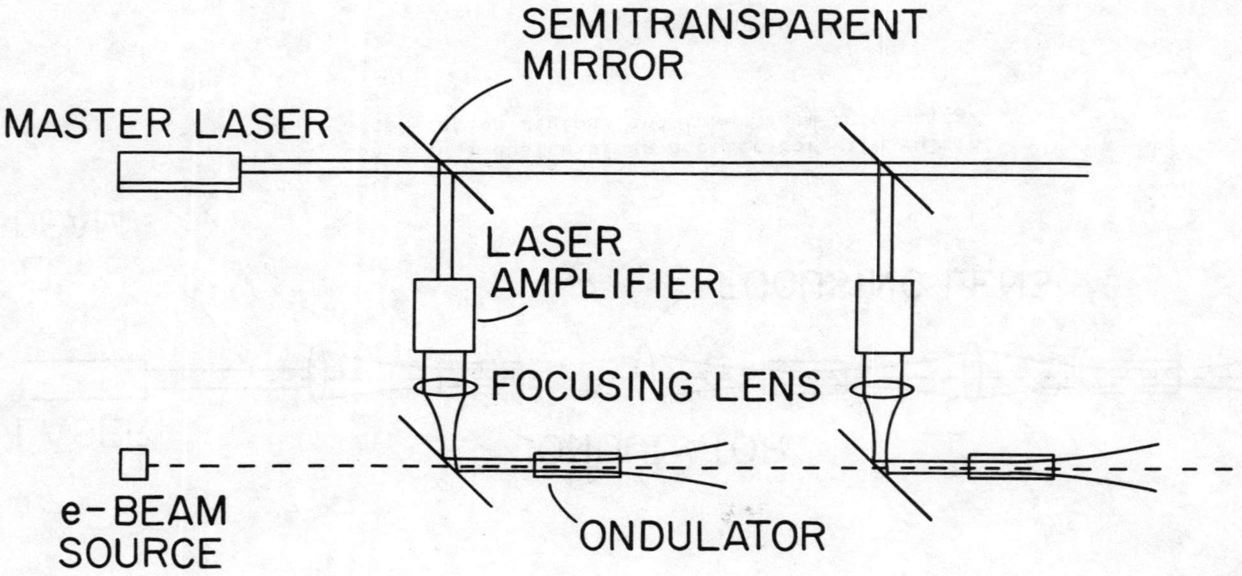

Fig. 3 Conceptual design of an accelerator using multiple laser beams and acceleration regions.

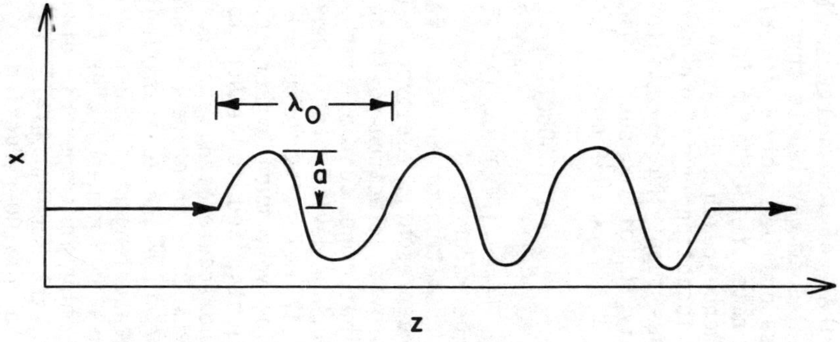

Fig. 4 Electron trajectory in ondulator. The period and amplitude of oscillation are λ_o and a.

THE FREE ELECTRON LASER AS A POWER SOURCE FOR A HIGH-GRADIENT ACCELERATING STRUCTURE*

Andrew M. Sessler

Lawrence Berkeley Laboratory
University of California
Berkeley, CA 94720

ABSTRACT

A two beam colliding linac accelerator is proposed in which one beam is intense ($\approx$1kA), of low energy ($\approx$ MeV), and long ($\approx$ 100 ns) and provides power at 1 cm wavelength through a free-electron-laser-mechanism to the second beam of a few electrons ($\approx 10^{11}$), which gain energy at the rate of 250 MeV/m in a high-gradient accelerating structure and hence reach 375 GeV in 1.5 km. The intense beam is given energy by induction units and gains, and looses by radiation, 250 keV/m thus supplying 25 J/m to the accelerating structure. The luminosity, L, of two such linacs would be, at a repetition rate of 1 kHz, $L = 4. \times 10^{32}$ cm^{-2} s^{-1}.

INTRODUCTION

A free electron laser (FEL) is a high-peak-power device which operates over a large range of frequencies and hence allows one to seriously consider operation of an accelerating structure at higher frequencies than are presently employed. The Stanford Linear Collider (SLC) will provide electron-positron collisions at 50 GeV x 50 GeV.[1] Beyond that, one contemplates a collider of (say) 300 GeV x 300 GeV.[2] Such a device, if it were to operate at the gradient of the SLC, 17 Mev/m, would be 18 km long. Thus one is driven to considering very much higher accelerating gradients. To achieve these gradients, and to reduce to a manageable level the energy stored in the accelerating structure one is driven to considering higher frequencies than are used in the SLC.

In this paper we take the accelerating structure to operate at 30 GHz; i.e., at a wavelength of 1.0 cm. Thus we consider just a factor of 10 increase in frequency over that in the SLC and, hence, a factor of 10 reduction in transverse dimensions of the accelerating structure. For the same accelerating gradient we would have, consequently, a factor of 100 reduction in stored energy.

For the accelerating gradient, at this high frequency, we believe we can achieve 250 MeV/m. Taking the accelerating structure to be 1.5 km long yields an energy of 375 GeV. One would also need

---

* This work was supported by the Director, Office of Energy Research, Office of Basic Energy Sciences, of the U. S. Department of Energy under Contract No. DE-AC03-76SF00098.

about 0.5 km of the present SLC accelerating structure, with its associated sources and damping rings, as an injector and hence the total length of one linac is 2.0 km. We note that two linacs just fit on the SLAC site (4 km) with the collision point half-way up the site.

Constructing and aligning the accelerating structure (if scaled from the present structure of the SLC it would have a radius of 3 mm and a beam hole of 1 mm!) is a formidable problem. We believe it is not, however, insuperable. Operation at 1 kHz, and with a bunch of 10^{11} particles, would give a luminosity of 4.0 x $10^{32}cm^{-2}sec^{-1}$.

The FEL for powering the structure can be designed to yield 1.0 GW/m which with a pulse length of 25 nsec is more than adequate to power the accelerating structure. The average beam power is 12.0 MW and the power from the mains is (about) 150 MW. Although the average power is high, and could be reduced by considering an even higher frequency, we believe the problems associated with smaller wavelengths are very difficult and unlikely to be solved in time for the next generation of collider. Beyond that, we will have to miniaturize and, hence research on these subjects is called upon.

The FEL should operate in "steady state"; i.e., the electron beam of (say) 3 MeV neither gains nor loses energy as it moves down the FEL. At an energy gain of 1 MeV/m, and a current of 1.0 kA, the FEL-beam is gaining 1.0 GW/m from the induction units, and radiating an equal amount due to the wiggler. The FEL-beam should have a length of 25 nsec so as to give 25 Joules/m.

Thus we are led to two beams traveling the length of the accelerating column (1.5 km). One is the FEL electron beam of 3 MeV, 1.0 kA, and length 25 nsec. The beam travels through induction units and a wiggler. The second beam consists of the 10^{11} electrons which are in a bunch of length 1 mm (so that the peak current is 480 A). This beam is taken to 375 GeV by traveling through the accelerating structure. The two structures are linked by pipes which carry the microwaves over from the FEL to the accelerating structure.

In this report we shall provide the reasoning and calculations behind the concept described in this Introduction. We shall end by suggesting experiments and theoretical studies which need to be done before one can feel confident about actually building the de vice proposed here. Fortunately, as will be seen, the requisite "proof-of-principle" work is rather modest in extent and cost.

ACCELERATING GRADIENT

In the present analysis we are motivated to obtain as large an accelerating gradient as possible. Just how high the gradient can be before sparkbreakdown occurs is unknown. In fact, this subject was considered by experts at this Workshop on the Laser Acceleration of Particles.[3] A gradient of 80 MV/m has been achieved by SLAC people and the Novosibirsk group has reported achieving

100 MV/m.[4] These gradients were at S-band and it is felt that one will do much better at 30 GHz. Estimates range from 150 MV/m through ("surely") 200 MV/m to ("possibly") 500 MV/m. We take 250 MV/m, but a larger gradient could shorten the device while only 200 MV/m would (in 1.5 km) still give 300 GeV x 300 GeV.

ACCELERATING STRUCTURE

In studying the accelerating structure we follow very closely the work by Wilson.[2] In fact, we really just use the methods of Wilson to study choice of parameters. We shall, therefore, use the notation of Ref. 2 and the reader may have to read that reference in order to understand this section.

Taking a charge bunch of 10^{11} particles, an RF wavelength of 1 cm, and an accelerating gradient of E_a = 250 MV/m, one, firstly, has to choose a structure type. We think a Jungle Gym is most suitable for this application and, hence, estimate the parameter k_0 ($E_a^2 = 4k_0 w_s$, where w_s is the stored energy per unit length) as 2.5 x 10^{15} V/C-m. One readily obtains that the structure energy, when it is excited, is 7.8 J/m and that the bunch gains an energy of 4.0 J/m.

The next thing to examine is the transverse wake field, which can be characterized by the parameter A. One readily finds that A is unacceptably large. One can increase the transverse focussing, or lower the frequency, or increase the beam-nearest structure size, (hole-size in a disk-loaded structure). All of these don't have to be done, but it seems convenient to (1) take λ_β = 10 m (It is 100 m in the SLAC.) and (2) take a = 2 mm. (It would be 1 mm if we "scaled-down" the SCL structure.) In this case, since A depends upon the parameter, a, as $a^{-3.5}$, we find an acceptable value of A.

Now, however, with the larger value of a, the structure is less efficient and thus $k_0 \approx$ 1.0 x 10^{15} V/C-m. The stored energy in the structure is now 19.5 J/m. (Alternatively one could decrease the frequency, but I think it is better to keep the frequency high, but make the accelerating structure less efficient.)

The luminosity one can obtain is dependent upon the repetition rate, the bunch length, and the beamstrahlung parameter. Taking δ = 0.05, which is close -- if maybe even beyond what experimentalists would find acceptable -- yields a bunch length, σ_z of 1 mm. (This is <u>not</u> what one would obtain by simply scaling SLC parameters for there the bunch length will be 1 mm so that scaling would yield 0.1 mm.)

The luminosity, with a repetition rate, f_r, of 1 kHz is 1.2 x 10^{32} $cm^{-2}s^{-1}$, but the disruption parameter, D, is 0.90. Thus the lumniosity is enhanced by about a factor of 3 to 4 x 10^{32} $cm^{-2}s^{-1}$. The crossing point β^* also has a reasonable value and is 1.04 cm. Finally, the beam transverse emittance ε_n = 3.0 x 10^{-3} cm, and is also reasonable.

With this long bunch length of 1 mm there will be an energy spread from acceleration of (about) 10%. Thus there is a large

(10%) energy spread in the present design. This is a difficulty with all high frequency colliders and there is no way to "get around the problem" and still have a good luminosity (since the luminosity varies as $f_r \sigma_z \delta$ and both σ_z and δ are restricted by the allowed energy spread) unless one contemplates either a higher rate, or a number of bunches per pulse.[5] The higher repetition rate is possible for an induction linac (greater than 1 kHz in possible) but makes the average power consumption probably unacceptably high. A pulse train is always a possibility -- as P: Wilson points out -- but is limited by the combination of resistive decay time in the accelerating structure and the cycle-time of particle detectors, so the number of bunches is probably limited to the number of distinct detectors on-line at any time.

FREE ELECTRON LASER

A free electron laser (FEL) can be used to generate the peak power needed to excite the accelerating structure. A rather extensive study has been made at 3 mm and one would expect not very large changes, from that work, at 1 cm.[6] In fact, the design will be easier at 1 cm so that one should be able to accommodate up to 2.0 kA of trapped particles.

On the other hand, induction modules should be able to be made compactly and stacked so that an acceleration of 1 MeV/m is readily obtainable. (Even a gain of 3 MeV/m should be possible.)

Thus, the beam would be given 2 GW/m. We have taken 250 MW/m so either there needs to be less energy gain to the electrons or a smaller current need be employed.

In steady-state, all of this energy will be radiated at 1 cm. Calculations are presently being done on wiggler wavelength, magnetic field, beam energy, current density, etc. Preliminary work, however, indicates that the parameters assumed, here, are quite reasonable.[7]

POWER REQUIREMENTS

The accelerated beam requires an energy of 4 J/m, while powering the structure requires 20 J/m. Thus the beam takes 20% of the energy out of the structure which is not so high as to give concern about instabilities or so low as to be grossly inefficient.

Let us assume an 80% efficiency for powering the structure, so that the FEL must produce 25 J/m. If the FEL is rated at 250 MW/m, then the FEL electron beam must have a duration of 100 ns which is just in the range of accessibility with ferrite cores on the induction linac. One can build such an induction linac with (about) a 50% efficiency from the mains to the beam. Hence, the over-all efficiency is (20%)(80%)(25%) = 8%. If the rep-rate is 1 kHz, and the length of two linace is 3 km then the beam power is 12 MW and the average power from the main is 150 MW.

The above estimate, which is conservative, may be quite conservative. We have designed an FEL which can fill the whole accelerating structure at the required 20 J/m. But the FEL beam, necessary to do this, is only 100 ns long. Thus we envision a 100 foot long, 1-2 kA, beam moving along and powering the accelerating structure. But in the accelerating structure the group velocity can be made large. In this case the pulse of energy can be arranged to move along just with the bunch of accelerated particles. Thus one has only to re-supply resistive losses and beam power losses and the power requirement is very much less than 150 MW.[8]

FURTHER WORK

In order to carry the suggestion made in this paper to the point where it is a serious contender for a linear collider project one must do experimental and theoretical work on a number of fronts:

1. Accelerating Structure

   Computer studies need to be made to see what structure is best at 1 cm and with a large beam hole (2 mm). Presumably one will find that it is a Jungle Gym structure. One needs to determine phase and group velocity so as to accurately evlauate power requirements of the structure. Tuning and tolerance requirements must also be determined.

   Most importantly, one must learn how to fabricate such a structure; first at all and then cheaply. It seems like a Swiss watch, but then we know such watches can be built....

2. Free Electron Laser

   The FEL, of Sessler and Prosnitz, at the ETA will be optimized at 3 mm. One would, probably, build a FEL optimized at 1 cm. Presumably such a device would use electrons of less energy than 4.5 MeV.

   Subsequently, one would want to build a "steady state" FEL of (say) 10 meters length. Such a device could then be used, with an injected beam in an associated accelerating structure, to give the beam (say) an energy gain of 2.5 GeV.

3. Beak-Down Studies

   Of great importance is just what gradient can be achieved before break-down. The FEL at the ETA allows one to do just such studies. They expect 500 MW for (say) 10 nsec which is 5 J. Thus this device can power 25 cm of accelerating structure to 250 MV/m. Of course, one can make shorter devices, etc...

4. Coupling Studies

   One needs to determine, theoretically, a good configuration for coupling the FEL to the accelerating structure. One imagines the FEL being slowly made twice as wide as usual (say) every meter and then the microwave energy is taken away in an over-moded pipe. This concept needs to be refined. Also, there is the very important subject of coupling into the accelerating structure.

   Once again, the FEL at the ETA could be employed for experimental studies.

ACKNOWLEDGMENT

The author has greatly profited from a conversation with Edward Knapp (LANL) and from numerous conversations with Perry Wilson (SLAC). Also helpful have been discussions with Andris Faltens (LBL), Edward Hartwig (LBL), Roger Miller SLAC), Philip Morton (SLAC), Donald Prosnitz (LLNL), Wolfgang Schnell (CERN), Donald Swenson (LANL), and Cha-Mei Tang (NRL).

REFERENCES

1. B. Richter, Proc. of the 11th Intl. Conf. on High Energy Accelerators, Birkhauser Verlag, Basel, 1980, p. 168.
2. P. B. Wilson, IEEE Trans. on Nuclear Science NS-28, 2742, 1981; and AATF/80/20, April 7, 1981, SLAC Internal Report (unpublished).
3. A Brief Report on the Workshop on the Laser Acceleration of Particles, Feb. 18-23, 1982, (unpublished).
4. A. N. Skrinsky, private communication.
5. For example, if the experimental requirement was that $\Delta E/E < 1$ then we could consider the (self-consistent) set of parameters: $\sigma_z = 0.1$ mm, $E_0 = 375$ GeV, $f = 1$ kHz, $N = 7.8 \times 10^9$, $\varepsilon_n = 3 \times 10^{-3}$ cm, $\beta^* = 0.5$ cm, $\delta = 0.01$, $D = 1.5 \times 10^{-2}$, $L = 2.5 \times 10^{30}$ cm^{-2} s^{-1}. In this case, compared to the example given in the text, there has been a one-order-of-magnitude reduction in $\Delta E/E$, but the luminosity has gone down by two-orders-of-magnitude.
6. D. Prosnitz and A. M. Sessler, "Multimeter Wave Generation by a Single-Pass, Compton Regime, Variable Parameter Free Electron Laser", to be published in the Proceedings of the Conference on Free Electron Lasers, Sun Valley, 1981; UCRL-86258.
7. Cha-Mei Kim and A. M. Sessler, private communication.
8. If, for example, the group velocity were $v_g = c/2$ (probably one would need a Disk and Washer Structure, rather than a Jungle Gym Structure, for this purpose.) then only 1/2 of the structure would need to be excited and the average power would be reduced from 150 MW to 75 MW.

GUIDING OF VERY INTENSE LIGHT PULSES FOR LASER ACCELERATORS

S. Solimeno
Università di Napoli, Italy

ABSTRACT

The field distributions and physical properties of several waveguides capable of transporting near infrared laser pulses are reviewed, with the aim of assessing the maximum transportable power and the limit fields in the laser beam.

INTRODUCTION

One of the main problems encountered in the conceptual design of laser accelerators is the periodic refocusing of light pulses having a peak fluence density of the order of some Tw/cm^2. In fact, in the so-called far-field acceleration schemes the electron is subject to the combined action of a slow-wave (wiggler field or microwave field) and a powerful laser beam along a straight trajectory.

For a acceleration gradient of 1 GeV/m, and an energy of 1 TeV, the laser photon bunches have to be guided over a distance of 1 Km with good collimation. The coherent interaction with the slow-wave and the electron can be only achieved by containing the deformation of the light wavefronts within the limits of fractions of wavelengths. If these requirements are quite difficult to meet when the fluence density is small, in case of very energetic pulses the damage to dielectrics and metals establishes a sharp threshold for the power density of the fields to be kept confined.

The fact that dielectrics and metals can withstand fluence densities which do not exceed the value of $10^{-3} \div 10^{-1}$ Tw/cm^2, does not automatically exclude the possibility of achieving the required guiding action on the beam by resorting to hollow dielectric or metallic guides. A way to concile higher fluence densities with relatively low damage threshold is to accurately shape the beam so as to obtain a fluence along the bunch trajectory much higher than that in the vicinity of the guiding structure. A second factor which could help is connected with the grazing incidence of the beam. While current data on laser damage have been taken by normal incidence irradiation of the sample, we can speculate that the threshold should increase for grazing incidence in view of the drastic reduction of the field inside the dielectric.

In the present contribution we intend to examine the problem of propagation through dielectric and metallic waveguides. We will address the problem from both the electromagnetic point of view and the laser damage of dielectric and metallic materials.

DIELECTRIC WAVEGUIDES

Basically, the guiding properties of these structures are associated with the process of total reflection, according to which a light beam traveling in a medium possessing an index of refraction n_1, can be totally reflected when impinging on a discontinuity surface separating the first medium from a second one having a refractive index $n_2 < n_1$. The waveguides most widely used for propagation over long distances have cylindrical symmetry (see Fig. 1).

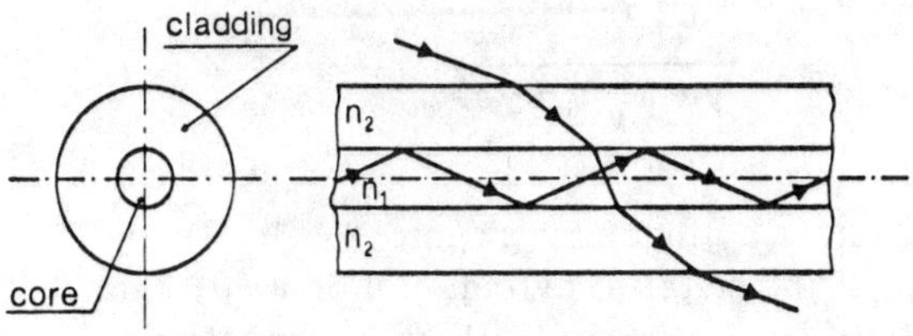

Fig. 1. - Cylindrical dielectric waveguide.-

The propagation of electromagnetic waves in these structures can be analysed by considering the field as a superposition of modes, characterized by a space dependence of the form exp $(-i\beta z)$, z being the axial coordinate.

From a physical standpoint the modes separate in two classes, guided modes and radiation modes according to whether the power carried along the propagation axis is conserved or not.

Solving Maxwell's equations, it can be shown[1] that the electric, E_z, and magnetic, H_z, axial components of a guided mode can be written as:

$$H_z(\varrho,\Phi) \propto E_z(\varrho,\Phi) \propto \begin{cases} J_\nu(x\varrho)\, e^{i\nu\Phi}, & \varrho \le a \\ \dfrac{J_\nu(xa)}{K_\nu(\gamma a)} K_\nu(\gamma\varrho)\, e^{i\nu\Phi}, & \varrho > a \end{cases} \qquad (1)$$

Here x and γ are related to the longitudinal propagation constant β through the relations:

$$\begin{aligned} \omega^2 \mu_o n_1 &= \beta^2 + x^2 \\ \omega^2 \mu_o n_2 &= \beta^2 - \gamma^2 \end{aligned} \qquad (2)$$

ν is an integer index, a is the radius of the inner region of the guide and K_ν is the modified Bessel function of order ν. The other components of the e.m. field can be obtained from Maxwell's equations. When the relative difference between the inner and outer refractive indices is much less than unity, a set of linearly polarized modes $(LP)_{\nu\delta}$ can propagate. The transverse electric components of these modes read:

$$(E_x)_{\nu\delta} \propto \begin{pmatrix} \sin \nu\Phi \\ -\mathrm{con}\nu\Phi \end{pmatrix} x \begin{cases} \dfrac{J_\nu(x\varrho)}{J_\nu(xa)}, & \varrho \leq a \\ \dfrac{K_\nu(\gamma\varrho)}{K_\nu(\gamma a)}, & \varrho > a \end{cases} \tag{3}$$

A similar equation holds for $(E_y)_{\nu\delta}$.

The constants β, x and γ depend on ν and δ through the characteristic equation:

$$\frac{J_\nu(xa)}{xa\, J_{\nu+1}(xa)} = \frac{K_\nu(a)}{\gamma a\, K_{\nu+1}(\gamma a)} \tag{4}$$

For $\nu = 0$, the fields of the dielectric cylinder break into TM (H_z=0) and TE (E_z=0) modes just as in the case of the metallic cylinder. For $\nu \neq 0$, hybrid modes, designated NE_m and EH_m, exist for which both E_z and H_z are non-zero. The designation HE and EH is given depending on whether or the largest contribution to the transverse field comes from H_z or from E_z.

An important mode parameter is the cutoff frequency, that is the frequency below which the mode becomes radiative. The cutoff conditions for the various mode types can be shown to be:

$$\left.\begin{matrix} EH_{\nu\delta} \\ EH_{1\delta} \end{matrix}\right\} J_\nu(xa) = 0$$

$$HE_{\nu\delta} \quad (n_1^2 + 1)\, J_{\nu-1}(xa) = \frac{xa}{\nu-1} J_\nu(xa) \quad \nu = 2, 3, \ldots \tag{5}$$

$$\left.\begin{matrix} TE_{o\delta} \\ TM_{o\delta} \end{matrix}\right\} J_o(xa) = 0$$

Note that the cutoff frequency of mode HE is zero. For frequency much larger than the cutoff frequency the characteristic equation simplifies notably into:

$$xa\, J_{\nu+1}(xa) = V\, J_\nu(xa) \tag{6}$$

where V is the so-called <u>normalized frequency</u> defined as:

$$V = a\,(x^2 + \gamma^2)^{1/2} = ak\,(n_1^2 - n_2^2)^{1/2} \tag{7}$$

By simple algebra it can be shown that eq. (4) yields:

$$xa = (xa)_{\infty} \left[1 - \frac{2\nu}{V} \right]^{1/2} \tag{8}$$

for $\nu \neq 0$ and

$$xa = (xa)_{\infty}\, e^{-1/V} \tag{9}$$

for $\nu = 0$, $(xa)_{\infty}$ being the zero of the equation $J_{\nu}(xa)=0$.

A plot of the normalized propagation constant β/K for a few of the low order modes is shown in Fig. 2.

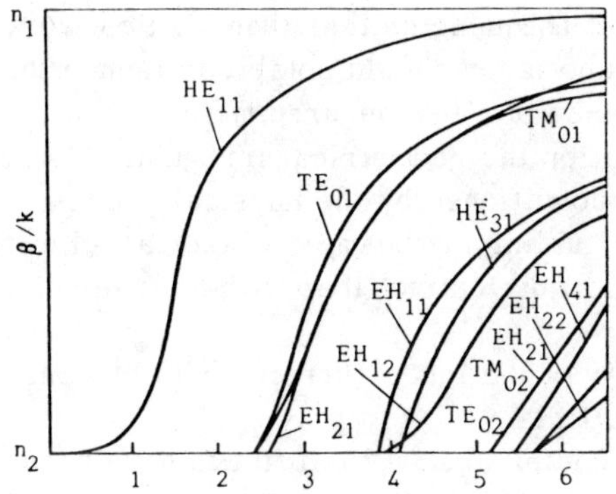

Fig. 2. - Propagation constant normalized frequency.-

For operating the waveguide single-made, V must be chosen less than 2.405.

As V increases, each mode becomes more strongly confined to the core and the fraction of power in the cladding approaches zero[2]. Far from cutoff Marcuse has obtained an asymptotic expression for P_{clad}/P,

$$\frac{P_{clad}}{P} = \left[(xa)_{\infty} \right]^{4} \frac{1}{V^{4}} \left(1 - \frac{2}{V}\right) \tag{10}$$

For the HE_{11} mode at V = 2.405, 84% of the power travels within the core. At V=10, the fraction of power reduces to 2.58‰.

It is also important to notice that for V=2.4, xa =1.68. Consequently, the field on the interface core-cladding is equal to:

$$\frac{E\,(a)}{E\,(o)} = J_o\,(1.68) \sim 0.4 \qquad (11)$$

Far from cutoff we have for the HE_{11} mode:

$$\frac{E\,(a)}{E\,(o)} = J_o \left[2.4\, e^{-1/V} \right] \xrightarrow[V\to 0]{} 0 \qquad (12)$$

Consequently, for reducing the field strength in the cladding it is necessary to use higher values of V. This requirement contrasts with the circumstance that small irregularities of the interface can scatter mode HE_{11} into the higher modes with cutoff frequencies less than V. However, there is no physical reason why this phenomenon should notably influence the propagation of a light pulse over a distance of a few hundred meters. While in optical fibers used for telecommunication the geometrical irregularities occur on a scale of a few microns and are conditioned by the physical process of pulling the fiber from molten glass, for guiding the beam of a laser accelerators it would be necessary to use a hollow dielectric fiber with an inner diameter of a few millimeters.

The optical quality of the inner surface could be controlled mechanically to obtain a good optical finish.

Various types of midinfrared optical fibers have been developed through intensive efforts to realize low-loss optical transmission of higher-power CO_2 laser beams. Sakuragi et al.[3] have recently developed a polycrystalline KRS-5 (thallium-bromide-iodide) optical fiber having a high-power transmission capability for cw CO_2 laser beams. A sample of fibers obtained by extrusion, 1 mm in diameter was found to remains free from damage up to a power flux of 30 KW/cm^2. Bridges et al.[4] have been able to grow fibers made with single-crystal AgBr. These infrared crystalline fibers have potentially extremely low losses. By reducing the metallic impurities by 4 orders of magnitude an attenuation of $\sim 10^{-3}$ dB/Km at 5 μm should be obtained. These fibers have the potential for high infrared power transmission.

Hidake et al.[5,6] have also proposed a new type of middle-infrared optical fiber in which oxide glass is used as the cladding material to define a hollow core. When the refractive index of the cladding material is smaller than unity, then there exists a critical angle for rays impinging on the core-dielectric interface from the hollow-core side. As already reported by Cleek[7], some materials have strong absorptions in the middle infrared region ($800 \div 1200\ cm^{-1}$) due to lattice vibrations. Consequently, on the high frequency side of the absorption land the real part of the refractive index is notably less than on the lower frequency side. For some materials, e.g. SiO_2, the reduction is so strong as to give a value of $n_r < 1$. Hidake et al. have remeasured accurately the reflectivity of fused silica, Pb glass and SF-6 glass and have found frequency intervals in the spectral region covered by CO_2 lasers, in which the refractive index is less than unity. They have also tested a hollow-core Pb glass fiber having an inner diameter of 1 mm. A loss factor of 7.7 dB/m was measured. It is worth noting that these losses compare well with those obtainable with a metallic silver waveguide having the same diameter. By making n_r small enough, the field remains confined within the core region and the oxide-glass-cladding will not be damaged.

HOLLOW DIELECTRIC WAVEGUIDES SUPPORTING RADIATIVE MODES

Hollow dielectric waveguides were first suggested by Marcatili and Schmellzer[8] as guiding media for gaseous lasers. Laser action in dielectric waveguides filled with He-Ne and CO_2 has been demonstrated by Smith[9], Bridges et al.[10], Burkhardt et al.[11], Jenson et al.[12], Degnan et al.[12], Chester et al.[14], Marcuse[15] and Laakmann et al.[16]. Different materials like BeO and Al_2O_3 in form of capillary tubes with bore diameters ranging between 1 and a few millimeters have also been used.

More recently several groups have built rectangular structures because of the ease of obtaining polished internal walls. The rectangular structures are fabricated from four polished slabs of dielectrics fitted together to form a hollow rectangular structure. By using slabs with corrugated surface these waveguides can be transformed into distributed feedback lasers in which the stationary wave pattern is obtained by coupling the two counterrining waves through the periodic wall corrugations.

Hollow dielectric waveguides do not support guided waves in the usual sense. In conventional dielectric waveguides guidance is achieved by total internal reflection. In hollow-core waveguides the "cladding" has a refractive index which is larger than that of the core (n = 1). This means that the critical angle is complex. Consequently, we can expect that the propagation constant β of the "radiative" modes supported by these structures is intrinsically complex, i.e. independent of possible losses in the dielectric. Marcuse[15] has calculated the loss coefficient of these leaky modes in parallel plates hollow dielectric waveguides. For TE_{oi} modes he has obtained:

$$\alpha_i^{TE} = \frac{(1+1)^2 \pi^2}{2\,(n^2-1)^{1/2}\,K^2 a^3} \tag{13}$$

while for the TM_{oi} modes it can be shown that:

$$\alpha_i^{TM} = \alpha_i^{TE}\, n^2 \tag{14}$$

More accurate calculations for rectangular waveguides have been carried out by Laakmann and Steier.

By limiting ourselves to the parallel plates case we have that the transverse distribution of the electric field of a TE_{oi} mode is of the form:

$$E(x) \propto \cos \varkappa x \tag{15}$$

where the generally complex transverse propagation constant K is related to the longitudinal propagation constant $K_z = \beta + i\alpha$ through the equation:

$$\varkappa^2 + \beta^2 - \alpha^2 + 2i\alpha\beta = K^2 \tag{16}$$

As a first approximation we can put:

$$\varkappa = \frac{\pi}{2a} - i\varkappa_i \tag{17}$$

where

$$\varkappa_i \sim \alpha_{01}^{TE} \quad \varkappa \frac{2a}{\pi} = \frac{4\pi}{(n^2-1)^{1/2}} \quad \frac{1}{Ka^2} \tag{18}$$

Consequently, the ratio between the field on the walls and on the center reads:

$$\frac{E_{wall}}{E_{centre}} = \sinh\left(Ka^2 \frac{2}{\pi} \alpha_{01}^{TE}\right) \sim \frac{2}{\pi} Ka \; a\,\alpha_{01}^{TE}$$

$$\sim \frac{4\pi}{(n^2-1)^{1/2}} \quad \frac{1}{Ka} \tag{19}$$

with a spacing of > 1 mm, we can easily obtain a reduction of the field on the wall of 10^{-2}.

Equation (18) also shows that as a rule this reduction is proportional to the factor a α^{TE}. Consequently, we can affirm that as a general rule a reduction of the transmission losses positively affects the effective field strength in the dielectric.

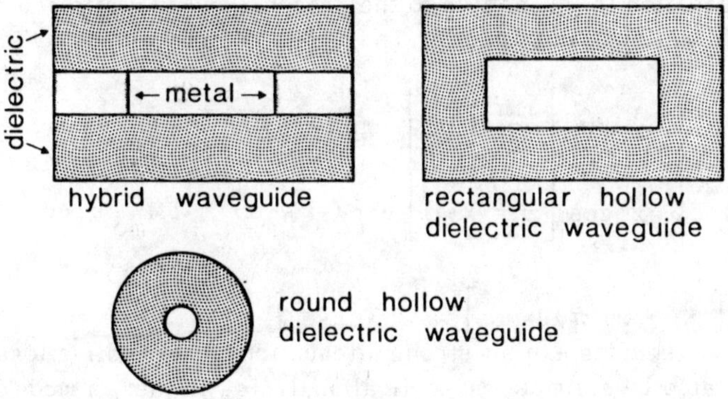

Fig. 3. - Hollow dielectric waveguides.-

METALLIC WAVEGUIDES

A number of metallic waveguides have been studied in the search for flexible systems able to deliver the output of CO_2 lasers. All these devices, the best understood being rectangular, helical and circular guides, share the feature of efficiently transmitting low-order modes.

The waveguide studied by the Center for Laser Studies, U.S.R. (see E.Garmire et al. Refs. 17-20) consists of two metallic strips separated by slimstocks whose sides form the sidewalls (see Fig. 4). It allows the propagation of TE and TM modes.

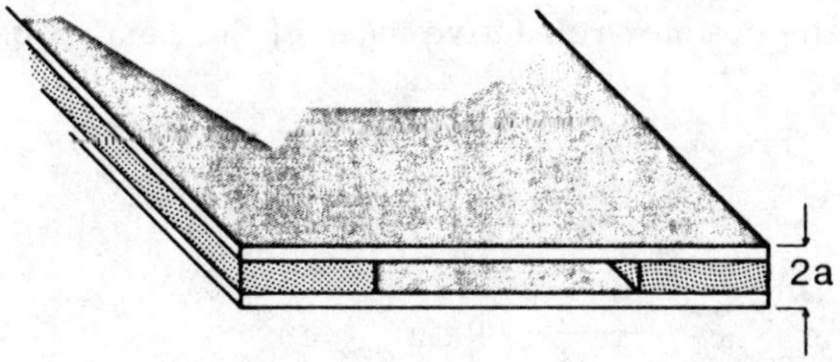

Fig. 4. - Metallic waveguide consisting of two metallic strips separated by slimstocks whose edges form the sidewalls.-

Customarily the waveguide height 2a is much smaller than the width so that the propagation is insensitive to the presence of sidewalls and occurs as through two parallel infinite plates. The modes are given by:

$$\underline{E} \propto \hat{y} \sin\left[\frac{m\pi}{2a}(x-a)\right] \exp(-i\,\beta_m z) \qquad TE_{mo} \text{ mode}$$

$$\underline{E} \propto \hat{x} \cos\left[\frac{m\pi}{2a}(x-a)\right] \exp(-i\,\beta_m z) \qquad TM_{mo} \text{ mode} \tag{20}$$

with $\beta_m = \left[K^2 - (m\,\pi/2a)^2 \right]^{1/2}$.

These waveguides exhibit strong attenuation. The relative loss factors can be calculated by a simple ray-optical analysis. In fact, a mode TE_{mo} or TM_{mo} is formed by two plane waves travelling at angle $\vartheta_m = \lambda_m\,\pi/2a$ with respect to the waveguide axis. Then, each ray travels a distance $d_m = a/\tan\vartheta_m$ between two reflections on upper and lower plates. Let $A(\vartheta)$ the loss per reflection at the incidence angle $i = \pi/2 - \vartheta_m$, the mode undergoes an attenuation per unit length equal to:

$$\alpha_m = \frac{m\,\lambda\,A(\vartheta)}{2\,a^2} \tag{21}$$

A being the absorptance of the metal at the grazing angle ϑ.

For grazing incidence $A(\vartheta)$ is given by (see f.i. Ref. 17):

$$A^{TE}(\vartheta) = 4\,\mathrm{Re}\,(n^{-1})\,\sin\vartheta \cong 4\,\vartheta\,\mathrm{Re}\,(n^{-1})$$

$$A^{TM}(\vartheta) = \frac{4\,\mathrm{Re}(n)\,\sin\vartheta}{1+2\,\mathrm{Re}(n)\,\sin\vartheta + |n^2|\,\sin\vartheta} \tag{22}$$

n being the generally complex refractive index of the metal. Consequently,

$$\alpha_m^{TE} \sim \frac{m^2\,\lambda^2}{a^3}\,\mathrm{Re}(n^{-1})$$

$$\alpha_m^{TM} \sim \frac{m^2\,\lambda^2}{a^3}\,\mathrm{Re}(n) \tag{23}$$

A transmission of more than 95% per meter in a straight alluminum waveguide having a cross section of .5x7 mm has been measured at $\lambda = 10\,\mu$.

It is noteworthly that α^{TE} scales with the inverse cubic power of the waveguide height. Consequently, using a waveguide 5 mm high the attenuation reduces to $5 \cdot 10^{-5}\ m^{-1}$.

Another factor to be considered is the ratio between the electric fields on the walls and in the middle of the waveguide for the TE_{10} mode, which exhibits the lowest losses,

$$\frac{E_{wall}}{E_{center}} = \frac{1}{2}|A| = \frac{1}{2}\ 4\ \mathrm{Re}(n^{-1})\frac{\pi\lambda}{2a} = \frac{\pi\lambda}{a}\ \mathrm{Re}(n^{-1}) \tag{24}$$

Since n oscillates between 60 and 70 for Al, Cu or Au, the above ratio is of the order of:

$$\frac{E_{wall}}{E_{center}} \sim \frac{0.5 \times 10^{-3}}{a} \tag{25}$$

with "a" expressed in mm. With a spacing of a few millimeters the field on the walls reduces to $10^{-3} \div 10^{-4}$ of the value in the middle. This reduction is of the same order of magnitude of the reduction which can be reasonably achieved in a hollow dielectric waveguide.

If we assume that E_{center} can reach values of the order of a few GV/m, a rough estimate of the field inside the metal yields E_{metal} MV/m. This field corresponds to a flux density of $\sim 10^5\ W/cm^2$. The same intensity can be reached by illuminating the metal almost at normal incidence with a flux density of $\sim 10\ MW/cm^2$, having assumed a reflectivity of $99 \div 98\%$. Good quality copper mirrors can withstand such high flux densities for a few nanoseconds.

If we illuminate a metallic waveguide with a well focused beam, we can excite a few modes in such a way as to keep the field on the walls below the damage threshold value. This means that the most critical stage of this system concerns the coupling of the input beam with the waveguide. In case of Gaussian beam, the percentage of power channeled through any waveguide mode can be calculated by assimilating the beam to that irradiated by a source located at a complex point. In so doing it is possible to apply the geometrical theory of diffraction (GTD) which can account very accurately for the geometry of the metal[21]. In Figs. 5a, 5b, 5c, 5d, 5e, 5f, 5g we have plotted the coupling coefficients for different parameters of a Gaussian beam. These plots show how critical the coupling with these structures is. A flared waveguide section or a compound parabolic concentrator can be used for reducing the sensitive of these guides to the misalignments of the beam and increasing the coupling to the fundamental mode[21] (see Fig. 6).

Recently, Marhic and Garmire[21] have reported the operation of a CO_2 laser over the TE_{01} and TE_{02} modes of a metallic waveguide having an inner diameter of 1.1 mm and a length of 3.4 cm.

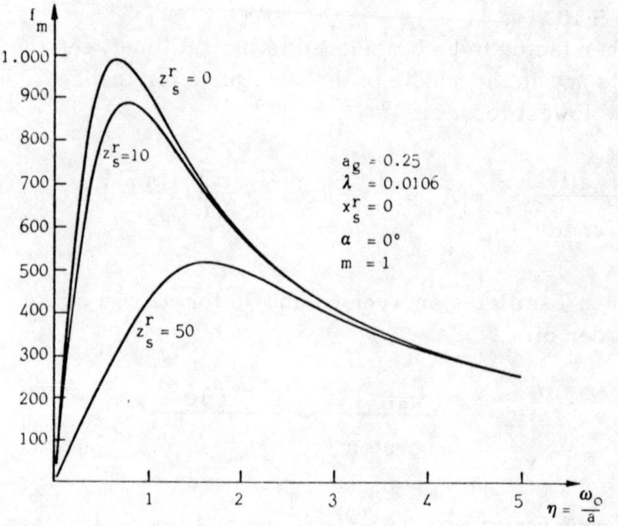

f_m
1.000
900
800
700
600
500
400
300
200
100
z^r_s = 0
z^r_s=10
z^r_s = 50
a_g = 0.25
λ = 0.0106
x^r_s = 0
α = 0°
m = 1
1
2
3
4
5
η = ω_o/a

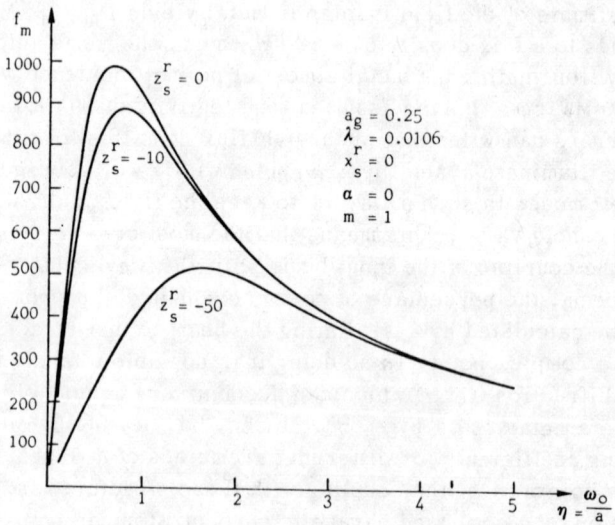

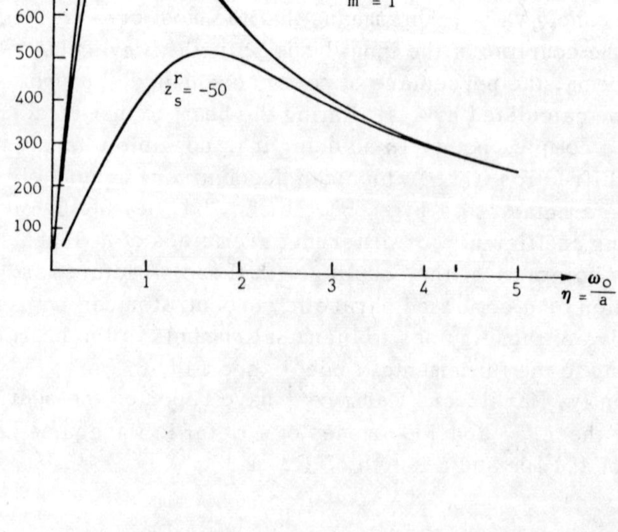

f_m
1000
900
800
700
600
500
400
300
200
100
z^r_s = 0
z^r_s = -10
z^r_s = -50
a_g = 0.25
λ = 0.0106
x^r_s = 0
α = 0
m = 1
1
2
3
4
5
η = ω_o/a

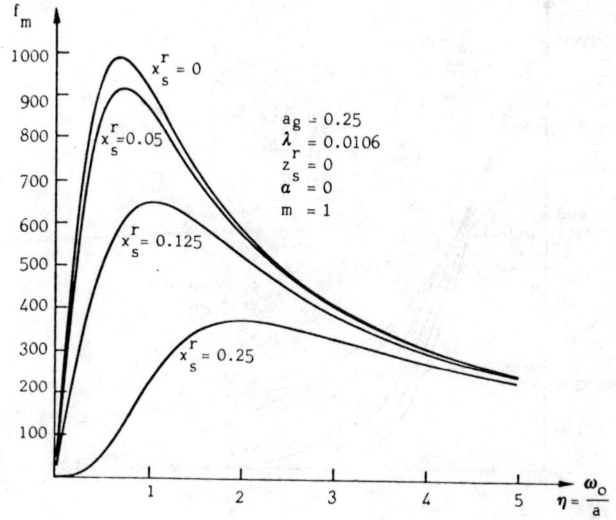

f_m
1000
900
800
700
600
500
400
300
200
100
$x_s^r = 0$
$x_s^r = 0.05$
$x_s^r = 0.125$
$x_s^r = 0.25$
$a_g = 0.25$
$\lambda = 0.0106$
$z_s^r = 0$
$\alpha = 0$
$m = 1$
1
2
3
4
5
$\eta = \frac{\omega_o}{a}$

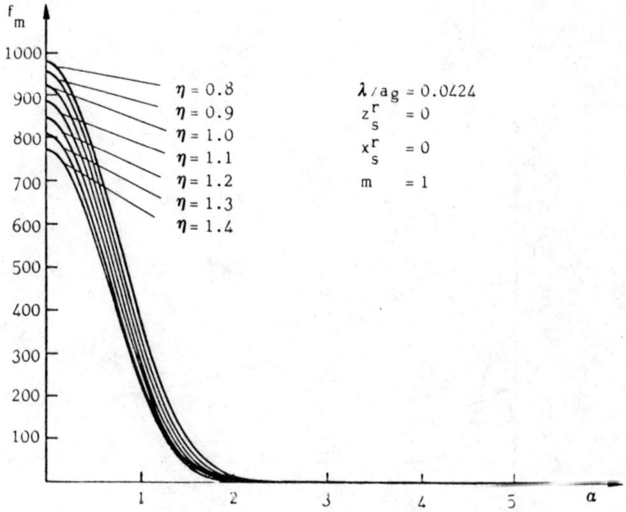

f_m
1000
900
800
700
600
500
400
300
200
100
$\eta = 0.8$
$\eta = 0.9$
$\eta = 1.0$
$\eta = 1.1$
$\eta = 1.2$
$\eta = 1.3$
$\eta = 1.4$
$\lambda / a_g = 0.0424$
$z_s^r = 0$
$x_s^r = 0$
$m = 1$
1
2
3
4
5
α

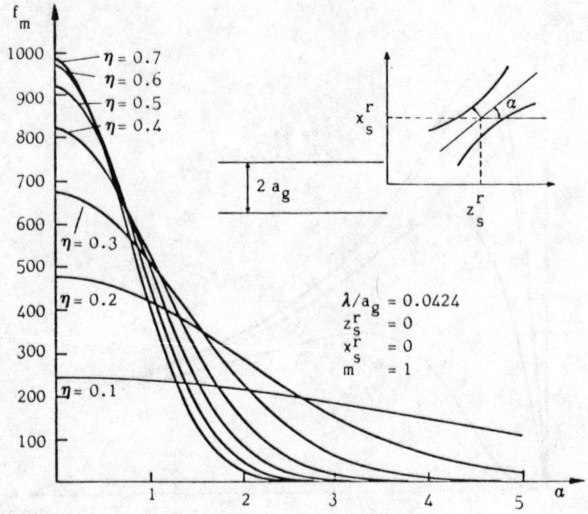
f_m
1000
900
800
700
600
500
400
300
200
100
η = 0.7
η = 0.6
η = 0.5
η = 0.4
η = 0.3
η = 0.2
η = 0.1
2 a_g
x_s^r
z_s^r
α
λ/a_g = 0.0424
z_s^r = 0
x_s^r = 0
m = 1
1
2
3
4
5
α

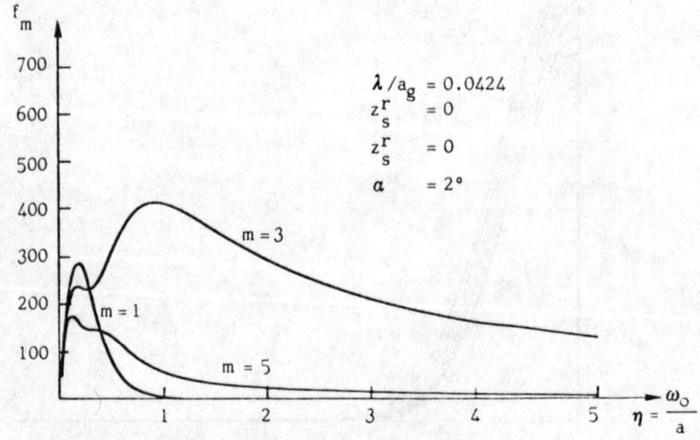
f_m
700
600
500
400
300
200
100
λ/a_g = 0.0424
z_s^r = 0
z_s^r = 0
α = 2°
m = 3
m = 1
m = 5
1
2
3
4
5
η = ω_o / a

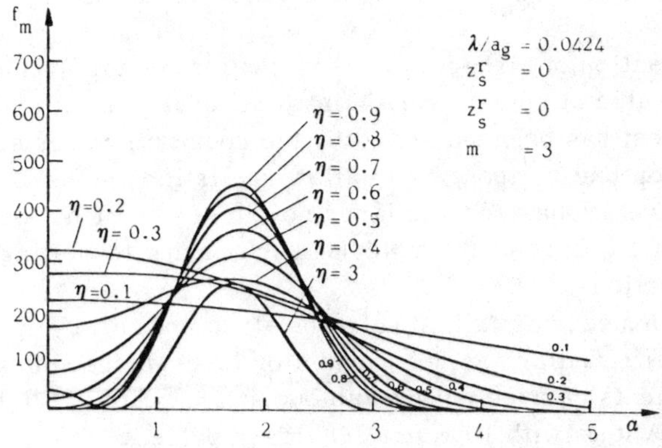

Figs. 5. - Coupling coefficient of a Gaussian beam with the fundamental (m=1) and third (m=3) modes of a metallic waveguide as a function of $\eta = \omega_0/a$, α and the coordinates x_s^r, z_s^r of the waist, α the angle formed by the propagation axis of the beam with the waveguide.-

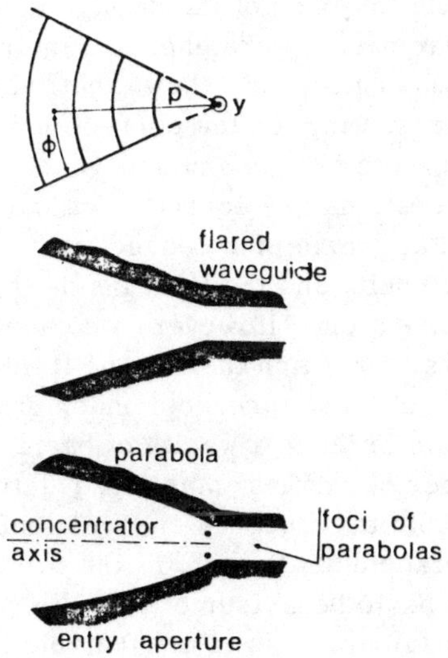

Fig. 6. - Geometry of flared waveguides and compound prabolic concentrators proposed as input sections for metallic guides.-

LASER INDUCED DAMAGE IN OPTICAL MATERIALS

The interaction of high-power lasers with material surfaces has received considerable attention over the past several years.

The interest has been motivated by the requirement for accurate damage thresholds for optical components to allow realistic designs of laser systems. Experiments have emphasized the laser power requirements necessary to cause catastophic failure of the material surface due to melting, fracture or slip plane distortions.

Breakdown mechanisms in insulators at dc and near-IR optical frequancies are very similar, as documented by laser-induced breakdown trheshold measurements carried out by Yablonovitch[22] who irradiated samples of alkali-halide crystals with 10.6 μm CO_2 laser pulses.

Crisp et al.[24,25] (see also Ref. 26) have observed that the nominal external intensity for damage at the rear surface of a plane parallel glass is about a factor 1.5 less than for the front surface. These results seem to indicate that local electric field strengths are of primary importance in determining surface damage either on the average energy density present at the surface or on the direction of the electric vector with respect to the surface normal.

Another indirect evidence of this dependence from electric field strength only is provided by the lowering of the damage threshold due to the electric field enhancement near pares, scratches and incipient cracks on the surface of a dielectric. Bloembergen[27] has shown that this reduction can be quantitatively explained by accounting for the correct depolarization factor depending on the geometry of surface cracks and grooves.

If these considerations are correct, we should observe an increase of the damage threshold at grazing incidence. In fact, in this case the reflectivity is almost equal to unity and the field inside the dielectric is a very small percentage of the incident one. However, we cannot disregard the role of the surface irregularities, for which there is no field reduction. In any case careful measurements of these threshold under grazing incidence and different surface finishing would produce a wealth of basic data.

The large number of experimental data relative to alcali-halides shows no variation attributable to frequency change alone, of the absolute magnitude of the breakdown threshold at 10.6, 1.06 and 0.694 μm. Thus, up to ω of almost $3 \cdot 10^{15}$ Hz the basic behaviour of bulk dielectric breakdown is quite similar to the dc behaviour. This is attributable to the fast hot electron collision process and to the large material bandgap as compared to the photon energy.

For shorter wavelengths multiphoton absorption processes become most relevant and produce a lowering of the damage threshold.

Since the breakdown process depends on thermal diffusion (particularly for long-duration pulses) or to plasma growth by electron diffusion from an initial location, we can expect a square-root functional form ($\sqrt{Dt}$) of the time-dependence of the breakdown evolution. This leads to predict that the breakdown threshold electric field will scale like $t^{-1/4}$.

Damage of metallic mirrors strongly depends on the increase with temperature of the absorptance A. This parameter can be approximated by the relation

$$A = A_o + A_1 T_s \tag{26}$$

T_s being the surface temperature. For damage at 100 ns in vacuum of clean-surface Portens et al.[28] have shown that the catastrophic failure of the material surface is due to melting.

A first principle calculation of the damage threshold based on the above equation and on the melting criterion for these materials by Sparks et al.[29,30], has given a value close to 45 J/cm^2 for Cu at 10.6 μm and 100 ns.

More complex phenomena characterize the damage induced by pulses lasting a few nonoseconds. Observations of the morphology of laser induced damage in copper mirrors irradiated with a succession of 1.7 ns CO_2 laser pulses, have been reported recently by Thomas et al.[31]. In particular they have observed that initial damage occurs at isolated sites due to random surface imperfections as for transparent dielectrics. After repeated irradiation the entire surface appears covered by an array of spheres with a rather uniform diameter of 1 μm. These alteractions of the surface structure of polished copper mirrors take place at a fluence level a factor 4 below that required for single-shot surface damage (11.2 J/cm^2).

While for transparent dielectrics the damage threshold depends on the strength of the electric field inside the material, the factors controlling the damage of metals seems to be the absorbed power. Consequently, a reduction of absorptance obtained by grazing incidence of the laser beam should produce a rather notable increase of the damage threshold. Accurate tests of metallic mirrors at grazing incidence would prove quite essential in assessing the order of allowed power leadings for different pulse lengths.

CONCLUDING REMARKS

The reader is justified in questioning how well founded is the suggestion of using dielectric or metallic waveguides for confining and guiding the very powerfull laser beams required by laser accelerators, over distances of some hundred meters. Inquiry of the laser damage litterature to date indicates that wide-bandgap insulators can withstand rms optical electric fields ranging from 1 MV/cm to 10 MV/cm depending on the pulse duration. These field breakdown thresholds correspond to flux densities of $\sim 10^{-3} \div 10^{-1}$ Tw/cm^2.

The experimental data refer to the case of normal incidence. In case of grazing incidence we can expect a notable increase of the damage threshold.

If we assume that the damage originates in the bulk of the dielectric the above quoted values of electric field will modifly as

$$\frac{E_{normal}}{E_{grazing}} = \frac{T_{graz}}{T_{nor}}^{1/2} \tag{27}$$

where $T_{nor} = 4\,n\,(n+1)^2$ is the transmission coefficient for normal incidence while T_{graz} refers to the case of grazing incidence at angle ϑ ($<<1$).

From Fresull's equations on reflectivity for optical quality surfaces we have at near incidence

$$T^{TE} = \frac{4\vartheta}{n}$$
$$T^{TM} = 4\vartheta n \tag{28}$$

So that we have

$$E_{threshold} \sim \vartheta^{-1/2}\ (1 \div 10)\ \mathrm{MV/cm} \qquad (TE)$$
$$\sim \frac{\vartheta^{-1/2}}{n+1}\ (1 \div 10)\ \mathrm{MV/cm} \qquad (TM) \tag{29}$$

If we assume that our dielectric waveguide is used to confine a gaussian beam having a spot-size ω_o, we obtain a rough estimate of the incidence angle in the above relation by equating ϑ to the far field aperture, i.e.

$$\vartheta \sim \frac{\pi\lambda}{\omega_o} \tag{30}$$

ω_o being the waist spot-size. So that,

$$E_{threshold} \sim \left(\frac{\omega_o}{\pi\lambda}\right)^{1/2} (1 \div 10)\ \mathrm{MV/cm} \tag{31}$$

If we consider that the field on the wall is equal to that on the axis times exp $(-a^2/w^2)$, we obtain a rough estimate of the highest rms field obtainable in the center of a hollow dielectric waveguide

$$E_{max} \sim \left(\frac{\omega_o}{\pi\lambda}\right)^{1/2} e^{a^2/\omega_o^2}\ /(1 \div 10)\ \mathrm{MV/cm} \tag{32}$$

As an example, if we choose a=1.5 mm, ω_0=1 mm, λ=10.6 μm, we obtain

$$E_{max} \sim 8 \div 80 \text{ GV/m} \tag{33}$$

These values are of the correct order of magnitude compared to laser fields considered for acceleration where useful gradients of 0.1 ÷ 1 GeV/m, are assumed.

ACKNOWLEDGEMENTS

Many interesting discussion with A. Renieri and S. Tazzari are gratefully acknowledged.

REFERENCES

1. D.Marcuse, Theory of dielectric optical waveguides,Accademic Press, N.Y. (1974).
2. D. Gloge, Appl., Opt. 10, 2252 (1971).
3. S.Sakuragi, M.Saito, Y.Kubo, K.Imagawa, H.Kotani, T.Morikawa and J.Shimada, Opt. Letters b, 629 (1981).
4. T.J.Bridges, J.S.Hasiak and A.R.Struad, Opt. Letters 5, 85 (1980).
5. T.Hidaka, T.Morikawa and J.Shimada, J.Appl.Phys. 52, 4467 (1981).
6. T.Hidaka, J.Appl.Phys. 53, 93 (1982).
7. G.W. Cleek, Appl. Opt. 5, 771 (1966).
8. E.A.J.Marcatili and R.A.Schmeltzer, Bell Sust. Techn. J. 43, 1783, (1964).
9. P.W.Smith, Appl. Phys. Lett. 19, 132 (1971).
10. T.J.Bridges, E.G.Burkhardt and P.W.Smith, Appl. Phys. Lett. 20, 403 (1972).
11. E.G.Burkhardt, T.J.Bridges and P.W.Smith, Opt. Commun. 6, 193, (1972).
12. J.J.Degnan, H.E.Walker, J.H.McElroy and N.McAvoy, IEEE J.Quantum Electron. QE-9, 489 (1973).
13. P.W.Smith, T.J.Bridges, E.G.Burkhardt and O.R.Wood, Appl.Phys. Lett. 21, 470 (1972).
14. A.N.Chester and R.L.Abrams, Appl.Phys.Lett. 21, 576 (1972).
15. D.Marcuse, IEEE J.Quantum Electron. QE-8, 661 (1972).
16. K.D.Laakmann and W.H.Steier, Appl. Opt. 15, 1334 (1976).
17. E.Garmire, T.McMahn and M.Bass, IEEE J. Quantum Electron. QE-16, 23 (1980).
18. L.W.Casperson and T.S.Garfield, IEEE J.Quantum Electron. QE-15, 491 (1979).
19. M.E.Marhie, L.I.Kwan and M.Epstein, Appl. Phys.Lett. 33, 874(1978).
20. M.E.Marhic and E.Garmire, Appl.Phys.Lett. 38, 743 (1981).

21. F.Crescenzi, P.Gay, A.Cutolo, I.Pinto, S.Solimeno, 1980 European Conf. on Optical Systems and Applications, Utrecht (1980), S.P.I.E. Vol. 236, p. 365.
22. E. Yablonovitch, Appl. Phys.Lett. 19, 495 (1971).
23. P.Salomon and N.Klein, Solid State Communications 17, 1397 (1975).
24. M.D.Crisp, N.L.Boling and G.Dubé, Appl. Phys. Lett. 21, 364 (1972).
25. N.L.Boling, M.D.Crisp and G.Dubé, Appl. Opt. 12, 650 (1973).
26. V.V.Lyubimov, I.A.Fersman and L.D.Khazov, Sov. J.Quantum Electronics 1, 201 (1971).
27. N.Bloembergen, Appl. Opt. 12, 661 (1973).
28. J.O.Portens, M.J.Soilean and C.W.Fountain, Appl. Phys. Lett. 29, 156 (1976).
29. M.Sparks, J.Appl. Phys. 47, 837 (1976).
30. M.Sparks and E. Loh, J.O.S.A. 69, 847 (1979); - 69, 859 (1979).
31. S.J.Thomas, R.F.Harrison and J.F.Figneira, Appl.Phys. Lett. 40, 200 (1982).

Near Field Accelerators

R.B. Palmer
Brookhaven National Laboratory
Upton, NY 11973

This is the report of the working group composed of Channell, P. Csonka, K. Kim, N. Kroll, J.D. Lawson, G. Loew, A. Luccio, P. Morton, R.B. Palmer, D. Prusnitz, D. Sutter, S. Tazzari, M. Tigner, T. Wangler, T. Weiland, and P. Wilson.

1. Introduction

We can conveniently divide near field accelerators into three types:

a) Conventional Linacs (Figure 1a)

These are periodic cavity structures with accelerating modes whose phase velocity matches the particles to be accelerated. If these are to be used for small (laser) wavelength, such a structure has to be reduced in scale rather drastically. This is certainly possible down to cm wavelength and might be possible even smaller using electron beam techniques. M. Tigner and B. Palmer had some ideas as to how this might be done.

b) Grating Linac (Figure 1b)

It has been shown by Palmer that a suitable grating can act as an accelerating cavity and behave in most respects the same as a conventional linac. In this case however we certaily know how to reduce its scale at least to 10 μm wavelength and possibly even below.

c) Dielectric Linac

Dielectric loaded non-periodic cavities have been known for a long time to support accelerating modes. If cylindrical or two sided symmetry is used (e.g. Lawson), then it is the dimension of the hole (Figure 1c) or gap (Figure 1d) between the dielectrics that must scale with the wavelength and could, certainly in the two sided case, be reduced to the 10 μ scale. There is also a one sided case (Figure 1e) which has no scale other than that the beam must be very close to the surface.

2. Factors Independent of Wavelengths (Tigner, Palmer)

Since from the point of view of construction the one sided structures (Figures 1b, 1e) are clearly easier to scale down, it is useful to ask if such structures are intrinsically less "efficient" than the more conventional cylindrically symmetric ones. Only in the grating case do we have a worked out example so we compare this with a "normal" iris loaded linac:

a) Power Loss. The Shunt Impedance $r = E_a{}^2/p$.

This relates the accelerating field gradient E_a with the power loss per cm p by resistive heating of the walls. r is about 20 times higher in a grating linac compared with the iris linac. This arises partly because a larger surface area is exposed to the field and partly became in the example the surface fields are higher:

b) Field Efficiency. $\kappa = E_a/E_s$.

Where E_s is the maximum surface field. κ is about 0.2 for the grating linac compared with about 0.5 for a conventional accelerator. A related factor is:

c) Stored Energy Factor. $S = E_a{}^2/J$.

Where J is the total stored electromagnetic energy in the cavities. S is about 10 times less for the grating case.

Thus we see that at least for the particular example known the one sided cavities is quite a bit less "efficient" than the closed cavity. Some of this difference certainly comes from the lack of development of the open cavity but some is almost bound to remain and would have to be made up for by its simplicity of construction or other advantages.

3. Scaling With Wavelength λ

For a number of criteria, using as standard a conventional Linac with λ = 10 cm, we consider

(i) A conventional Linac with λ = 1 cm
(ii) A conventional Linac with λ = 10 μm
(iii) A grating Linac with λ = 10 μm

a) Power Loss. (Tigner, Lawson, Loew)

Here we ask for the relationship between the RF power loss P and final beam energy ε.

$$\varepsilon = E_a \, \ell \propto (P \, \ell \, r)^{1/2}$$

$$\varepsilon \propto \left(P \, \ell \, r_o\right)^{1/2} \rho^{1/2} \lambda^{-1/4}$$

where ℓ = linac length
r = shunt impedance
r_o = shunt impedance at fixed λ
ρ = electrical conductivity of surface.

Thus for fixed P:

(i)	1 cm Linac	$\varepsilon = 1.8\ \varepsilon_o$
(ii)	10 μ Linac	$\varepsilon = 10\ \varepsilon_o$
(iii)	10 μ Grating	$\varepsilon = 2.5\ \varepsilon_o$

i.e., we gain only very slowly with decreasing λ and lose most of the gain if a one sided cavity is used.

b) <u>RF Energy</u>. W_{RF}

$$W_{RF} \propto E_a^2\ \lambda^2/S$$

where S is the stored energy factor (see above)

$$\varepsilon \propto \lambda^{-1}\ S^{1/2}$$

thus for fixed RF energy W_{RF}

(i)	1 cm Linac	$\varepsilon = 10\ \varepsilon_o$
(ii)	10 μ Linac	$\varepsilon = 10{,}000\ \varepsilon_o$
(iii)	10 μ Grating	$\varepsilon = 3{,}000\ \varepsilon_o$

c) <u>Cavity "Q" and Fill Time τ</u>

$$Q \propto W_{RF}/(P\ \lambda)$$

$$\propto \lambda^2\ S^{-1}/(\lambda^{3/2}\ r_o^{-1})$$

$$\propto \lambda^{1/2}\ r_o/S$$

$$\tau \propto Q\ \lambda \propto \lambda^{3/2}\ r_o/S$$

e.g.,	(i)	1 cm Linac	$Q = Q_o/3$	$\tau = \tau_o/30$
	(ii)	10 μ Linac	$Q = Q_o/100$	$\tau = \tau_o/10^6$
	(iii)	10 μ Grating	$Q = Q_o/200$	$\tau = \tau_o/2\ 10^6$

4. Details of a Grating Linac

a) <u>Transverse Stability in Grating Linac.</u> (Kroll, Kim)

Palmer had shown that in the presence of a horizontal fixed field a grating linac shows strong vertical focusing with one phase and strong vertical defocussing with the other. It was shown at the workshop however that the strong vertical focus is accompanied by strong horizontal defocussing and visa versa. A solution proposed is to use alternating phase which can be easily generated in a grating by periodically displaying the grating lines. The focusing/-defocussing strength β for fixed phase φ is given by

$$\beta = \frac{2\pi}{\kappa}\left(\gamma \frac{\partial\gamma}{\partial z} \tan\phi\right)^{1/2}$$

where κ is a parameter of the mode $\kappa = E_a/E_s$ and $\partial\gamma/\partial z$ is the acceleration. For $\phi = 45°$, $\gamma = 1$ GeV, $\partial\gamma/\partial z = 1$ GeV/m:

$$\beta = 4 \text{ cm}$$

If the phase is alternated every 2 cm

$$\beta(\text{effective}) = 10 \text{ cm}$$

which is satisfactory.

b) Tolerances in the Grating. (Kroll)

Since the particles become longitudinally very stiff the energy spread due to grating spacing errors δS become small:

$$\left\langle \frac{\Delta\varepsilon}{\varepsilon} \right\rangle = \frac{\delta S}{S} N^{-1/2}$$

where N is the number of lines (very large)
The transverse situation is more complex but was found, rather surprisingly, to be independent of the length, energy or anything else about the accelerator.:

$$\left(\frac{\Delta x}{\lambda}\right)^2 = \frac{16}{\pi}\left(\frac{\delta S}{S}\right)^2$$

If we require $\Delta x/\lambda \approx 1/8$ then $\delta S/S \leqslant 1/16$ and $\delta S \leqslant 0.5\mu$ for $\lambda = S = 10\mu$ which is okay.

c) Loading. (P. Wilson, M. Tigner)

The fraction of the stored RF power that can be given to the beam is limited by effects due to the so called "wake fields" that are generated by the particles themselves and effect other particles later in the bunch (in the wake).

We define $\eta = P_{beam}/P_{stored}$

The total RF power $P_{TOT} = T_{beam} + P_{stored} + P_{loss}$

For maximum P_{beam}: P_{stored} P_{loss}

$$P_{TOT} = P_{beam}\left(1 + \frac{2}{\eta}\right)$$

If $\eta \ll 1$

$$P_{TOT} \approx P_{beam} \cdot \frac{2}{\eta}$$

η is limited by both transverse and longitudinal effects: In both cases the effects were seen to be independent of wavelength λ, the accelerating gradient, and not obviously much different in the grating compared with the conventional case. The maximum value of η that might be acceptable would be of the order of 10%.

5. Accelerating Field Limits (Proznitz, Tigner)

a) Electrical Breakdown

Conventional cavities are limited by such breakdown which occurs for λ = 10 cm at about 80 MeV/m. For a field efficiency κ of 0.5 this limits the accelerating fields to 40 MeV/m. Breakdown is believed to be linear with λ because it depends on the maximum energy attained by an electron before the field reverses. Thus at λ = 1 cm a gain of 10 yielding 400 MeV/m may be expected. Shorter wavelengths would yield higher gradients but for:

b) Surface Distruction by Heating

This has been calculated by Takeda and Matsui and should occur with a copper surface when the energy absorbed is 0.07 Joules/cm^2 for a pulse deviation of τ = 1 n sec. For other durations they predict a $\tau^{-1/2}$ dependency so long at τ is long enough that conduction cooling is effective. For long wavelengths these calculations are probably exact. They are also found to be correct for reflection off plane mirrors at 10 μ wavelength. Gratings, however, are in practice damaged at levels about 7 times lower, presumably due to grating imperfections. We will ignore such imperfections in the following calculations.

If we can keep the pulse duration to the ideal fill time then as shown above

$$\tau \propto \lambda^{3/2}$$

$$\text{energy absorbed} \propto E_a^2 \, \lambda^{-1/2}$$

and this for conduction cooling is limited by $\tau^{-1/2} \propto \lambda^{-3/4}$. From which we obtain

$$E_a(\max) \propto \lambda^{-1/8}$$

i.e., a very slow gain with decreasing λ

λ = 10 cm	$E_a(\max) \approx 0.4$ GeV/m, τ = 1 μsec
λ = 1 cm	$E_a(\max) \approx 0.5$ GeV/m, τ = 30 nsec
λ = 10 μm	$E_a(\max) \approx 1.3$ GeV/m, τ = 1 psec

At 3 psec $E_a(max) = 1.0$ GeV/m. If this field is exceeded the surface is destroyed.

c) Plasma Growth. (P. Channel)

At first the thought of destroying the grating seems clearly unacceptable but one should think some more. One notes that it is a very thin surface layer (~ 2 μm) that is destroyed, in fact evaporated. Various methods suggest themselves for restoring the surface. The grating could consist of the ripples on a mercury surface which would automatically reform. Another idea is to use a thin layer of low melting point material (e.g., water) evaporated onto a copper substrate. The material could be blown off but could easily and quickly be replaced between pulses. Once this limitation is removed we must look to a further limit: when the surface plasma grows so fast that the grating is destroyed before it can be used as an accelerator. One notes in passing that plasmas have very high conductivity and that a plasma close to the surface should make a very good cavity wall.

A conservative estimate of the distance t of plasma growth is:

$$t \propto E_s \lambda \left(\frac{\tau}{\lambda}\right)^{1/2} = E_s \lambda^{1/2} \tau^{1/2}$$

If we require $t < \lambda/8$, $\lambda = 10\ \mu$, $\tau = 100$ psec

$$E_s \leq 5 \text{ GeV/m}$$

At the ideal filling time of ≈ 1 psec $E_s \leq 50$ GeV/m and $E_a \leq 10$ GeV/m for $\kappa = 0.2$.
If this ideal filling time ($\propto \lambda^{3/2}$) is used we get the general scaling low

$$E_a(max) \propto \lambda^{-1/4}$$

$\lambda = 10\ \mu$ $\quad E_a(max) = 10$ GeV/m
$\lambda = 1\ \mu$ $\quad E_a(max) = 17$ GeV/m

It must be noted that the calculation used here is primitive at best and could be wrong by an order of magnitude either way.

d) Summary of Accelerating Field Limits

The limits discussed here are plotted in Figure 2. It is seen that a big gain is gradient maybe obtained dropping the wavelength from 10 cm to 1 cm. After that the gain is very slow limited by electrical heating of the cavities walls. Bigger gains are made if the wall is allowed to become a plasma provided the very short required pulses are available.

6. Examples

For illustration only we give the parameters of five 1 TeV accelerators. Number 1 and #2 are more or less conventional but with the second using 1 cm wavelength; #3 to #5 are grating accelerators in #3 the grating survives in #4 and #5 it does not. Number 5 is really pushing things but even the possibility of 1 TeV in 60 m is worth recording.

Example	1	2	3	4	5
Wavelength	10 cm	1 cm	10 μm	10 μm	1 μm
Gradient MeV/m	30	300	300	3000	300
Length	30 km	3 km	3 km	300 m	30 m
Pulse Deviation	1 μsec	30 msec	30 psec	30 psec	1/30 psec
Power Watts	$5\ 10^{10}$	$1.7\ 10^{12}$	$2\ 10^{12}$	$2\ 10^{13}$	$2\ 10^{14}$
Energy Joules	50,000	60,000	60	600	7
Particles/Bunch	10^7	10^7	10^4	10^5	10^3

7. Conclusions

The Workshop has made enormous gains in our understanding.

Many scaling laws of linac behavior with wavelength have been defined; some probably for the first time. A comparison has been made between a conventional and grating linac, also probably for the first time.

A problem in transverse stability of the grating accelerator was exposed but also solved at the workshop with a great net gain in understanding. Loading of linacs appears not to be independent on wavelength, it had not previously been studied for a laser accelerator.

The limits on accelerating field as a function of wavelength have been studied. A gain of a factor of 10 is likely if the wavelength can be reduced to the 1 cm range. After that the gains are small unless it is acceptable to destroy the structure. If this is allowed, as could be the case for a grating consisting of liquid ripples, then plasma development will limit the fields. A first calculation suggests that very high gradients may be attainable.

The most remarkable feature of laser driven grating acceleators are the low RF energies required to attain high beam energies for a few particles. One example is of a 1 TeV accelerator driven by a 60 Joule laser!

What is needed now is:

a) theoretical work to design a more efficient grating
b) more work on the filling of a grating
c) experimental work on grating survival
d) work on 1 cm wavelength sources
e) another workshop.

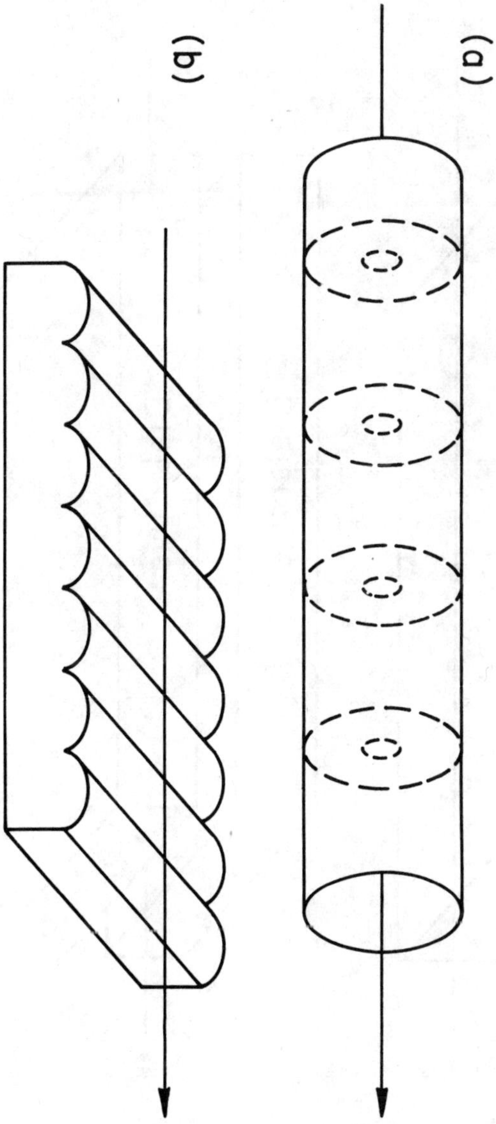

Fig. 1

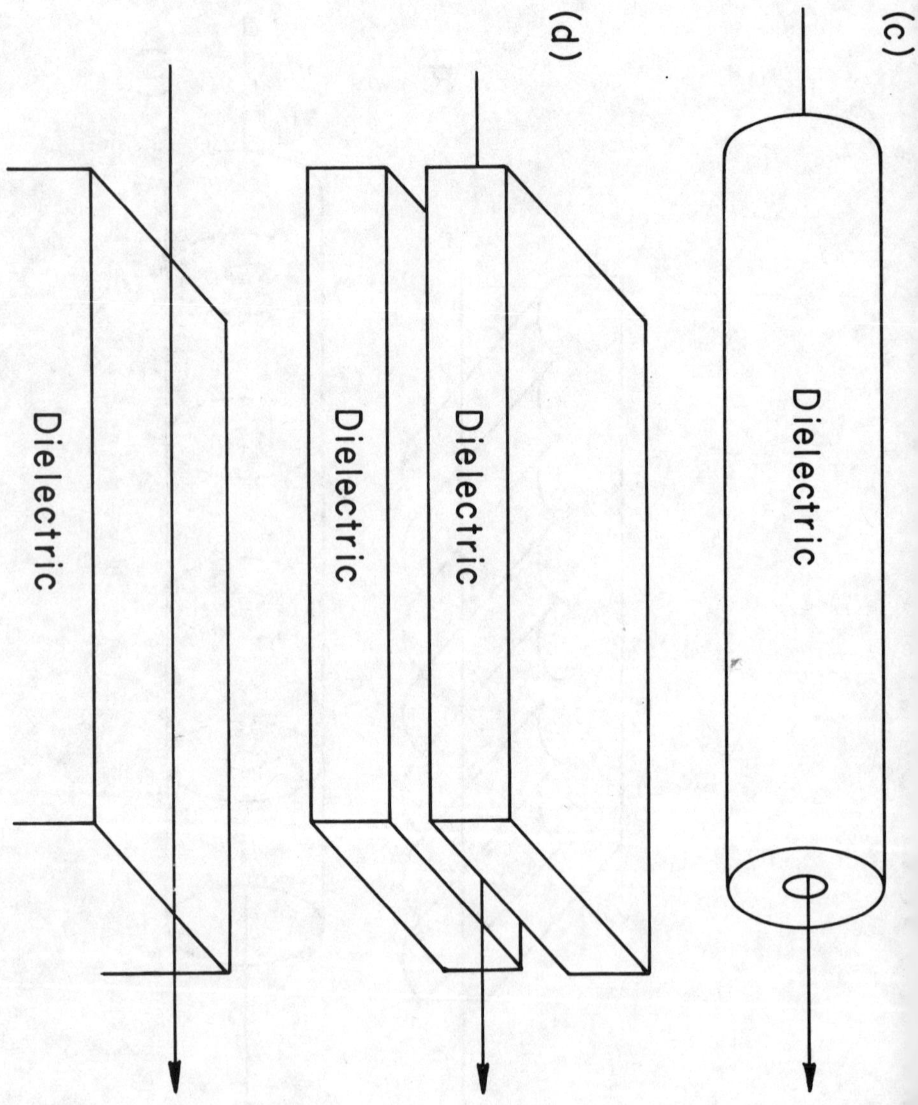

Fig. 1

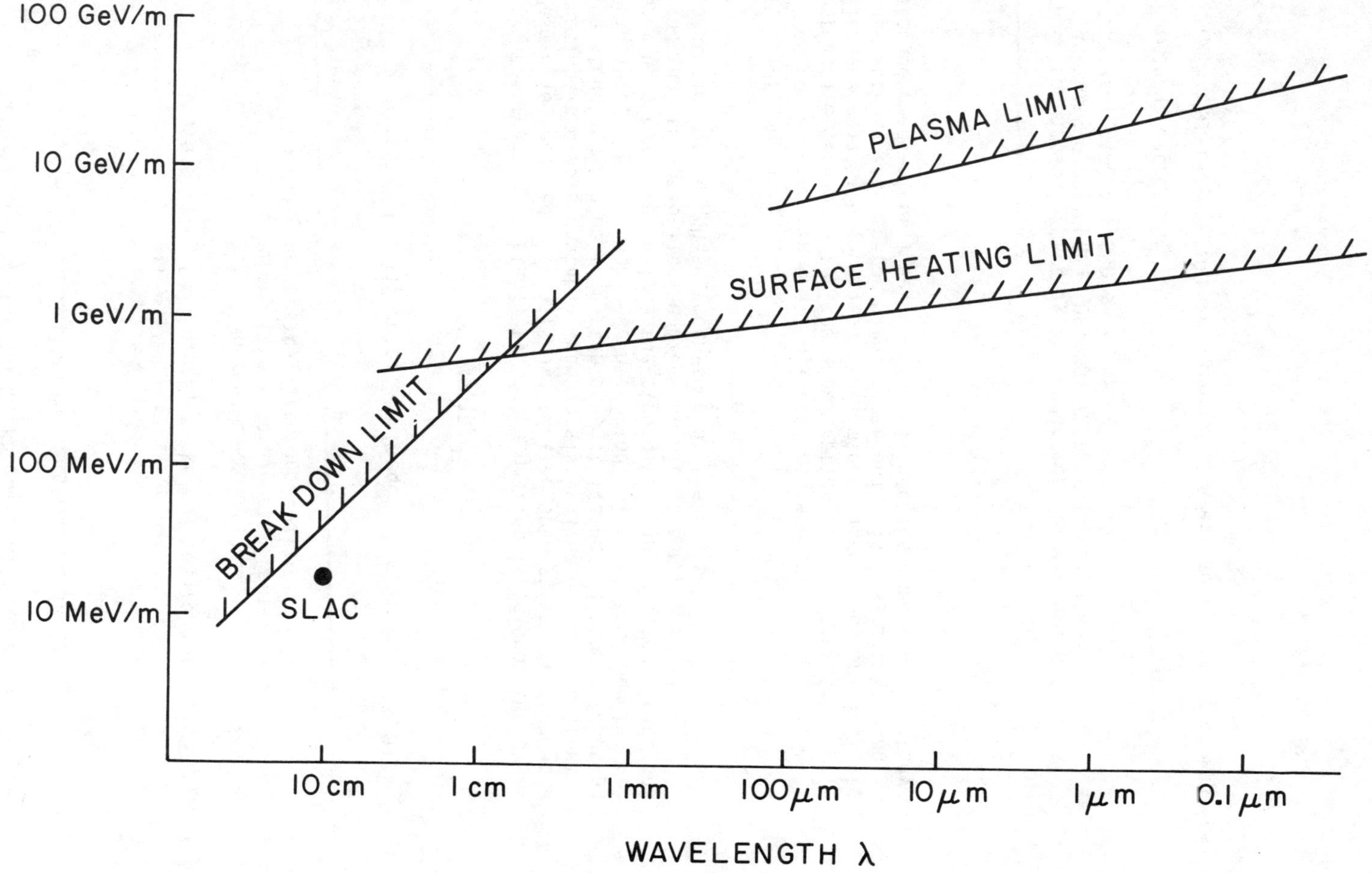
100 GeV/m
10 GeV/m
1 GeV/m
100 MeV/m
10 MeV/m
PLASMA LIMIT
SURFACE HEATING LIMIT
BREAK DOWN LIMIT
SLAC
10 cm
1 cm
1 mm
100 μm
10 μm
1 μm
0.1 μm
WAVELENGTH λ

SOME EFFECTS OF THE TRANSVERSE STABILITY REQUIREMENT ON THE DESIGN OF A GRATING LINAC*

Kwang-Je Kim
Lawrence Berkeley Laboratory
University of California, Berkeley, California 94720

Norman M. Kroll
Standford Linear Accelerator Center
Stanford University, Stanford, California 94305

and

University of California, San Diego+, San Diego, CA 92093

ABSTRACT

The transverse stability of the grating linac proposed by Palmer is analyzed. It is shown that an open structure such as a grating is always unstable transversly as long as it is uniform. The structure can be made stable by utilizing the strong focusing principle. This is achieved by periodically interrupting the grating shape. We analyze the strong focusing grating linac, and find that the stability requirement places a non-trivial constraint on the phase acceptance of the system.

I. Introduction

The laser driven grating linac[1] has emerged as an attractive candidate for the acceleration of electrons to ultra-relativistic energies. Preliminary investigations indicate that such an accelerator might be technically feasible with an average gradient of 1 GeV/m. Thus it becomes important to study

---

*Work supported in part by the Department of Energy, contracts DE-AC03=76SF00098 (LBL), DE-AC03-76SF00515 (SLAC) and DE-AT03-81ER40029 (UCSD).

+Permanent address.

the beam dynamics in such a system. In this paper we analyze the transverse motions of the particles in order to determine possible restrictions on the design of a grating linac.

The geometry of a grating linac differs fundamentally from that of a conventional linac in that the former is an open structure in which the accelerating field decreases rapidly away from the grating surface, while the latter is a closed system with a cylindrical symmetry around the axis. For a conventional linac in extreme relativistic limit, it is well known that the transverse force is negligible and the focusing requirement is not severe. The situation is entirely different in the case of a grating linac; indeed, we find that the transverse force is large and unstable in such a structure. Therefore, the grating linac as originally proposed will not work.

However, the transverse force is similar to the field inside a quadrupole: thus if the force is defocusing in one direction, it is focusing in the other direction. Furthermore, the focusing or defocusing effect depends on the phase of the particle under consideration. Therefore, an obvious solution of the problem is to employ the concept of the strong focusing principle[2] by changing the phase periodically so that a particle sees the focusing and the defocusing field alternatively. Our analysis here shows that such a scheme works in principle if some non-trivial requirements are met. In particular we find that the net focusing requirement in both directions implies a limited acceptance in particle phase.

Sec (II) discusses the transverse stability from a general point of view. Sec (III) deals specifically with the grating linac. In Sec (IV), we analyze the transverse stability of a strong focusing grating linac by means of a simple model and find the stable phase acceptance region.

II. General Discussion

Consider an extremely relativistic electron traveling along the z-direction through a general accelerating structure. The force acting on the paritcle is (M.K.S. unit)

$$\underset{\sim}{F} = -e\,(\underset{\sim}{E} + c\,\hat{z} \times \underset{\sim}{B}) + \ldots.. \tag{1}$$

In the above, terms which are smaller by a factor $1/\gamma = \sqrt{1 - v^2/c^2}$ are neglected. The Panofsky-Wenzel theorem [3] states that, for a traveling wave with a phase factor $\exp i(\omega t - kz)$, the transverse part of the force is simply related to E_z

as follows:

$$\underset{\sim}{F}_{\perp} = -e \frac{i}{k} \underset{\sim}{\Delta}_{\perp} E_z \tag{2}$$

In the above, the phase velocity of the wave must match the particle velocity which is almost the light velocity c. Thus

$$\frac{\omega}{k} = c + 0\left(\frac{1}{\gamma^2}\right) \tag{3}$$

The fields then satisfy the two dimensional Laplace equation:

$$\underset{\sim}{\Delta}_{\perp}{}^2 E_z = \left(\left(\frac{\partial}{\partial x}\right)^2 + \left(\frac{\partial}{\partial y}\right)^2\right) E_z(x,y) = 0. \tag{4}$$

For a system which is uniform along the z-direction, Eqs (2) and (4) imply that the tranverse motion is unstable if E_z has an appreciable transverse variation. To see this, assume that the motion is stable around (x_0, y_0). Eq (2) implies that the point must be a minimum (or a maximum depending on the phase) of the function E_z (x,y). However, E_z is a solution of the Laplace equation (4). Thus an extremum of E_z cannot be a minimum or a maximum but can only be a saddle point. Therefore, if the motion is stable in the x-direction, it must be unstable in the y-direction and vice versa.

In general F contains a constant force which must be cancelled out by an external force. This amounts to changing E_z by

$$E_z \rightarrow E_z' = E_z - ax - by. \tag{5}$$

Since E'_z is still a solution of the Laplace equation, our conclusion remains valid.

In a conventional linac with cylindrical symmetry, $\Delta_{\perp} E_z$ vanishes on the axis and the defocusing force is negligible in the extreme relativistic limit. However, in a one-sided open structure like a grating linac, E_z vanishes necessarily rapidly away from the grating surface. The transverse defocusing effects are then quite large. The difficulty can be overcome by shifting the grating ruling periodically so that the phase of a particle alternates between the focusing and defocusing region. The transverse motion can then be made effectively focusing in both x and y direction. This is known as the strong focusing principle[2].

III. Application to Grating Linac

We shall now work with an explicit field solution of the

grating linac treated by Palmer[1]. The n = 1 accelerating component which is in phase with particle is given by

$$E_z = -E_0 \cos\phi \cos py\, e^{-px}, \quad c B_z = -E_0\, k \cos\phi \sin py\, e^{-px},$$

$$E_x = E_0 \frac{k}{p} \sin\phi \cos py\, e^{-px}, \quad c B_x = E_0\, p \sin\phi \cos py\, e^{-px}, \tag{6}$$

$$E_y = 0, \quad c B_y = E_0\left(\frac{k^2}{p} - p\right) \sin\phi \cos py\, e^{-px}.$$

Here

$$\phi = \omega t - kz + \text{const.} \tag{7}$$

In the above, x is the direction pointing away from the grating surface, y is along the grating ruling and z is the particle direction. The electric fields are given in the Palmer's paper[1]. The magnetic fields are then determined from the Maxwell's equation. It is straightforward to check that the above field satisfy eqs (2) and (4).

Taking z as the independent variable, the equations of motion are

$$mc^2 \frac{d\gamma}{dz} = e\, E_0 \cos\phi \cos py\, e^{-px}, \tag{8a}$$

$$mc^2 \frac{d}{dz} \gamma \frac{dx}{dz} = -e\, E_0 \frac{p}{k} \sin\phi \cos py\, e^{-px}, \tag{8b}$$

$$mc^2 \frac{d}{dz} \gamma \frac{dy}{dz} = -e\, E_0 \frac{p}{k} \sin\phi \sin py\, e^{-px}. \tag{8c}$$

Assuming that the particle moves around the point x = 0, y = 0, we expand eq (8) near this point. Introducing a constant magnetic field along y-direction to cancel the constant deflecting force in the x-direction, on obtains

$$\frac{d^2x}{dz^2} = -\frac{p}{\gamma k} \sin\phi \ (1 - px) + \frac{p}{\gamma k} \sin\phi_e, \tag{9a}$$

$$\frac{d^2y}{dz^2} = -\frac{p}{\gamma k} \sin\phi\ y \tag{9b}$$

The last term in (9a) represents the external force, and we have introduced the quantity $\mathscr{L}$ via

$$\mathscr{L} = \frac{mc^2\gamma}{eE_0} \qquad (10)$$

In deriving eq (9), we have assumed that γ is a slowly varying function of z.

In a laser driven grating linac, eE_0 is typically 1 GeV/m. At the injection point $mc^2\gamma \sim 1$ GeV, thus $\mathscr{L} \sim 1$ m. In the following, we shall work with the following values of parameters specified by Palmer[1].

$$\mathscr{L} = \frac{\gamma}{\gamma_{in}} \cdot (m), \ \lambda = \frac{2\pi}{k} = 10\mu, \ \ p = .2\ k, \qquad (11)$$

where $mc^2\gamma_{in}$ is the injection energy. Taking $\sin\phi \sim .5$, an external magnetic field of ~ 3.3 KG will produce the last term in Eq (9a).

From Eq (9), one sees that the coefficient of the linear force for the x and y motions are equal in magnitude but opposite in sign. Thus the motion is stable in x and unstable in y or vice versa depending on the sign of $\sin\phi$, in agreement with our general discussion. The focusing or the defocusing strength is characterized by a distance $2\pi\sqrt{\mathscr{L}k/p^2} \sim 4$ cm for the case (11) at injection point. This is quite serious for an accelerator of 1 Km long.

The structure can be made focusing in both directions by introducing the strong focusing principle. One shifts the grating rulings so that the phase ϕ changes by $\pm\,\Delta$ periodically. The external field represented by ϕ_e must also change in order to maintain an appropriate relation with the changing ϕ. Ideally, one would like to have $\phi = \phi_e$ for all particles but this is impossible because different particles have different phases. A detailed analysis of the focusing properties taking into account this effect is presented below.

IV. A Strong Focusing Grating Linac and Phase Acceptance

In this section, we consider a simple model of the strong focusing grating linac in which the phase ϕ changes discontinuously. ϕ_e (z) is a periodic function of period L and is given by

$$\phi_e(z) = -\Delta/2, \quad - L/2 < z < 0 \qquad (12)$$
$$= \Delta/2 \qquad 0 < z < L/2.$$

ϕ differs from ϕ_e by a constant phase ϕ_0 which characterizes different particles:

$$\phi(z) = \phi_e(z) + \phi_0 \qquad (13)$$

Eq (9a) becomes

$$\ddot{x} = \frac{1}{p}(q_1{}^2 - k^2) - q_1{}^2 x, \quad -L/2 < z < 0, \qquad (14)$$
$$= -\frac{1}{p}(q_2{}^2 - k^2) + q_2{}^2 x, \quad 0 < z < L/2,$$

while Eq (9b) becomes

$$\ddot{y} = q_1{}^2 y, \ -L/2 < z < 0 \qquad (15)$$
$$= -q_2{}^2 y, \qquad 0 < z < L/2$$

In the above, $\ddot{x} = d^2x/(dz)^2$ and

$$K^2 = \frac{p^2}{\lambda k} \sin\frac{\Delta}{2}, \ q_1{}^2 = K^2 \frac{\sin(\Delta/2 - \phi_0)}{\sin \Delta/2}, \ q_2{}^2 = K^2 \frac{\sin(\Delta/2 + \phi_0)}{\sin \Delta/2}. \qquad (16)$$

Eqs (14) and (15) are linear differential equations with periodic coefficients, and can be solved by the standard matrix method[2]. The discussion of the stability of the motion involves two parts: First, Eq (14) is inhomogenious. Therefore, we have to find a periodic solution to define the equilibrium orbit. If the equilibrium orbit deviates too much from x = 0, the particle will either hit the grating surface, while if it deviates too much from x = 0 the acceleration becomes small. Second, we have to consider the stability of the homogeneous betatron motion around the equilibrium orbit. We consider the homogeneous part first

The betatron motion is stable if

$$|\cos \mu_x| < 1, \ |\cos \mu_y| < 1, \qquad (17)$$

where

$$\cos \mu_x = \cosh \theta_2 \cos \theta_1 - \frac{1}{2}\left(\frac{q_1}{q_2} - \frac{q_2}{q_1}\right) \sinh \theta_2 \sin \theta_1, \tag{18}$$

$$\cos \mu_y = \cosh \theta_1 \cos \theta_2 + \frac{1}{2}\left(\frac{q_1}{q_2} - \frac{q_2}{q_1}\right) \sinh \theta_1 \sin \theta_2 .$$

Here we have defined

$$\theta_i = q_i L/2 \tag{19}$$

in the above, q_i and θ_i could be imaginary if $\Delta < 0$. The above formula still apply with the replacement $\sin i\theta = i \sinh \theta$, etc. In addition to (18), we must require

$$-\pi/2 < \pm \Delta/2 + \phi_0 < \pi/2, \tag{20}$$

since otherwise the particle will be decelerated.

The inequalities (18) and (20) are analyzed numerically to find the stable region of the parameter space. It is convenient to present the results in terms of Δ, ϕ_0 and

$$\psi = KL = L\sqrt{\frac{p^2}{\lambda k}} \sin (\Delta/2) \tag{21}$$

We wish to determine the range of ϕ_0 corresponding to stable motion for given ψ and Δ. We find

(i) for all values of Δ, the maximum phase acceptance (range of ϕ_0) is obtained near $\psi = \pi$,

(ii) setting $\psi = \pi$, the value $\Delta = .7\ \pi$ corresponds to the maximum phase acceptance $|\phi_0| \leqslant .11\ \pi$,

(iii) for comparison, the phase acceptance for the case $\psi = \pi$, $\Delta = .5\ \pi$ is $|\phi_0| \leqslant .06\ \pi$.

If the injected beam is uniformly distributed in phase, case (ii) accepts 11 percent of the beam while case (iii) accepts only 6 percent. The ratio of the accelerating field to the peak available field is roughly $\cos(\Delta/2)$, which is .16 for case (ii) and .71 for case (iii). Thus case (ii) accepts more particle at the expense of less accceleration. Given Δ and ψ, the period length L is determined from Eq (21). Using the values given by Eq (11), one obtains

$$L = 2.1 \frac{\gamma}{\gamma_{in}} \text{ cm for case (ii).} \tag{22}$$

Next we consider the equilibrium orbit by the x-motion, which is the periodic solution of the inhomogeneous equation. One obtains

$$x(z) = f_1 + G\,(a_1 \cos q_1 z + b_1 \sin q_1 z)\;,\; -L/2 < z < 0,$$

$$= f_2 + G\,(a_2 \cosh q_2 z + b_2 \sinh q_2 z),\; 0 < z < L/2. \tag{23}$$

Here

$$f_1 = \frac{1}{p}\left(1 - \frac{\sin \Delta/2}{\sin(\Delta/2 - \phi_0)}\right),\quad f_2 = \frac{1}{p}\left(1 - \frac{\sin \Delta/2}{\sin(\Delta/2 + \phi_0)}\right),$$

$$G = \frac{1}{2(1-\cos\mu)}\,(f_1 - f_2),$$

$$a_1 = \cosh\theta_2 - \cos\theta_1 - 1 + \cosh\theta_2 \cos\theta_1 + \frac{q_2}{q_1}\sinh\theta_2 \sin\theta_1 \tag{24}$$

$$b_1 = \sin\theta_1\,(1 - \cosh\theta_2) - \frac{q_2}{q_1}\sinh\theta_2 (1 - \cos\theta_1),$$

$$a_2 = \cosh\theta_2 - \cos\theta_1 + 1 - \cosh\theta_2 \cos\theta_1 + \frac{q_1}{q_2}\sinh\theta_2 \sin\theta_1,$$

$$b_2 = \frac{q_1}{q_2}\sin\theta_1 (1 - \cosh\theta_2) - \sinh\theta_2\,(1 - \cos\theta_1).$$

In the above , we have assumed $\Delta > 0$. If $\Delta < 0$, the orbit can be obtained by setting $\Delta \to -\Delta > 0$ and displacing z by half a period.

Qualitative behavior of the equilibrium orbit can be seen directly from Eq (14). Without loss of generality, we may assume $\Delta > 0$. The inhomogeneous terms of Eq (14) are negative when ϕ_0 is positive, and vice versa. Thus, one may expect that x will generally be negative when $\phi_0 > 0$, and vice versa. When $\phi_0 > 0$ and increases, the focusing in the region $-L/2 < z < 0$ becomes weaker while the defocusing in the region $0 < z < L/2$ becomes stronger. Thus the orbit deviation will become large and negative until ϕ_0 reaches the limit of the homogeneous stability region where it blows up. This is also apparent from the factor $(1 - \cos \mu)^{-1}$ in the expression of G in Eq (24). On the other hand, when ϕ_0 is negative and decreases, the focusing becomes stronger and the defocusing weak, and we expect that the equilibrium orbit will behave better in that region.

We need to specify the maximum tolerable excursion of the

equilibrium orbit. We shall take

$$|x| \lesssim \frac{1}{p} = \frac{5\lambda}{2\pi} \tag{25}$$

We have evaluated Eq (23) numerically, and find that (25) implies

$$\begin{aligned} -.11\,\pi < \phi_0 < .05\,\pi \quad &\text{for case (ii)},\\ -.06\,\pi < \phi_0 < .02\,\pi \quad &\text{for case (iii)}. \end{aligned} \tag{26}$$

The phase acceptance region is reduced in the $\phi_0 > 0$ region as expected.

The limit of the phase acceptance region for negative ϕ_0 in (26) is due to the linear stability of the y-motion. There is a good reason to believe that this could be extended substantially if one takes into account the full non-linear equation (8). The equilibrium x-orbit, when $\phi_0 < 0$, tends to have a maximum in the region $-L/2 < x < 0$ and a minimum in the region $0 < z < L/2$. Therefore, the factor e^{-px} in Eq (8c) tends to neutralize the mismatch of the focusing and defocusing effects of the factor $\sin\phi$. This point needs to be studied further.

We conclude, therefore, that the strong focusing scheme for the grating linac works in principle. However, the phase acceptance region derived for the linear stability consideration is rather small. One hopes that the analysis of the full non-linear equation improves this limit substantially.

Acknowledgement

One of us (KJK) thanks Lloyd Smith for kindly providing him with his original notes on the transverse force in the grating linac. We are grateful to Robert Palmer's remark that the strong focusing can be achieved by simply shifting the phase. We also thank John Lawson and other workshop participants for helpful discussions.

References

1. R. B. Palmer, Particle Accelerators 11, 81 (1980), and contribution to this workshop.

2. E. D. Courant and H. S. Snyder, Ann. Physics 3, 1 (1958).

3. W. K. H. Panofsky and W. A. Wenzel, Rev. of Sci. Inst. 27, 967 (1956).

BEAM LOADING AND EMITTANCE GROWTH FOR A DISK-LOADED STRUCTURE SCALED TO 10 μm*

Perry B. Wilson
Stanford Linear Accelerator Center
Stanford University, Stanford California 94305

ABSTRACT

Beam loading and transverse emittance growth are studied in a disk-loaded accelerating structure which has been scaled to a wavelength of 10 μm. The resulting limitations on the charge per bunch which can be accelerated in such a scaled structure should provide a crude estimate of the charge per bunch which can be accelerated in a laser driven grating accelerator operating at the same wavelength. For an accelerator 100 m in length delivering an energy of 500 GeV, it is found that the number of particles per bunch that can be accelerated is on the order of $10^5 - 10^6$.

INTRODUCTION

Laser driven grating accelerators have been proposed which are capable of achieving very high energy gradients at a wavelength on the order of 10 μm. The charge per bunch which can be accelerated in such a device, as determined by beam loading and transverse emittance growth considerations, is also of importance. Beam loading effects and transverse emittance growth are readily calculated if the longitudinal and transverse wake potentials for an accelerating structure are known. These wake potentials have been computed[1] for the SLAC disk-loaded structure, and experimental measurements have verified that these computed wake potentials are at least approximately correct. In the absence of similar detailed calculations for the grating geometries proposed for a near-field laser driven accelerator, we can obtain a rough estimate of the charge per bunch that can be accelerated by studying beam loading and emittance growth in a disk-loaded structure scaled to the same wavelength.

BEAM LOADING

Associated with each mode in an accelerating structure is a quantity k_n, called the loss parameter, defined by

$$k_n \equiv \frac{E_a^2}{4w} \qquad (1)$$

*Work supported by the Department of Energy, contract DE-AC03-76SF00515.

0094-243X/82/910199-04$3.00

where E_a is the accelerating gradient and w is the energy stored per unit length for that mode. Note that k_n scales as ω^2. The above quantity is called the loss parameter because the energy loss per unit length of structure to a point charge q is given by $\Delta U_n = k_n q^2$. Also, the average beam loading gradient per particle is $\bar{E}_{bn} = k_n q$. Let k_o be the loss parameter for the accelerating mode. The energy loss and average beam loading gradient per particle due to higher-order modes can be taken into account by the beam loading enhancement factor B, such that the total loss to all modes is given by

$$\Delta U = B(\sigma)k_o q^2$$

$$\bar{E}_b = B(\sigma)k_o q \quad .$$

Note that B is a function of the bunch length σ. For the SLAC disk-loaded structure (λ = 10.5 cm), B = 3.0 for a bunch length σ = 1 mm. The value of k_o is 2×10^{13}V/C-m.

In addition to an average energy loss per particle, there will be an energy spread within the bunch due to beam loading. The particles at the front of the bunch will experience very little energy loss, while the particles at the back of the bunch will see a beam loading gradient on the order of 2 $\bar{E}_b$. If the accelerating field is E_a, the total energy spectrum width will be on the order 2 $\bar{E}_b/E_a$. This energy spread can be reduced by letting the bunch ride ahead (in space) of the accelerating wave crest so that the more heavily loaded particles at the rear of the bunch pick up more energy from the accelerating wave. Let us ignore this complication and compute the average energy loss per particle normalized to the peak accelerating field,

$$\frac{\bar{E}_b}{E_a} = \frac{B\,k_o q}{E_a} \quad .$$

Rewriting this expression in term of the number of particles that can be accelerated for a given value of $\bar{E}_b/E_a$, using also Eq. (1),

$$N = \frac{E_a}{e\,B\,k_o}\left(\frac{\bar{E}_b}{E_a}\right) \quad . \tag{2}$$

If the SLAC structure is scaled to λ = 10 μm, k_o increases by the ratio (10 cm/10 μm)^2 = 10^8 to $k_o = 2 \times 10^{21}$ V/C-m. Using also E_a = 5 GeV/m, $\bar{E}_b/E_a$ = 0.1 and B = 3, we obtain N = 5×10^5. The total relative energy spread in this example would be on the order of 20%. This could be reduced to a few per cent by moving the bunch ahead of the accelerating wave crest, but at the expense of an additional loss in average particle energy.

The efficiency for transfer of energy stored in the structure to the bunch is also of interest. This efficiency is given by

$$\eta = \frac{q(E_a - \bar{E}_b)}{w} = \frac{4}{B}\left(\frac{\bar{E}_b}{E_a}\right)\left(1 - \frac{\bar{E}_b}{E_a}\right) \quad . \tag{3}$$

The maximum efficiency is seen to be $\eta = 1/B$ at $\bar{E}_b/E_a = 1/2$. For the preceding example with $\bar{E}_b/E_a = 0.1$ and $B = 3$, the efficiency is 12%.

EMITTANCE GROWTH

The presence of dipole (deflecting) modes in a disk-loaded accelerating structure can lead to single bunch emittance growth.[2] Assume that a bunch moves off axis at the beginning of an accelerator, executing betatron oscillations with a wavelength $\lambda_\beta = 2\pi/k_\beta$ and an initial amplitude x_0. The bunch can be modeled by a simple two-charge approximation, a head charge $q/2$ at $z' = +\sigma$ and a tail charge $q/2$ at $z' = -\sigma$, where z' is the coordinate with respect to the bunch center. The head charge excites a transverse deflection wake, which drives the betatron oscillations of the tail charge. After traveling length z, it is easy to show[3] that the amplitude of the oscillation of the tail charge is

$$x(z) = \frac{q\,x_0\,z\,w_\perp(2\sigma)}{4\,V_0\,k_\beta} \quad ,$$

where eV_0 is the beam energy and $w_\perp$ is the dipole wake potential. If $A \equiv x(z)/x_0$, then the number of particles that can be accelerated is

$$N = \frac{4\,A\,V_0\,k_\beta}{e\,z\,w_\perp(2\sigma)} \quad . \tag{4}$$

The transverse wake for the SLAC structure at $\lambda = 10.5$ cm and $2\sigma = 2$ mm is[1] $w_\perp = 3 \times 10^{15}$ V/C-m^2. The dipole wake scales[3] as ω^3; at $\lambda = 10$ m it therefore becomes $w_\perp = 3 \times 10^{27}$ V/C-m^2. Assuming $\lambda_\beta \approx 100$ m for SLAC, then $\lambda_\beta = 1$ cm at the shorter wavelength. As a numerical example, again assume a gradient of 5 GeV/m, a total energy of 500 GeV and a length of 100 m. Equation (4) is strictly valid only for V_0 constant, but take $V_0 = 250$ GeV as an approximation for uniform acceleration to 500 GeV. Let the amplitude of the oscillation of the tail particle grow by a factor of 10. Inserting these values in Eq. (4), we obtain $N = 1 \times 10^5$.

DISCUSSION

The luminosity in a linear collider is given by $\mathscr{L} = N^2 f_r/\left(4\pi\sigma_x^*\sigma_y^*\right)$, where f_r is the repetition rate and σ_x^* and σ_y^* are the transverse beam dimensions at the collision point. To be interesting as a machine for high energy physics, an accelerator must be capable of achieving a luminosity 10^{30} cm^{-2}s^{-1}, at the very least. Assuming transverse beam dimensions of 0.1 μm and a repetition rate of 1 kHz, 10^9 particles per bunch are required to achieve this luminosity. The preceding analysis indicates that it will be very difficult to reach this high a charge per bunch in a grating accelerator for a beam with $\sigma_x^* \approx \sigma_y^*$. If the charge per bunch is lower, the repetition rate must increase above 1 kHz as N^{-2} to maintain a given luminosity with the above beam dimensions. It is difficult to believe that beams with a transverse dimension much less than 0.1 μm can be made to collide. Using a wide flat beam ($\sigma_x^* >> \sigma_y^*$), or many beams in parallel, is then the only remaining possibility. The total beam width in this case would have to be on the order of 1 cm to achieve the above luminosity.

REFERENCES

[1]K. Bane and P. Wilson, 11th Int'l Conf. on High Energy Accelerators (Birkhäuser Verlag, Basel, 1980) p. 592.

[2]A. W. Chao, B. Richter and C. Y. Yao, 11th Int'l Conf. on High Energy Accelerators (Birkhäuser Verlag, Basel, 1980) p. 597.

[3]P. B. Wilson, 1981 Summer School on High Energy Particle Accelerators. To be published in AIP Conf. Proceedings.

A THIN LAYER DIELECTRIC NEAR FIELD LASER ACCELERATOR

T. Weiland

DESY, Hamburg 52, Notkestr.85, Germany

ABSTRACT

A thin layer dielectric laser accelerator is proposed and analytical expressions are given for accelerating wave structures. All necessary ingredients for a full study of the beam dynamics including longitudinal and transverse collective effects are calculated. Expressions are found for the coupling impedance between the bunched beam and the accelerator as well as for the total parasitic energy loss. Integrals are given for the wake potentials inside the bunch. Dielectric strip line technique may be used to feed the structure with laser light and to focus and steer the light towards the beam.

INTRODUCTION

The very intense electromagnetic field produced by a laser beam gave rise to various ideas of how to use the field for acceleration of charged particles to very high energies. We will consider here only the so called "near field accelerators". A material or a grating[1-9] is used to slow down the phase velocity of the laser wave to the particle velocity and to turn the electric field vector of the transversely polarized wave towards the direction of the beam axis. The partciles travel near to the surface of the material or grating. In specific there exists the idea of using a half space filled with a dielectric medium and laser light coming from the inside towards the surface to the vacuum having a total reflection[3]. This scheme has some practical problems but is the easiest one to describe mathematically. The idea of using gratings[2] seems to be more realistic but needs further studies of all the collective effects which are very likely to limit strongly the maximum number of particles which can be accelerated in one bunch. A calculation of the beam induced electromagnetic fields on a grating can only be done by means of numerical computations.

In this paper we will describe a dielectric near field accellerator which consists of a thin layer on a meatllic surface. The system has two main components: the feeding in part and the beam line. For the electromagnetic fields on the beam line simple expressions are given. The coupling impedance between the beam and the fields are calculated as well as the fundamental beam loading[10]. Integrals describe the beam induced deccelrating and deflecting forces (wake potentials) and the total parasitic energy loss. Thus all essential ingredients are available to study the beam dynamics including collective effects such as bunch lengthening or shortening and transverse instabilities and beam break up.

For the feeding in device a mathematical method is described which enables the calculation of coupling coefficients between the incoming light wave and the accelerating wave. A dielectric strip line technique may be used to concentrate the fields in the beam region and to steer the light towards the beam line. Since a dielectric layer and a grating have almost the same wave structure the beam dynamics should also be very similiar in both accelerators.

THE SURFACE WAVES

The particles to be accelerated travel at the speed $v=\beta c \approx c$ at distance d above the surface of a thin dielectric layer on an infinitely conducting plate of metal, see Fig. 1. In order to

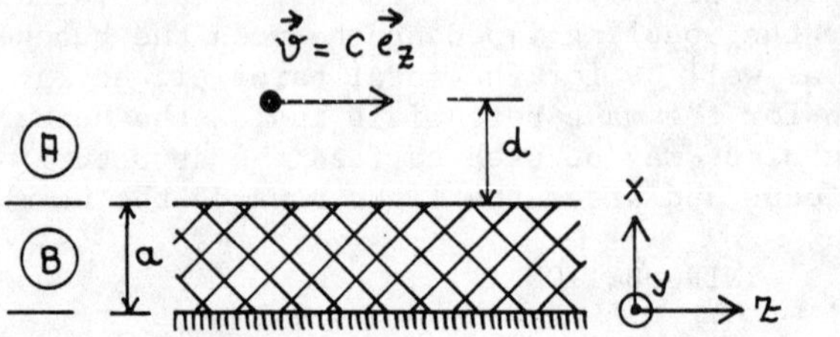

Fig. 1. A thin layer of a dielectric on a metal and particles moving in z direction at distance d above the surface. The thickness of the plate is a, the dielectric constant is ε_r.

accelerate the partciles efficiently we have to find travelling waves in +z direction with phase velocity equal to c. The waves also should carry a finite energy in the fields (evanescent in y direc.). General expressions for such fields may be found in any text book on micro wave theory. For the wave numbers in medium A **and** B we find :

$$\text{medium A}: \quad k_o^2 = k_x^2 + k_y^2 + k_z^2, \quad k_z = k_o, \quad k_x^2 = -K^2, \quad k_y^2 = K^2, \quad K \text{ real+pos} \tag{1}$$

$$\text{medium B}: \quad \varepsilon_r k_o^2 = k_x^2 + K^2 + k_o^2, \quad k_x = \xi, \quad (\varepsilon_r - 1)k_o^2 = K^2 + \xi^2 \tag{2}$$

The field components may be derived from the two vector potentials:

$$\vec{A}_I = \vec{e}_x E_I (1/KZ_0) \cos Ky \exp(-jk_o(z-ct)) \cdot \begin{cases} \exp(-K(x-a)); & a \le x \\ \cos\xi x/\cos\xi a; & 0 \le x \le a \end{cases} \tag{3}$$

$$\vec{H}_I = \text{curl}\, \vec{A}_I$$

$$\vec{A}_{II} = -\vec{e}_x E_{II} (1/K) \sin Ky \exp(-jk_o(z-ct)) \cdot \begin{cases} \exp(-K(x-a)); & a \le x \\ \sin\xi x/\sin\xi a; & 0 \le x \le a \end{cases} \tag{4}$$

$$\vec{E}_{II} = \text{curl}\, \vec{A}_{II}$$

E_I and E_{II} are the amplitudes of the electric longitudinal field at the surface. Wave type I has no magnetic component in x direction, type **II** has no electric field perpendicular to the surface. For an arbitrary initial phase of the particles $-\phi = k_o(z-ct)$ the electric fields of wave I in the vacuum read as:

$$\vec{E}_{In} = E_{In} \exp(-K_n d) \Big(\vec{e}_x ((K_n/k_o) + (k_o/K_n)) \sin\phi \cos K_n y + \vec{e}_y (K_n / k_o) \sin\phi \sin K_n y + \vec{e}_z \cos\phi \cos K_n y \Big) \tag{5}$$

For a given frequency k_o the K_n have to be solutions of :

$$K_n \cdot \varepsilon_r = \xi \tan \xi a \; ; \quad (\varepsilon_r - 1) k_o^2 = K_n^2 + \xi^2 \; ; \xi > 0 \; , \; K_n > 0 \tag{6}$$

The two curves are illustrated in Fig. 2. The given frequency corresponds to the big circles and it can be seen that only a finite number of solutions (small circles) exist. The cut-off frequencies for the modes are given by poles of the tan-function as :

$$a k_{Ic,n} = n \pi / \sqrt{\varepsilon_r - 1} \quad ; \; n=0,1,2,... \tag{7}$$

For the second set of waves we find the electric fields above the dielectric at a initial phase ϕ to be:

$$\vec{E}_{IIn} = E_{IIn} \exp(-K_n d) \quad (-\vec{e}_y \cdot (k_o/K_n) \sin\phi \; \sin K_n y + \vec{e}_z \qquad \cos\phi \; \cos K_n y \;) \tag{8}$$

The K_n are now solutions of the two equations :

$$K_n = -\xi \cot \xi a \; ; \quad (\varepsilon_r - 1) \; k_o^2 = K_n^2 + \xi^2 \quad ; \xi > 0 \; , \; K_n > 0 \tag{9}$$

These equations are illustrated in Fig.3. The cut-off frequencies are found to be :

$$a k_{IIc,n} = ((2n+1)/2) \pi / \sqrt{\varepsilon_r - 1} \; ; \; n=0,1,2.. \tag{10}$$

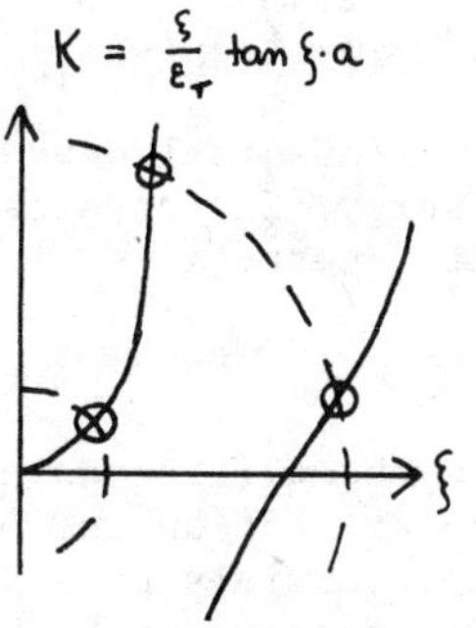

Fig. 2 The two curves giving the eigen solutions K_n of waveI

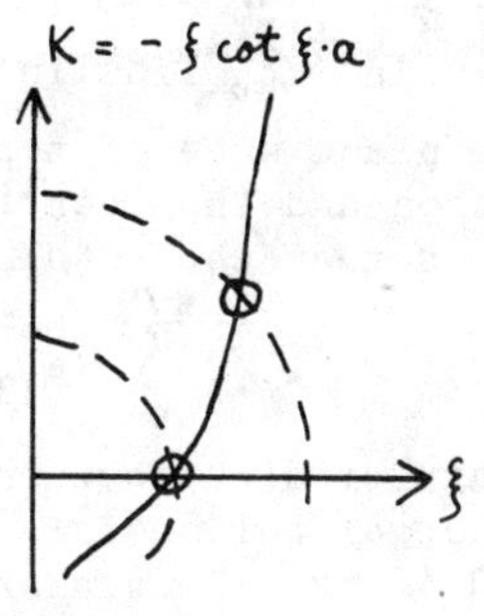

Fig. 3 The two curves giving the eigen solutions K_n of wave II

Both types of waves are able to accelerate partciles at the speed of light. The wave numbers K must be real and positiv to keep the energy in the field concentrated in a region near the surface. The y dependence of the field follows with the same wave numbers just by solving the boundary value problem and by asking for waves with phase velocity equal to the speed of light. Since for any given frequency only a finite number of eigen waves are found there must exist other field types on the structure. An arbitrary field cannot be described by a finite set. In order to find such fields we have to give up the condition for the phase velocity.

THE PLANE WAVE PACKETS

Any plane wave falling onto the dielectric surface from the medium A will result in a travelling wave, see Fig. 4. Again there exist two types of waves, called III and IV, depending on whether the electric or magnetic field vector is parallel to the x-z plane. Both waves always fulfill the boundary conditions for any incident angle φ and at any frequency. There may also be an angle ψ between the direction of the wave and the z direction. Since we are interested in waves travelling in +z direction we restrict the angle φ to $[0,\pi/2]$ and assume for any nonzero ψ a pair of waves at $+/-\psi$ with the same amplitude resulting in a sinusoidal dependence in y. Without writing down the rather lengthy expressions for the field we conclude that for a given frequency and a given wave number in y the fields are functions of the parameter φ. An arbitrary field on the dielectric plate may now be written as a sum over the surface waves plus integrals over the plane waves (see for example ref. 11,13) :

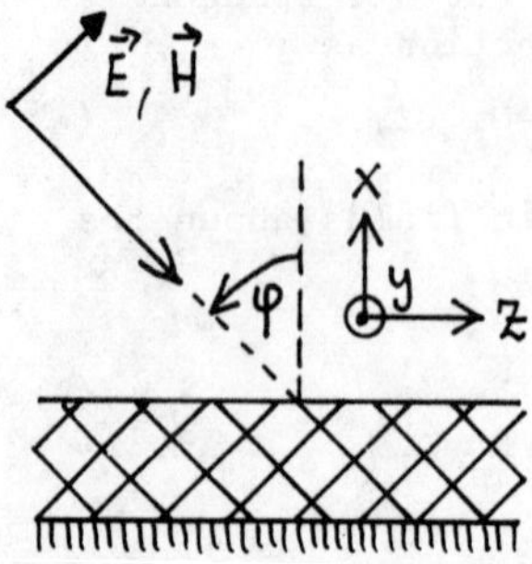

Fig. 4 The two possible polarisations of plane waves falling on the dielectric plate.

$$\vec{E} = \sum_{n=0}^{\infty} a_{In}\vec{E}_{In} + \sum_{n=0}^{\infty} a_{IIn}\vec{E}_{IIn} + \int_0^{\pi/2} a_{III}(\varphi)\vec{E}_{III}(\varphi)d\varphi + \int_0^{\pi/2} a_{IV}(\varphi)\vec{E}_{IV}(\varphi)d\varphi \tag{11}$$

Each single plane wave of type III or IV has a sinusoidal dependence in x direction and thus carries an infinite energy. The physical constraint is that the total field is of finite energy :

$$\int_0^{\pi/2} (a^2_{III}(\varphi) + a^2_{IV}(\varphi))\, d\varphi < \infty \tag{12}$$

Thus the angular distributions a_{III} and a_{IV} may be written as a sum over orthonormal polynomials which form a complete set on the interval $[0,\pi/2]$ [12,14]. The scalar product of the polynomials p_n and the fields $E_{III}(\varphi)$, $E_{IV}(\varphi)$ yields a set of wave packets with the running index according to p_n :

$$e_{III,IV\,n} = \int_0^{\pi/2} p_n(\varphi)\, E_{III,IV\,n}(\varphi)\, d\varphi \tag{13}$$

Any field on the structure may now be written as a sum :

$$\vec{E} = \sum_{n=0}^{N_I} \vec{E}_{In} + \sum_{n=0}^{N_{II}} \vec{E}_{IIn} + \sum_{n=0}^{\infty} \vec{e}_{IIIn} + \sum_{n=0}^{\infty} \vec{e}_{IVn} \tag{14}$$

The summing index of the two first sums is limited by the number of surface waves at the given frequency. The wave packets have distributions over the phase velocity which extends from c to infinity. These waves never couple to a surface wave if the structure is perfect. These waves will be used to feed the accelerator with light as will be shown later on.

BEAM DYNAMICS

The surface waves I and II are the only ones which determine the dynamical behaviour of the particles. For both mode types we find that the accelerating forces fall off exponentially from the dielectric surface as :

$$F_{znI,II} = e\, E_{I,IIn} \exp(-K_n d) \cos\phi \cos K_n y \tag{15}$$

The transverse forces due to both waves read as :

$$F_{xnI,II} = e\, E_{I,IIn} \exp(-K_n d) \sin\phi \cos K_n y \tag{16}$$

$$F_{ynI,II} = e\, E_{I,IIn} \exp(-K_n d) \sin\phi \sin K_n y \tag{17}$$

The force in x direction tries to push the particles away from the dielectric surface. It has been suggested[3] to compensate this force by a static magnetic field. Such a field on the dielectric surface could easily produced by a dc current in the metallic plate, Fig.5.

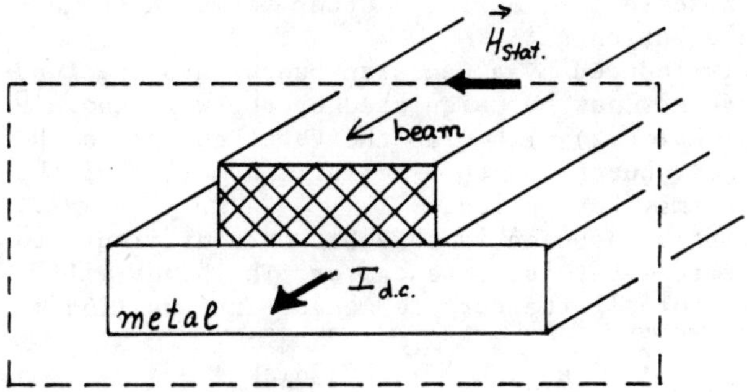

Fig. 5 A magnetostatic field produced by a dc current in the metal plate for compensating the deflecting force in x direction

The betatron wave numbers are then found to be :

$$-k_{\beta y}^2 = k_{\beta x}^2 = eE_{I,II} \cdot K_n^2 \cdot \sin\phi \cdot \exp(-K_n d) \,/\, (k_o \cdot E_0) \text{ , with } Z_0 = \sqrt{\mu_0/\varepsilon_0} \tag{18}$$

$$H_{y,stat.} = E_{I,II} \cdot K_n \cdot \sin\phi \cdot \exp(-K_n d) \,/\, (k_o \cdot Z_0) \text{ , } E_0 = \text{nom. energy}$$

The forces in y direction are always defocussing. This fact may be overcome by switching the initial phase of the partciles with respect to the wave from $+\phi$ to $-\phi$ by changing the wave phase as suggested by Kroll and Palmer for grating structures.[17] In this case the dc field has to alternate also and thus cannot be provided by a dc current in the plate anylonger. Phase switching provides an alternating gradient focussing in the transverse plane but whether a focussing is necessary at all can only be decided for a specific set of parameters.

The forces induced by the beam itself are proportional to the number of particles per bunch. Since the surface waves I and II are the only ones which can accelerate the bunch these waves are also the only ones which may take energy away from the partciles. The collective effects may be described using the formalism of loss-parameters and wake potentials[10]. For the lossparameters per unit length we find:

$$k'_{nI} = \frac{4\,c\,Z_0}{\pi\,a^2\,M}\,\exp(-K_n d)\cdot\frac{a^2K_n^2}{(1+a^2k_o^2/a^2K_n^2)\cdot(1+aK_n/\cos^2\zeta_n a)} \tag{19}$$

$$k'_{nII} = \frac{4\,c\,Z_0}{\pi\,a^2\,M}\,\exp(-K_n d)\cdot\frac{a^2K_n^2}{(1+a^2k_o^2/a^2K_n^2)\cdot(1+aK_n/\sin^2\zeta_n a)} \tag{20}$$

M is the number of half wave lengths in y direction taken into account for the integration of the total stored energy in the modes. The accelerating voltage is integrated at y=0 and at distance d above the surface. Both lossparameters fall off exponentially from the dielectric surface and both scale like a^{-2}. The other terms in the expressions depend not on a but on ak_o!

The voltage which is induced by a Gaussian bunch into the fundamental accelerating mode may now be calculated by the well known[10] formula $V_b = 2\,q\,k_n\,\exp(-k_o^2\sigma^2/2)$ with q as the total charge in the bunch and with σ the r.m.s. bunch length. The full formalism of the conventional accelerators may be applied.

The total energy lost by a bunch is somewhat more difficult to evaluate since a dielectric layer is no resonator. The bunch will couple to all frequencies of all the surface modes. Thus we find :

$$k'_{tot}(\sigma) = \sum_{n=0}^{\infty}\int_{ak_{Ic,n}}^{\infty} k_{In}(ak_o)\,\tilde{j}(ak_o)\,d(ak_o) + \sum_{n=0}^{\infty}\int_{ak_{IIc,n}}^{\infty} k_{IIn}(ak_o)\,\tilde{j}(ak_o)\,d(ak_o) \tag{21}$$

$\tilde{j}(ak_o)$ is the normalized frequency spectrum of the bunch power spectrum which corresponds to the bunch form factor $\exp(-k_o^2\sigma^2)$. The total parasitic losses can now be obtained by subtracting the loss into the accelerating mode from the above expression. The wake forces inside the bunch depend only the the partciles infront of the position s where the potential is to be calculated. Due to causaltiy a partcile at position s can see fields only from earlier particles. Thus we obtain :

$$w(s) = 2\cdot q\left\{\sum_{n=0}^{\infty}\int_{ak_{Ic,n}}^{\infty} k_{In}(ak_o)\,\tilde{j}(ak_o,s)\,\cos(sk_o)\,d(ak_o) + \sum_{n=0}^{\infty}\int_{ak_{IIc,n}}^{\infty} k_{IIn}(ak_o)\,\tilde{j}(ak_o,s)\,\cos(sk_o)\,d(ak_o)\right\} \tag{22}$$

$\tilde{j}(ak_o,s)$ is the frequency spectrum of the bunch current cut off at s inside the bunch.

Since the transverse deflecting forces and the longitudinal forces are due to the same waves on the accelerator the full beam dynamics including collective effects may be investigated using the equations given above.

FEEDING THE ACCELERATOR WITH LASER LIGHT

The wave structure on the dielectric accelerator described above splits into two groups of fields: one which couples to the beam and one which does not couple to the beam having a phase velocity always greater than the speed of light. Unfortunately the waves which do not interact with the beam are the only ones which are able to feed in energy from above the dielectric layer. Thus we need a device which couples the two types of waves. A simple example of

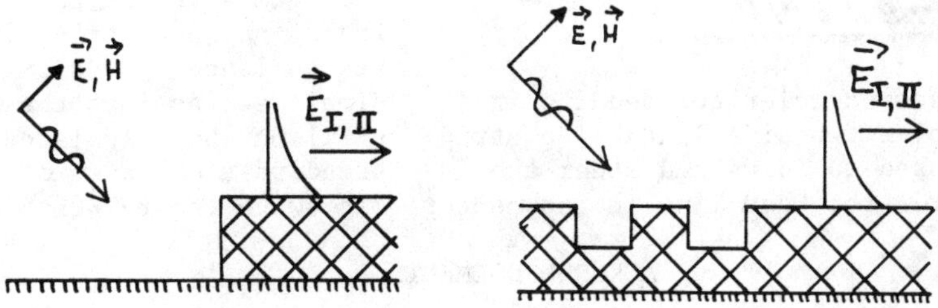

Fig. 6 Two simple structures which transfer energy from the incoming laser wave to a surface wave of type I or II

such a structure is given in Fig.6 . Both devices are able to transfer energy from an incoming laser wave to surface waves travelling to the right hand side. The frequency may be chosen such that only the surface wave $E_{I,1}$ is excited. The mathematics to solve this kind of problem is reduced to the problem of Fig.7. By standart eigenmode technique using the incoming field ($\vec{1}$) plus two sets of forward and backwards travelling waves ($\overleftarrow{\Sigma}$, $\vec{\Sigma}$) on the left hand side the field may be matched to a sum over waves running to the right hand side containing the accelerating mode. After inverting a matrix the transmission coefficient from the incoming laser wave to accelerating wave is known. Since the surface waves must have a y dependence a complementary set of two laser waves is used with the same amplitudes. This fact also solves the problem of keeping the feeding-in device away from the beam axis. The most simple arrangement which could do the job

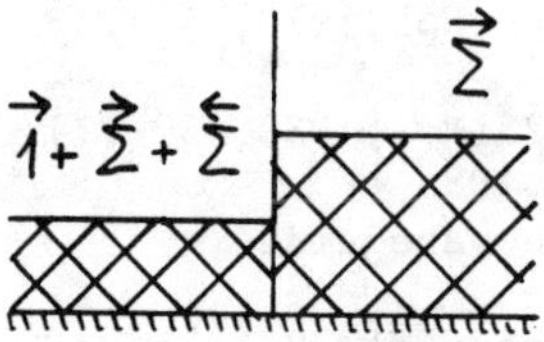

Fig. 7 The basic problem to be solved for calculating the coupling

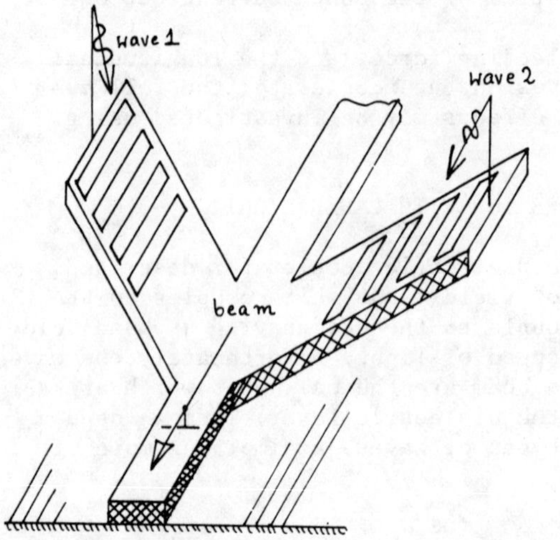

Fig.8 A laser accelerator feeding in the light over two side lines. The strip lines are used to focus and steer the light towards the beam line in the center

is shown in Fig. 8. Instead of a dielectric plate a dielectric stripline device is used. The side lines may be adjusted to the light aperture and may be tapered in addition in order to focus the incoming wave towards the beam line. The center strip may be kept small in order to concentrate the fields near the beam. Phase switching can be done by interupting the the dielectric layer under the beam.Although the fields on such strip lines are no longer given by simple expressions they still may be calculated by standard techniques of microwave theory [15,16].

ACKNOWLEDGEMENTS

The author wishes to thank A.Chao and H.Henke for helpful discussions.

LITERATURE

1. Y.Takeda,I.Matsui,Nucl.Inst.a.Meth. 62(1968) 306-310
2. K.Shimoda, Appl. Optics 1,33(1962)
3. S.A.Keihfets, 8.Conf.on High Energy Acc., 1971 CERN,597
4. K.Mizemo,S.Ono,O.Shimoe, Nature,Vol 253 Jan.1975
5. S.J.Smith,E.M.Purcell, Phys.Rev. 92(1969),1053
6. R.Rossmanith, Nucl.Instr,a.Meth.,154(1978),29-39
7. J.D.Lawson, IEEE Trans.on.Nucl.Science,VolNS26,No5,1979, 4217
8. R.Palmer,Particle Acc.,198o,Vol11,81-9o
9. Peng Huan,Zhuang Jiejia, Scienta Sinica ,Vol23,No2,1980,159-171
10. P.B.Wilson,KEK Lectures,KEK Accelerator 79-7,198o
11. V.V.Shevchenko,Continuous transitions in open waveguides, Golden Press, Boulder,Colorado 1971
12. F.M.Samir,J.C.Beal,IEEE Trans.on Micr.Wave.Theor.MTT23,1975
13. D.Marcuse,Theory of dielectric optical waveguides,Acad.Press,74
14. T.Weiland, waves on dielectric plates,TH-Darmstadt 1975,unpubl.
15. T.Weiland, Archiv fuer Elektronik u.Übertragungstechnik,33,170-4
16. T.Weiland, Kleinheubacher Berichte 22(1979),435-443
17. N.Kroll,R.Palmer,private communication,this workshop

A NOTE ON CYLINDRICAL WAVES WHICH PROPAGATE AT THE VELOCITY OF LIGHT*

Norman M. Kroll
Stanford Linear Accelerator Center
Stanford University, Stanford, California 94305

and

University of California, San Diego†
San Diego, California 92093

INTRODUCTION

The continuous linear acceleration of ultra relativistic particles in free space by an electromagnetic field requires the presence of a cylindrical wave component with phase velocity that differs negligibly from c and with non-vanishing electric field component in the direction of propagation. Lawson and Woodward have pointed out the fact that certain geometries proposed for laser driven acceleration fail to satisfy these requirements.[1] On the other hand, complex wave number plane wave fields which do satisfy these requirements have been constructed by Palmer, who also points out that any cylindrical wave with the required properties can be formed from superposition of plane waves of the form which he has obtained.[2]

The situation is analogous to that which occurs in standard waveguide theory. There it is also true that any waveguide mode can be constructed by superposition of plane waves. Nevertheless, the study of the general properties of cylindrical waves has proved to be a very powerful tool for the analysis of waveguides and similar structures. Because this may also prove to be the case for fields with propagation velocity c we present a brief study of their properties below.

A BRIEF REVIEW OF CYLINDRICAL WAVES

By definition, cylindrical waves with sinusoidal time dependence are solutions of Maxwell's equations in which the fields have the general form

$$\begin{aligned} \vec{E} &= \vec{E}(x,y)\ \mathrm{expi}(kz-\omega t) \\ \vec{B} &= \vec{B}(x,y)\ \mathrm{expi}(kz-\omega t) \end{aligned} \quad . \qquad (1)$$

* Work supported in part by the Department of Energy, contracts DE-AC03-76SF00515 (SLAC) and DE-AT03-81ER40029 (UCSD).
† Permanent address.

Since each field satisfies the vector wave equation we have

$$\nabla^2\vec{E} + \frac{\omega^2}{c^2}\vec{E} = \nabla^2\vec{B} + \frac{\omega^2}{c^2}\vec{B} = 0$$

and hence the transverse variation of each field component satisfies the two dimensional Helmholtz equation

$$\nabla^2 f(x,y) + K^2 f(x,y) = 0 \tag{2}$$

where

$$K^2 = \frac{\omega^2}{c^2} - k^2 \tag{3}$$

writing

$$\begin{aligned} \vec{E}(x,y) &= \vec{E}_T + \hat{z}E_z \quad , \quad \hat{z}\cdot\vec{E}_T = 0 \\ \vec{B}(x,y) &= \vec{B}_T + \hat{z}B_z \quad , \quad \hat{z}\cdot\vec{B}_T = 0 \end{aligned} \tag{4}$$

and substituting in Maxwell's equations yields[3] (Gaussian units)

$$ik\hat{z}\times\vec{E}_T - \hat{z}\times\vec{\nabla}E_z = i\frac{\omega}{c}\vec{B}_T \tag{5a}$$

$$ik\hat{z}\times\vec{B}_T - \hat{z}\times\vec{\nabla}B_z = -i\frac{\omega}{c}\vec{E}_T \tag{5b}$$

$$\vec{\nabla}\times\vec{E}_T = i\frac{\omega}{c}B_z\hat{z} \tag{6a}$$

$$\vec{\nabla}\times\vec{B}_T = -i\frac{\omega}{c}E_z\hat{z} \quad . \tag{6b}$$

From Eq. (3) one obtains

$$\vec{E}_T = \frac{ik}{K^2}\vec{\nabla}E_z - \frac{i\omega/c}{K^2}\hat{z}\times\vec{\nabla}B_z \tag{7a}$$

$$\vec{B}_T = \frac{ik}{K^2}\vec{\nabla}B_z + \frac{i\omega/c}{K^2}\hat{z}\times\vec{\nabla}E_z \quad . \tag{7b}$$

The standard TE (TM) modes are obtained by setting E_Z (B_Z) equal to zero and choosing a suitable set of solutions of (2) for B_Z (E_Z). The transverse fields are then obtained from (7).

THE SPECIAL CASE $k = \omega/c$

For waves which propagate at velocity c, the quantity K^2, which appears in the denominator of Eq. (7) vanishes, so that the previously described procedure fails. Setting $E_z = B_z = 0$ provides an obvious way out of the difficulty with Eq. (7) and leads to the usually discussed TEM modes. There are additional possibilities, however, which we discuss below.

Setting $K^2 = 0$ in (2) we obtain:

$$\nabla^2 f = 0 \quad . \tag{8}$$

Setting $k = \omega/c$ in (5) and carrying out a little vector algebra yields

$$\vec{\nabla} B_z = -\hat{z} \times \vec{\nabla} E_z \quad . \tag{9}$$

Equation (8) tells us that B_z and E_z satisfy Laplaces equation, and (9) tells us that they are related by the Cauchy-Riemann equations. Thus if we write

$$W(x+iy) = B_z(x,y) + iE_z(x,y) \quad , \tag{10}$$

Equations (8) and (9) are satisfied wherever W is analytic. We note that in the general case the modes are neither TE nor TM as Eq. (9) implies that non-vanishing $\vec{\nabla} E_z$ implies non-vanishing B_z.

An additional set of useful restrictions on the transverse fields is obtained by taking the divergence of (5), which when combined with (6) yields

$$\vec{\nabla} \cdot \vec{E}_T = -ikE_z \tag{11a}$$

$$\vec{\nabla} \cdot \vec{B}_T = -ikB_z \tag{11b}$$

Equations (5), (6), (10), and (11) are a set of necessary relations which the fields must satisfy. The relations are not, however, independent since satisfaction of a suitable subset implies the others. It is sufficient, for example, to satisfy (6a), (11a), (5a) and either (10) or the two dimensional <u>vector</u> Laplace equation for $\vec{E}_T$. The simplest procedure for generating a solution is to begin by specifying any solution of the two dimensional vector Laplace equation. If one identifies this solution with $\vec{E}_T$, then (6a), (11a), and (5a), provide explicit formulas for B_z, E_z, and $\vec{B}_T$, respectively. Alternatively, if one identifies this solution with $\vec{B}_T$, then (6b), (11b), and (5b) provide explicit formulas for E_z, B_z, and $\vec{E}_T$. Thus each solution of the vector Laplace equation generates two independent solutions.[4]

SOME SIMPLE APPLICATIONS

A solution of the type applied by Palmer to the grating problem[2] is obtained by setting

$$E_x = E \cos q y e^{-qx}$$

$$E_y = 0 .$$

Then from (6a)

$$B_z = \frac{i}{k}\frac{\partial E_x}{\partial y} = -\frac{iq}{k} E \sin q y e^{-qx}$$

from (11a)

$$E_z = \frac{i}{k}\frac{\partial E_x}{\partial x} = -\frac{iq}{k} E \cos q y e^{-qx}$$

and from (5a)

$$\vec{B}_T = \hat{z} \times \left(\vec{E}_T + \frac{i}{k}\vec{\nabla}E_z\right)$$

$$= E\left[\left(1 - \frac{q^2}{k^2}\right)\hat{y} \cos qy + \frac{q^2}{k^2}\hat{x} \sin qy\right] e^{-qy}$$

The cylindrically symmetric solution suitable for acceleration is obtained by setting

$$E_z = \text{constant}$$

$$H_z = 0$$

Then from (6a) and (11a)

$$\vec{E}_T = -\frac{ik\vec{r}}{2} E_z$$

and from (5a)

$$\vec{B}_T = -\frac{ikr\hat{\theta}}{2} E_z .$$

Finally, we recall that Lawson and Woodward[1] showed that any physical solution independent of y and valid in free space for all $x > 0$ must have vanishing E_z. This result is easily obtained in the present context from the fact that for y independent fields (11a) and (8) imply that E_z is constant and $E_x = ikE_z x + \text{constant}$.

REFERENCES

1. J. D. Lawson, Rutherford Laboratory report RL-75-043 (1975); IEEE Transactions on Nuclear Science, NS-26, 4217 (1979); P. M. Woodward, Journal IEE93, Part IIIA, 1554 (1947).

2. R. B. Palmer, Particle Accelerators 11, 81-90 (1980).

3. For a more detailed discussion see, for example, J. D. Jackson, Classical Electrodynamics (Wiley, New York, 2nd Edition 1975), Ch. 8.

4. Related TEM modes are obtained by setting $\vec{E}_T = \vec{\nabla}(\vec{\nabla}\cdot\vec{F})$, $\vec{B}_T = \hat{z}\times\vec{E}_T$ (or $\vec{B}_T = \vec{\nabla}(\vec{\nabla}\cdot\vec{F})$, $\vec{E}_T = -\hat{z}\times\vec{B}_T$) where $\vec{F}$ is the solution to the vector Laplace equation referred to above. Note further that although $\hat{z}\times\vec{F}$ is also a solution of the vector Laplace equation, the solutions which it generates by the same procedure are linear combinations of those obtained with $\vec{F}$.

"NEAR FIELD" LASER ACCELERATORS

Paul L. Csonka
Institute of Theoretical Science and Department of Physics
University of Oregon, Eugene, Oregon 97403

ABSTRACT

One particular laser accelerator is discussed in detail: the Template Laser (LT) Accelerator. According to one design, it employs mirrors placed periodically along the particle trajectory. According to another design, the mirrors are omitted, and additional electromagnetic fields are employed instead. The former design is a near field accelerator, but the latter is not. It is shown that contrary to intuition, a multiplicy of Fourier components will accelerate more efficiently than if the radiation is concentrated into only one component. Radiation damage on thin mirrors can be reduced by allowing appropriately phased additional radiation to impinge on the backside of the mirror. The RLT accelerator makes use of a partially reflecting cylinder to "stack" photons. A second type of resonance can further enhance the accelerating field (R2LT accelerator).

INTRODUCTION

I am the fourth speaker today. My talk follows three excellent review lectures. On the one hand, that puts me in a somewhat difficult position, because, to avoid repetition, I will have to omit a fair fraction of the material which I prepared for discussion today. On the other hand, since many of the obligatory general topics have already been adequately covered, I can afford the luxury to select, and elaborate on a few of my favorite topics. That is what I will do.

Before coming to those topics though, let me make some introductory remarks.

First of all, I think it is clear that this workshop will not teach us anything we do not already know. After all, Maxwell's Equations and the far field properties of electrons and protons have been known for decades. This meeting is really like a Rorschach test: everything we will come to recognize, we already see.

How can we speed up this process of recognition? Like in a Rorschach test, a new look, a new point of view may help. Here are some suggestions:

1. Think in terms of Fourier components as opposed to space time coordinates (God surely does).

2. Think of photons as the building blocks, as the material "stuff" of which the accelerator is to be built. Magnets,

waveguides, etc. are merely tools to shape the photons. They play the same kind of auxiliary role as screwdrivers, pliers, and other instruments used to fashion the accelerator. The accelerator is the photon field.

3. Design principle: maximize the number of parameters to enable us to accommodate constraints imposed by conservation (and other) laws.

WHY HIGH ENERGY ACCELERATORS?

Without them, there will be no progress in elementary particle physics. The reason was basically given by Heisenberg, who realized that $\Delta x \gtrsim \hbar/\Delta p$, so that in order to achieve small space resolution Δx, large momenta p are required. There is no way around this.

To distinguish details of 10^{-16} cm, one needs momenta of order $\gtrsim 10^3$ GeV/c.

Now, it is not hard to build an accelerator which will produce not only this much momentum, but even a thousand times more. For example, I myself can do it by throwing this piece of chalk. But that does not seem to be satisfactory. Or could it be? I will not answer that question here. If you do not know the answer, you may want to think about it for a moment.

LASER ACCELERATORS

Figure 1 shows a classification of possible laser acceleration schemes.

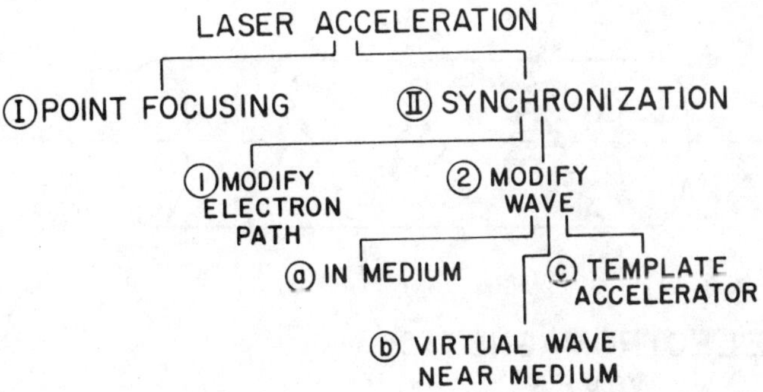

Fig. 1 Classification of possible laser acceleration schemes.

LASER TEMPLATE (LT) ACCELERATOR

I would like to take time to talk about one particular proposed laser accelerator, the "Laser Template Accelerator" (LT accelerator) outlined in Fig. 2a. As you will see, according to one proposed design it is a near field accelerator, but, according to another design it is not.

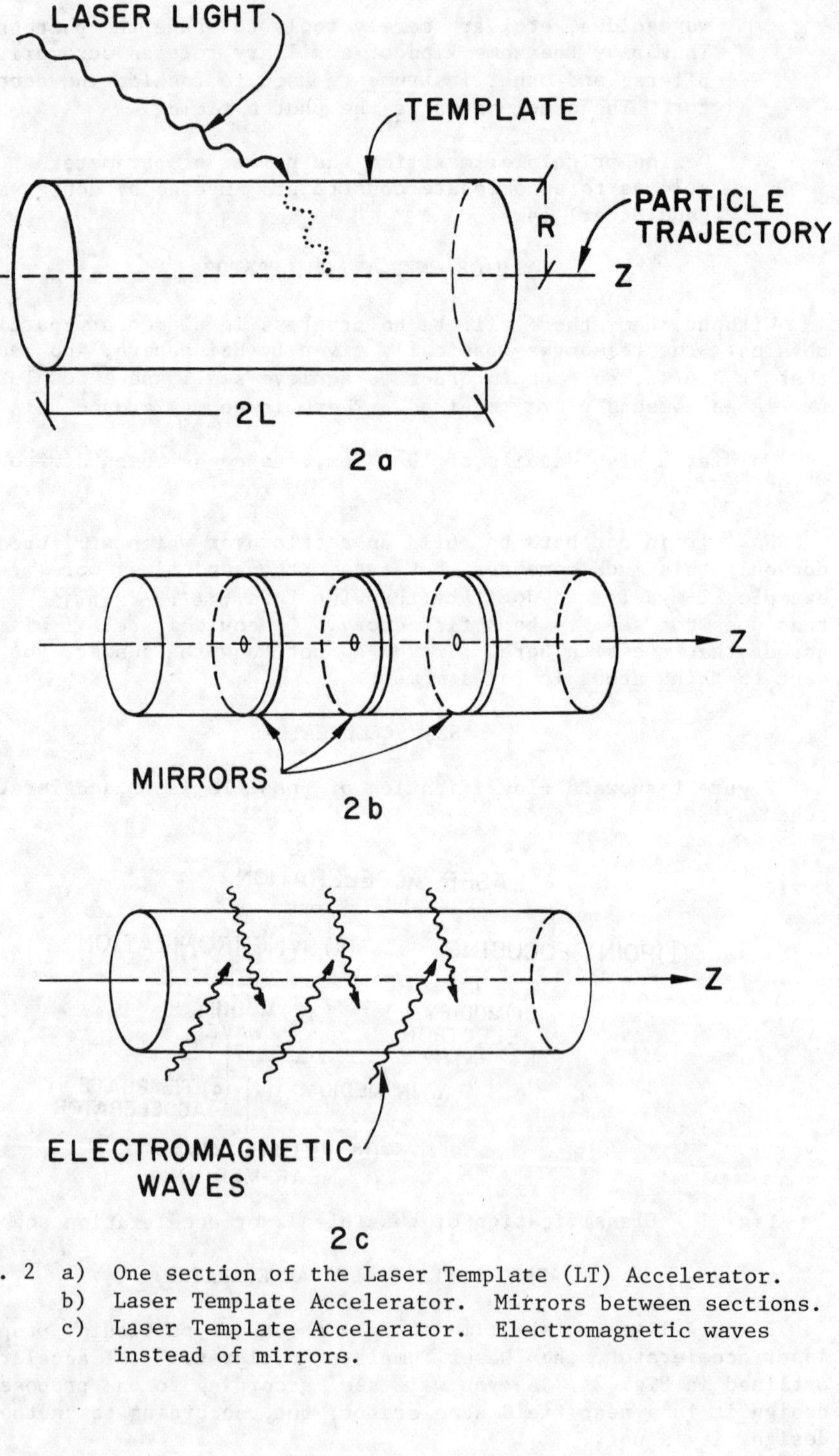

Fig. 2 a) One section of the Laser Template (LT) Accelerator.
b) Laser Template Accelerator. Mirrors between sections.
c) Laser Template Accelerator. Electromagnetic waves instead of mirrors.

The basic principle is simple: coherent radiation (e.g. laser light) converges on a cylinder called the "template". A pattern (e.g. a sequence of light and dark lines) is imprinted on this cylindrical template. This pattern is a holograph, i.e. a pattern which has the property that when the coherent radiation passes through it, it will produce a hologram along the cylinder axis. The hologram, in turn, is a field configuration (remember, a hologram determines not only the intensity, but also the phase) designed to accelerate an electron moving along the cylinder axis.

Before exploring some details, let me point out that the template is the most general optical element, because it controls both the phase and the amplitude for all waves passing through the template, and does this for each point on the template surface. You may now think that it must be very hard to make templates. Actually, it is not hard in general. For example a grating (a simple sequence of dark and light lines) is one particular template which is easy to make. Generally, one can calculate the template properties if the field configuration near the cylinder axis is given. One can then inscribe it with - say - electron beams. Alternatively, one can produce the hologram by simply recording the holograph of the desired field configuration along the axis. (One can eliminate the so called "twin image", but I do not want to take time to explain that now.)

What are the expected advantages of the template accelerator?

1. The fields converge cylindrically on the axis. Therefore, the field intensity at the axis will be of order $\sqrt{R/\lambda_r}$ times more than at the template. Here R is the cylinder radius, and λ_r is the "radial wavelength" of the field configuration (i.e. the period of the appropriate cylindrical Bessel function. See below.) When λ, the wavelength of the coherent radiation, is $\ll R$, the field near the axis can be much higher than it is on the cylinder surface.

2. Since the particle trajectory is in empty space, we can have fields of arbitrary intensity there (except, if mirrors are used: see below).

3. The very high accelerating fields will produce strong acceleration. The accelerating field configuration can have typical dimensions of order λ. In conventional accelerators the corresponding dimensions are of the order of centimeters. When λ is in the visible range, the resultant reduction of typical dimensions is of order 10^5.

4. The cylindrical geometry is very favourable (as opposed to, say, a planar geometry, as near a plane grating) because, first, it focusses in two dimensions (not just one). In particular, beam stability along both the x and y directions is easily achieved. Second, the cylindrical Bessel functions are well known, many results are immediate.

5. Once a suitable template is available, another one can be produced by simply copying the first one. If λ is not too small, the copying can be done by photographic techniques. (That will produce twin images, but they will generally not impair the operation of the accelerator.) Of course, the accelerator contains other components besides the template (for example a laser. Also, perhaps mirrors, see below). Nevertheless, it is significant, that at least the most important element can be manufactured by photography. In this sense one can say that if one has one template accelerator, one can produce a second one by simply photographing the first.

So far everything looks promising. Now let us look at the proposed template accelerator in more detail, as well as the difficulties one will encounter, and ways to overcome them.

OPTIMAL FIELD DISTRIBUTION

What field distribution will best accelerate the particle?

Focus Field on a Spot

The highest field intensity can, naturally, be obtained when the radiation is focussed on the smallest possible area, i.e. an area whose diameter is of the order of λ. Several people studied this possibility, and it was shown that if you can produce a stepfunction-like field distribution, existing lasers could accelerate electrons to a few times 10^7 eV energy.[1] Actually, stepfunction like field distributions cannot be produced by a laser, so the energy attainable is much less.[2]

But, in any case, is it a good idea to focus laser radiation on a small spot? I will now show that if the total energy output of the laser is limited (and it always is) and if you want the particle to gain the highest possible energy, then focusing on a small spot is <u>not</u> what you should do. Here is why:

Denote the kinetic energy gained by the particle by ΔK. It is proportional to $E_{\parallel}$ the accelerating field component along the particle trajectory, times the distance ℓ, over which the particle moves while the field is accelerating it:

$$\Delta K \sim E_{\parallel}\, \ell \sim E\, \ell \, .$$

The E, on the other hand, is proportional to the square root of U, the energy density in the electromagnetic field. The U itself is inversely proportional to the volume within which the radiated electromagnetic energy is distributed. Assume that this volume is roughly cylindrical, its length is ℓ, while its cross section is A. Denote the total electromagnetic energy radiated by the laser by $\mathcal{E}$. Then one can write so that $E \sim \sqrt{U} \sim \sqrt{\frac{\varepsilon}{A\ell}}, \qquad \Delta K \sim \sqrt{\frac{E}{A}}\ \sqrt{\ell}.$

Clearly, to get large ΔK, one should increase ℓ i.e. spread out the radiation along a line instead of focusing it on a spot. (Of course, ℓ can not be made arbitrarily large, because the coherence length of the radiation is finite, and repeated optical manipulation will gradually degrade beam quality.) Also, the radiation should be concentrated in a narrow cylinder (i.e. reduce A). It is now clear why the cylindrical template is so well suited to our problem: it concentrates the radiation near the cylinder axis, while spreading it out along the same axis.

Focus Cylindrically on a Line

Try to imagine, for a moment, the accelerating field as a superposition of plane wave Fourier components. Inside the template cylinder, near the particle trajectory, there is no medium (except perhaps mirrors . Those will be discussed later.) All electromagnetic plane waves travel in vacuum with velocity c. That is, the phase velocity for any plane wave along any straight line $c_\phi = \frac{c}{\cos\theta} \geqslant c$, where the = sign holds only when the straight line is parallel to the direction of wave propagation ($\theta = 0$), as is clear from Fig. 3. The particle to be accelerated, of course, always travels with velocity $v < c$ along the cylinder axis. Therefore, any Fourier component travels faster than the particle. Sooner or later, the electromagnetic force exerted on the particle by any particular Fourier component will switch sign. If originally it accelerated the particle, then after a while it will decelerate it. One Fourier component may still accelerate, while others are already decelerating.

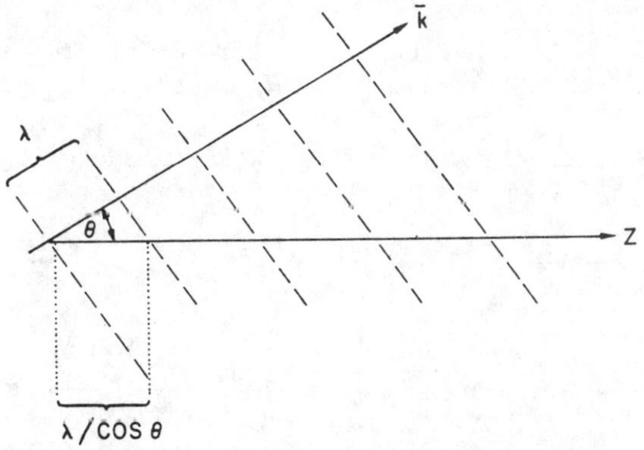

Fig. 3 Diagram showing that the phase velocity of an electromagnetic plane wave in vacuum along any straight line is $c_\phi = c/\cos\theta$.

To prevent net deceleration of the particle, the cylindrical volume surrounded by the template can be divided into sections by circular mirrors, whose plane is perpendicular to the cylinder axis

(Fig. 2b). A hole is drilled in the center of each mirror, to allow the accelerated particles to pass. The radius of these holes is sufficiently small to insure that only little radiation will leak thorugh them from one section to the next. (That little bit does not matter.) The radiation reflected from the mirrors will set up a standing wave pattern in each section.

The most general electromagnetic field pattern can be easily written down once the boundary conditions are given. Let us assume that we want to use mirrors as explained above. For simplicity, let us also assume that all mirrors are perfectly reflecting, and are plane. Here I will not write down the most general field, but consider only cylindrically symmetric configurations. (This restriction is not necessary, the most general configuration was given in Ref. 3). The result is displayed in Fig. 4.

$$E_z(t,\bar{x}) = i\,\frac{\omega}{c}\,e^{-i\omega t}\sum_k c_k\,e^{ikz}\,J_0(\kappa r)$$

$$E_r(t,\bar{x}) = i\,\frac{\omega}{c}\,e^{i\omega t}\sum_k c_k\,e^{ikz}\,\frac{(-ik)}{\kappa}\,J_1(\kappa r)$$

$$B_s(t,\bar{x}) = e^{-i\omega t}\sum_k c_k\,e^{ikz}\left(\frac{\omega}{c}\right)^2\frac{1}{\kappa}\,J_1(\kappa r)$$

$$E_s(t,\bar{x}) = B_z(t,\bar{x}) = B_r(t,\bar{x}) = 0$$

$$E_z(t,\bar{x}) \xrightarrow[\kappa r \to 0]{} \frac{i\omega}{c}\,e^{-i\omega t}\sum_k c_k\,e^{ikz}\left[1 - \frac{(\kappa r)^2}{8} \pm \dots\right]$$

$$E_r(t,\bar{x}) \xrightarrow[\kappa r \to 0]{} \frac{i\omega}{c}\,e^{-i\omega t}\sum_k c_k\,e^{ikz}\,(-ik)\left[\frac{r}{2} - \frac{\kappa^2}{23}\,r^3 \pm \dots\right]$$

$$B_s(t,\bar{x}) \xrightarrow[\kappa r \to 0]{} e^{-i\omega t}\sum_k c_k\,e^{ikz}\left(\frac{\omega}{c}\right)^2\frac{1}{\kappa}\left[\frac{r}{2} - \frac{\kappa^2}{32}\,r^3 \pm \dots\right]$$

$$E_z(t,\bar{x}) \xrightarrow[\kappa r \to \infty]{} \frac{i\omega}{c}\,e^{i\omega t}\sum_k c_k\,e^{ikz}\left(\frac{2}{\pi\kappa r}\right)^{1/2}\cos\left(\kappa r - \frac{\pi}{4}\right)$$

$$E_r(t,\bar{x}) \xrightarrow[\kappa r \to \infty]{} \frac{i\omega}{c}\,e^{-i\omega t}\sum_k c_k\,e^{ikz}\left(\frac{2}{\pi\kappa r}\right)^{1/2}\left(\frac{-ik}{\kappa}\right)\sin\left(\kappa r - \frac{\pi}{4}\right)$$

$$B_s(t,\bar{x}) \xrightarrow[\kappa r \to \infty]{} e^{-i\omega t}\sum_k c_k\,e^{ikz}\left(\frac{2}{\pi\kappa r}\right)^{1/2}\frac{\left(\frac{\omega}{c}\right)^2}{\kappa}\sin\left(\kappa r - \frac{\pi}{4}\right)$$

Fig. 4 Cylindrically symmetric electric and magnetic field components in one section of a Laser Template Accelerator. The notation used in Fig. 4 is as follows. The coherent radiation has wavelength

Fig. 4 continued – λ, and circular frequency ω. The velocity of light in vacuum is c. The cylinder axis is the z axis. The distance from it is denoted by r, and $\hat{z} \times \hat{r} = \hat{s}$ defines $\hat{s}$. The components of the electric field along $\hat{z}$, $\hat{r}$ and $\hat{s}$ are E_z, E_r and E_s respectively. The B_z, B_r and B_s are defined analogously. The c_k are coefficients which determine the amplitude of the various Fourier components. The zeroth and first order cylindrical Bessel functions are denoted by J_0 and J_1 respectively. The κ is defined as

$$\kappa = [(\frac{\omega}{c})^2 - k^2]^{1/2} .$$

An interesting feature of these fields should be noticed: Along the cylinder axis the electromagnetic field has no transverse components, nor any magnetic components (neither transverse, nor longitudinal), but has a longitudinal electric component travelling along the axis.

Now, let us ask ourselves what is the best accelerating field? More precisely: if the total energy of the electromagnetic field is given, then how should we choose the coefficients c_k in the formulae given in Fig. 4, such that the particle acquires the most kinetic energy, ΔK, while crossing one section of the accelerator. Of course, we already know that it would be wrong to focus the radiation near one spot; one should spread it out along the axis. But exactly how?

The obvious answer is: concentrate all available energy in a single wave, namely that one, which oscillates as exp(ikz) as a function of z, where k is chosen just right. How to choose this k? So that the phase of this wave at the position of the accelerated particle should shift by $\leqslant \pi$ as the particle moves from one end of the section to the other end. The reason is, that for such a k, the acceleration need not turn into deceleration at any time. If the particle is relativistic, its velocity $v \approx c$ will hardly change while moving the distance $2L_s$ from one end of the section to the other end. The shift in phase can then be written quite simply:

$$\Delta\phi \simeq 2\pi \frac{1}{\lambda} (c_\phi - v) \frac{2L_s}{v} = 2\pi(\frac{c_\phi}{v} - 1) \frac{2L_s}{\lambda} . \quad (1)$$

Now remember that we want as much acceleration as possible, i.e. as much longitudinal electric field as we possibly can have. For given $|\bar{E}|$, the longitudinal component is given by $|\bar{E}| \sin\theta$ which is larger for larger θ ($<\frac{\pi}{2}$). So it is obvious that we want the largest θ consistent with our condition $\Delta\phi \leqslant \pi$. Since

$$k = \frac{\omega}{c_\phi} = \frac{\omega}{c} \cos\theta \quad , \quad (2)$$

we want the smallest k for which $\Delta\phi \leqslant \pi$, i.e., we choose

$$k = (\omega/v)\ 2L_s/(2L_s + \lambda/2).$$

This result is intuitively obvious. Nevertheless, this apparently obvious result is quite wrong. You may find the correct result surprising. To derive it, first of all one has to write down ΔK, the kinetic energy gained by the accelerated particle for a given field distribution, then vary the field and find that one for which ΔK is maximum, subject to the condition that the total field energy in the template is constant.

The correct answer: The total field energy, ε, inside a cylindrical template section $2L_s$ long and with radius R, is the integral of the field energy density $u = \frac{1}{8\pi}(E^2 + B^2)$ over the cylinder volume. In the limit $R \gg \frac{1}{\kappa}$ the result is particularly simple:

$$\varepsilon = \frac{2L_s R}{2\pi} \sum_k |c_k|^2 \frac{(\omega/c)^4}{\kappa^3} . \qquad (3)$$

Next, we write down the kinetic energy gained by the accelerated particle, ΔK. For purposes of this discussion, let me consider the simple case when the particle trajectory is a straight line along the cylinder axis, which is the z axis. Assume that the particle is relativistic, its velocity remains essentially constant over the distance $2L_s$, i.e. $v \approx v_0$. The electric field component along z is E_z, it can be found from Fig. 4, so that for a particle of charge q,

$$\Delta K \approx q\int_{-L_s}^{L_s} dz\, E_z(z) \approx q \sum_k \frac{\sin(k_z - \frac{\omega}{v_0}) L_s}{\frac{c}{v_0} - \frac{\lambda}{\lambda_z}} \, 2 \, \mathrm{Im}\, c_k, \qquad (4)$$

where c_k are the coefficients in Fig. 4, and $\lambda_z \equiv \frac{2\pi}{k}$ is the wavelength along the z axis of the field whose amplitude is proportional to c_k. If ΔK is to be maximum as the c_k are varied while ε is left unchanged, then

$$\frac{\partial}{\partial c_k}\{\Delta K - \Lambda\, \varepsilon\} = \frac{\partial}{\partial c_k^*}\{\Delta K - \Lambda \varepsilon\} = 0 \qquad (5)$$

has to hold. Here Λ is a lagrange multiplier. Substituting ε and ΔK in Eq. (5) and solving for c_k gives

$$c_k = \frac{-i}{\Lambda} q \frac{\lambda}{2RL_s} \left[1 - \left(\frac{\lambda_0}{\lambda_z}\right)^2\right]^{3/2} \frac{\sin(k - \frac{\omega}{v_0})L_s}{\frac{c}{v_0} - \frac{\lambda}{\lambda_z}} (-1) . \qquad (6)$$

Now let me assume that $L_s k = n\pi$, with n: integer. (This will hold, for example, if the cylindrical section under discussion is closed at both ends by perfect mirrors.) Introduce the notation $\Delta\lambda = \lambda_z - \lambda$, and $\gamma_0 = [1 - v_0^2/c^2]^{1/2}$. Then, if $\frac{\omega}{v} L_s = \Pi\,(1/_2 + \text{integer})$, and in the limit $v_0 \to c$ (i.e. when $\gamma_0^2 \gg \lambda/\Delta\lambda$), we have

$$c_k = \begin{cases} \xrightarrow[\lambda_o/\lambda_z \to 0]{} \dfrac{-i}{\Lambda}\, q\, \dfrac{\lambda}{2RL_s} (\sin \dfrac{\omega}{v_o} L_s)\, (-1)^n & \text{(6a)} \\ \xrightarrow[\lambda_o/\lambda_z \to 1]{} \dfrac{-i}{\Lambda}\, q\, \dfrac{\lambda}{2RL_s} (\sin \dfrac{\omega}{v_o} L_s)\, (-1)^n \sqrt{8}\, [\sqrt{\Delta\lambda/\lambda} - \dfrac{1}{2\gamma_o^2}\, \dfrac{1}{\sqrt{\Delta\lambda/\lambda}}\,]. \end{cases}$$

Substituting Eqs. (6) in Eq. (4), one gets, again assuming $k2L_s = n\pi$ and choosing $L_s = \frac{v_o}{\omega}\pi\, (1/4 + \text{integer})$, in the limit $\gamma_o^2 \gg \lambda/\Delta\lambda$:

$$\Delta K = \text{constant} \cdot \sum_k [1 - (\frac{\lambda}{\lambda_z})^2]^{3/2} \frac{1}{[\frac{c}{v_o} - \frac{\lambda}{\lambda_z}]^2}$$

$$= \begin{cases} \xrightarrow[\lambda/\lambda_z \to 0]{} \text{Constant} \cdot \sum_k 1 \\ \xrightarrow[\lambda/\lambda_z \to 1]{} \text{Constant} \cdot \sum_k \sqrt{8}\, [(\Delta\lambda/\lambda)^{-1/2} - \dfrac{1}{\gamma_o^2} (\Delta\lambda/\lambda)^{-3/2}] \end{cases} \qquad (7)$$

Equation (6) demonstrates that the "intuitively obvious" solution is quite wrong. Indeed, the c_k is not like a delta function δ_{k,k_o}, rather, the c_k is quite broad as a function of k. (This is analogous, for example, to a linear accelerator which would work best when a multitude of simultaneous modes is generated in each accelerating cavity). On the other hand, it is clear from Eq. (7) that the largest contribution to ΔK comes from that Fourier-Bessel field component, for which k is such that $\Delta\lambda$ is the smallest, i.e. from that wave which moves most with the particle. That is in agreement with common intuition.

The improvement which can be achieved by choosing c_k correctly, is far from trivial. A factor of 10 or more in effective acceleration is quite realistic. That corresponds to an effective reduction of required laser power (while still achieving the same particle acceleration) by a factor of 100 or more.

MIRRORS - HOW TO PROTECT THEM FROM RADIATION DAMAGE

As I mentioned a little while ago, any electromagnetic wave in vacuum (e.g. near the particle trajectory) will sooner or later overtake the particle. If it originally accelerated the particle, then after a while it will start to decelerate it. To avoid such deceleration, one has to intervene somehow and change the phase of the wave relative to the particle. One way to achieve that, is to intercept the wave before it would start to decelerate, by placing a mirror in its way. The plane of the mirror may be chosen to be perpendicular to the particle trajectory. Behind the mirror another wave takes over, its phase is so chosen that it will continue to accelerate the particle until the next mirror is reached, beyond which yet another wave will take over, etc. A small hole is drilled at the center of each mirror, which will allow the accelerated particles to pass through the mirror, while reflecting most of the electromagnetic wave. (One can think of an alternative method: allow the particle to interact with an appropriately chosen electromagnetic field which will temporarily displace the particle. I will come to this possibility later.)

If one intends to use mirrors in the manner just described, an important problem has to be faced: The mirror will be exposed to the electromagnetic fields which accelerate the particles. Near the cylinder axis these fields are, of course, very intense, and could cause damage to the mirrors.

Damage due to electromagnetic radiation is primarily the result of two distinct mechanisms.

The first mechanism is caused by the electric field near the mirror surface: electrons are pulled out of the material, and form a dilute plasma which may cover the surface. When the electromagnetic wave is exactly or almost normally incident, the electric field is exactly or almost parallel to the surface: this mechanism is unimportant. At any rate, the phenomenon is reversible: when the fields are reduced or reversed, the electrons re-enter the material. Furthermore, such a plasma film covering the mirror is mostly harmless, because a plasma, even if dense, is also a mirror, so that in effect it at most causes only a change in effective mirror thickness.

The second mechanism is the result of Joule heating. As the electromagnetic wave is reflected by a metal, electron currents are set up near the surface which will heat, melt or even evaporate the material. The damage is irreversible. For our purposes, this second mechanism is the more devastating one.

I will now describe a method which can be used to reduce heat damage to a mirror. The idea is simple: Let not only one wave be incident on the mirror, but two waves, one from each side of a thin metal film. Choose the phases of the waves such that the two fields tend to cancel inside the mirror. That will, in turn, reduce the electric currents flowing, and therefore, the Joule heat produced. In other words, I will show that if one incident electromagnetic wave burns up a thin mirror, then shining an appropriately chosen second wave of approximately equal intensity on the other side of the same mirror will not only <u>not</u> cause it to burn up twice as fast, but can be so chosen as to prevent heat damage to the mirror altogether, even if both wave intensities are then increased, under certain circumstances by several orders of magnitude.

Since this is a review talk, I am sure you do not want to look at complicated formulae. Therefore, I will explain the basic idea for the simple case of a scalar field incident on a mirror. The mathematics which describes this case is quite transparent. At the end I will show how to derive from these the more complicated formulae which apply to the electromagnetic field (mass zero vector field).

SCALAR WAVE INCIDENT ON POTENTIAL BARRIER

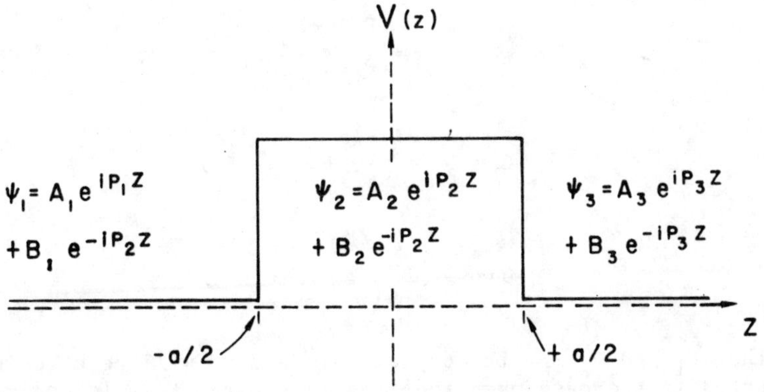

Fig. 5 Potential barrier for a scalar field. The waves are ψ_1 and ψ_3 outside the barrier, and ψ_2 inside it.

Figure 5 shows a potential barier. For $z < -a/2$, the wave has wavelength (along z) $\lambda_1 = 2\pi/p_1$; inside the barrier, i.e. for $-a/_2 < z < a/_2$, it has wavelength $\lambda_2 = 2\pi/P_2$; while for $z > a/2$, the wavelength is $\lambda_3 = 2\pi/p_3$. At both boundaries of the barrier the wave and its derivative have to be continuous. These boundary conditions allow one to express, for example A_2 and B_2 in terms of A_1 and B_1 as follows:

$$A_1 e^{-i P_1 a/_2} = \frac{P_1 + P_2}{2 P_1} A_2 e^{-i P_2 a/_2} + \frac{P_1 - P_2}{2 P_1} B_2 e^{i P_2 a/_2}$$

$$B_1 e^{i P_1 a/_2} = \frac{P_1 - P_2}{2 P_1} A_2 e^{-i P_2 a/_2} + \frac{P_1 + P_2}{2 P_1} B_2 e^{i P_2 a/_2} \tag{8a}$$

We will be interested in the case when

$$P_1 - P_3 \ .$$

In that case, using Eqs. (8a), and a similar set which is valid at the boundary at $z = a/_2$, one obtains

$$\begin{bmatrix} A_3 e^{i P_1 a/_2} \\ B_3 e^{-i P_1 a/_2} \end{bmatrix} = \underbrace{\begin{bmatrix} C - \frac{1}{2}(\alpha + \frac{1}{\alpha})D & \frac{1}{2}(\alpha - \frac{1}{\alpha})D \\ -\frac{1}{2}(\alpha - \frac{1}{\alpha})D & C + \frac{1}{2}(\alpha + \frac{1}{\alpha})D \end{bmatrix}}_{=M} \cdot \begin{bmatrix} A_1 e^{-i P_1 a/_2} \\ B_1 e^{i P_1 a/_2} \end{bmatrix} \tag{8}$$

The matrix M is a "transfer matrix". Knowing it, one can calculate A_3 and B_3, from A_1 and B_1. Here we used the notation

$$\alpha = \frac{-P_1}{P_2} = \frac{-P_3}{P_2}$$

$$C = \frac{1}{2}\left(e^{i P_2 a} + e^{-i P_2 a}\right) \tag{8b}$$

$$D = \frac{1}{2}\left(e^{i P_2 a} - e^{-i P_2 a}\right)$$

Note that P_1 and P_2 may be complex. Indeed we will be especially interested in the case when the imaginary part of P_2 is non-zero; that term will be responsible for energy absorption in the barrier.

Let us define the eigenvectors of M:

$$\phi^{(+)} = \begin{bmatrix} \phi_1^{(+)} \\ \phi_2^{(+)} \end{bmatrix}, \quad \text{with eigenvalue } \lambda^{+} = e^{i P_2 a},$$

and (9)

$$\phi^{(-)} = \begin{bmatrix} \phi_1^{(-)} \\ \phi_2^{(-)} \end{bmatrix}, \quad \text{with eigenvalue } \lambda^{-} = e^{-i P_2 a},$$

$$M\, \phi^{(\pm)} = \lambda^{\pm}\, \phi^{(\pm)} . \tag{10}$$

Here $\phi_2^{(+)} = \rho\, \phi_1^{(+)}$, where $\rho = -\frac{1+\alpha}{1-\alpha}$,

(11)

$$\phi_2^{(-)} = \frac{1}{\rho}\, \phi_1^{(-)} .$$

(The $\phi^{(+)}$ and $\phi^{(-)}$ are not orthogonal in general, because M need not be unitary.)

The energy flowing into the barrier from the left is proportional to $|A_1|^2$, while that from the right is proportional to $|B_3|^2$. Similarly, the energy flowing out of the barrier is proportional to $|B_1|^2$ and $|A_3|^2$. The net energy flow into the barrier is, therefore, proportional to

$$F = |A_1|^2 - |B_1|^2 - \left[|A_3|^2 - |B_3|^2\right] . \tag{12}$$

To evaluate F, we first make use of the fact that any vector (A,B) can be expanded in terms of $\phi^{(+)}$ and $\phi^{(-)}$. In particular, we choose $\phi_1{}^{(+)} = \phi_1{}^{(-)} = 1$, and then write

$$= \begin{bmatrix} A_1 e^{-i P_1 a/2} \\ B_1 e^{i P_1 a/2} \end{bmatrix} = a\phi^{(+)} + b' \phi^{(-)} = a\,\phi^{(+)} + \rho\, b\, \phi^{(-)} = \begin{bmatrix} a + \rho b \\ \rho a + b \end{bmatrix} \tag{13a}$$

Next, make use of Eq. (10) to express (A_3, B_3) as

$$= \begin{bmatrix} A_3 e^{i P_1 a/2} \\ B_3 e^{-i P_1 a/2} \end{bmatrix} = a\,\lambda^+ \phi^{(+)} + \rho\, b\, \lambda^- \phi^{(-)} = \begin{bmatrix} \lambda^+ a + \rho\,\lambda^- b \\ \rho\,\lambda^+ a + \lambda^- b \end{bmatrix} . \tag{13b}$$

It is convenient to use instead of a and b two other complex constants, u and h defined by

$$a = hu \quad ; \qquad h = \exp(\eta + i\xi) \ , \qquad b = \frac{1}{h}\, u \ . \tag{14}$$

Substituting Eqs. (13a), (13b) and (14) in Eq. (12), one can expres F in terms of $|u|$, η and ξ. I will write down here the result which is valid in the special case of interest to us, namely when the imaginary and real parts of P_1 and P_2 satisfy:

$$\operatorname{Im} P_1 = 0 \quad , \qquad \operatorname{Im} P_2 = + \operatorname{Re} P_2 \ . \tag{15}$$

In particular, the second of Eqs. (15) holds to a very good approximation when the barrier is an energy absorbing "good conductor", and the first of Eqs. (15) holds when the barrier is located in vacuum. In this special case one finds

$$F = |u|^2 \ \frac{8\sqrt{2}\ |\alpha|}{|1 - \alpha|^2} \ \{[\operatorname{ch}(\varepsilon + 2\eta)]\operatorname{sh}\varepsilon + [\cos(\varepsilon + 2\xi)]\sin\varepsilon\} \tag{16}$$

where we defined

$$\varepsilon = a \operatorname{Re} P_2 \ .$$

The quantities ε and α are determined by the height and width of the barrier. Once the barrier is given, the ε and α are determined. The quantities $|u|$, η and ξ characterize the wave. We want to find that wave, for which the energy flow into any given barrier is smallest. Therefore, we vary η and ξ and determine those η and ξ values for which F is a minimum. At

$$\eta = -\frac{\varepsilon}{2} \ , \quad \xi = -\frac{\varepsilon}{2} \pm \frac{\pi}{2} \ , \tag{17a}$$

F assumes its minimum value of

$$F_{min} = |u|^2 \frac{\sqrt{2}\ 16\ |\alpha|}{|1-\alpha|^2} (1/2) \{\text{sh}\ \varepsilon - \sin\varepsilon\}$$

$$\xrightarrow[\varepsilon \to 0]{} |u|^2 \frac{\sqrt{2}\ 16\ |\alpha|}{|1-\alpha|^2} \frac{1}{6} \varepsilon^3 , \tag{17b}$$

and here

$$\begin{bmatrix} A_1 e^{-i P_1 a/2} \\ B_1 e^{i P_1 a/2} \end{bmatrix} = v \begin{bmatrix} \text{ch}\,\epsilon + \alpha & \text{sh}\,\epsilon \\ -\text{ch}\,\epsilon + \alpha & \text{sh}\,\epsilon \end{bmatrix}$$

$$\begin{bmatrix} A_3 e^{i P_1 a/2} \\ B_3 e^{-i P_1 a/2} \end{bmatrix} = v \begin{bmatrix} \text{ch}\,\epsilon - \alpha & \text{sh}\,\epsilon \\ -\text{ch}\,\epsilon - \alpha & \text{sh}\,\epsilon \end{bmatrix} \tag{18}$$

$$\epsilon = \frac{\varepsilon}{2}(-1 + i) , \quad v = \pm i \frac{2u}{1-\alpha} .$$

In the definition of v, the upper and lower sign holds depending on whether the upper or lower sign is chosen in Eq. (17a).

Let us study Eqs. (16) and (17). Equation (16) shows that the energy deposited in the barrier layer is generally proportional to ε, whenever $\varepsilon \ll 1$. This is not surprising. For small ε the barrier thickness is small compared to the skin depth, so that "surface currents" fill the whole volume. Therefore, the Joule heat produced is proportional to the volume, i.e. ε On the other hand, Eqs. (17) say that one can arrange A_1 and B_1 so, that a special wave cancellation occurs inside the barrier, and reduces the energy absorption there. Indeed, by choosing η and ξ appropriately, one can insure that in the expression for F, not only the coefficient of ε, but also that of ε^2 vanish, so that the leading term will be proportional to ε^3, which results in a total reduction of absorbed energy by a factor $(1/6)\varepsilon^2$.

The physical interpretation of this phenomenon is illustrated in Fig. 6. The waves ψ_1 and ψ_3 are arranged such that they cancel exactly at the center of the barrier. Elsewhere inside the barrier they cannot cancel, because the absolute value of one of them increases with increasing z, while the other decreases. In the first approximation, the difference between them at the edges is proportional to the barrier width. The current in the barrier is proportional to this difference, the joule heat energy absorbed per volume element is proportional to its square, therefore, the total absorbed energy flux is proportional to its cube, i.e. to ε^3.

Remember, a metal mirror is a good reflector even when $\varepsilon \ll 1$, because $|\alpha| \ll 1$. In other words, there will be only little radiation leaking through the mirror even for $\varepsilon \ll 1$ (as long as the atoms behave as they do inside a bulk metal, i.e. as long as ε is of order $\gtrsim 10^2$ Å).

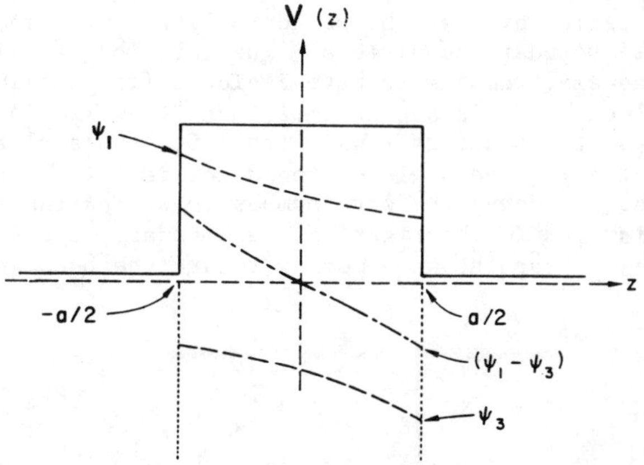

Fig. 6 Field cancellation inside a potential barrier reduces Joule heating.

This is a very interesting result. But, by itself, it does not help us much. We are interested not only in the energy deposited, but also in the shift between the phase of the waves on both sides of the barrier. From Eq. (18) we find the difference between the wave amplitudes

$$\begin{bmatrix} A_3 e^{i P_1 a/2} - A_1 e^{-i P_1 a/2} \\ B_3 e^{-i P_1 a/2} - B_1 e^{i P_1 a/2} \end{bmatrix} = v \begin{bmatrix} -2\alpha \text{ sh } \varepsilon \\ -2\alpha \text{ sh } \varepsilon \end{bmatrix} \xrightarrow[\varepsilon \to 0]{} v \begin{bmatrix} -2\alpha\varepsilon \\ -2\alpha\varepsilon \end{bmatrix} \quad . \qquad (19)$$

For real metals, $|p_2| \gg |p_1|$, therefore $|\alpha| \ll 1$. The difference given by Eq. (19) is very small indeed. Next we see if we can increase that difference while still keeping the absorbed energy small, by varying η and ξ by $d(\eta + i\xi)$. We find

$$d \begin{bmatrix} A_3 e^{i P_1 a/2} - A_1 e^{-i P_1 a/2} \\ B_3 e^{-i P_1 a/2} - B_1 e^{i P_1 a/2} \end{bmatrix} \xrightarrow[\varepsilon \to 0]{} v \begin{bmatrix} 2\varepsilon \\ -2\varepsilon \end{bmatrix} d(\eta + i\xi) \quad ,$$

$$\frac{dF}{d(\eta + i\xi)} = 0 \quad , \qquad (20)$$

$$dF \xrightarrow[\varepsilon \to 0]{} |u|^2 \frac{\sqrt{2}\ 8\ |\alpha|}{|1 - \alpha|^2} \varepsilon \left[(d\eta)^2 + (d\xi)^2\right] \quad .$$

Clearly, if $|d\eta + id\xi| \lesssim \varepsilon/3$, one has $dF \lesssim F_{min}$, so that F at most doubles its value compared to what it is at F_{min}. On the other hand, the difference between waves is multiplied by a factor $d(\eta + i\xi)/\alpha$, which can be large, since one can have $\alpha \approx 10^{-2}$ in the visible range for electromagnetic waves, and several orders smaller for microwaves. Therefore, a very large gain in phase shift can be so obtained, at the price of hardly any change in the energy absorbed by the barrier.

Electromagnetic Waves

Electromagnetic waves at the boundary between two media have to satisfy certain boundary conditions. The notation is illustrated in Fig. 7. In the electromagnetic wave incident from the left, the oscillating electric field has an amplitude E_1, while in the wave moving away from the boundary toward the left, the amplitude is E_{1r} (r for "reflected"). The angle of incidence is θ_1. In the medium to the left of the boundary, the wave number is k_1 and the magnetic permeability is μ_1. To the right of the boundary, the corresponding quantities are E_2 (for the wave moving toward the boundary), E_{2r}, k_2, and μ_2.

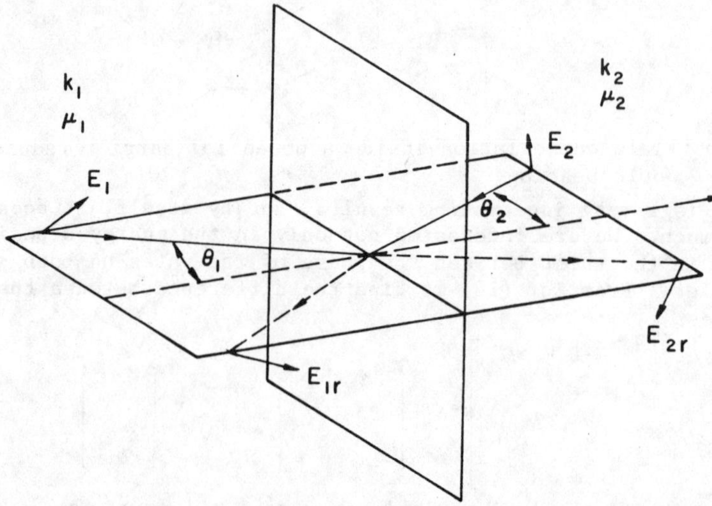

Fig. 7 Electromagnetic waves at the plane boundary of two media. In the first medium the wave number is k_1, and the magnetic susceptibility is μ_1. The k_2 and μ_2 refer to the second medium.

The waves have to satisfy the well known conditions at the boundary:

$$E_2 = \frac{2\dfrac{k_1}{\mu_1}\cos\theta_1}{\dfrac{k_1}{\mu_1}\cos\theta_i + \dfrac{k_2}{\mu_2}\cos\theta_j} E_1 ,$$

$$E_{1r} = \frac{\dfrac{k_1}{\mu_1}\cos\theta_i - \dfrac{k_2}{\mu_2}\cos\theta_j}{\dfrac{k_1}{\mu_1}\cos\theta_i + \dfrac{k_2}{\mu_2}\cos\theta_j} E_1 , \tag{21}$$

where $i = 1$, $j = 2$ if E_1 is perpendicular to the plane of incidence, and $i = 2$, $j = 1$, if E_1 is parallel to that plane.

It turns out that Eqs. (21) can be obtained directly from Eqs. (8a), if in the latter equations we substitute as follows:

If E_1 is perpendicular to the plane of incidence:

$$A_1 = E_1, \quad B_1 = E_{1r}, \quad A_2 = E_{2r}, \quad B_2 = 0$$

$$P_1 = \frac{k_1}{\mu_1} \cos \theta_1 \quad , \quad P_2 = \frac{k_2}{\mu_2} \cos \theta_2 \tag{22a}$$

If E_1 is parallel to the plane of incidence:

$$A_1 = E_1 \cos \theta_1, \quad B_1 = -E_{1r} \cos \theta_1, \quad A_2 = E_{2r} \cos \theta_2, \quad B_2 = 0$$

$$P_1 = \frac{k_1}{\mu_1 \cos \theta_1} \quad , \quad P_2 = \frac{k_2}{\mu_2 \cos \theta_2} \tag{22b}$$

These same substitutions will take ψ_1 and ψ_2 into waves which will describe the propagation of electromagnetic waves along z. Therefore, with these very same substitutions we can transform conditions (16) - (20) into expressions which are applicable when the waves under study are electromagnetic. In this manner we conclude that when electromagnetic radiation is incident from both sides of a film of good conductor, the energy absorbtion per unit surface can be reduced by a factor of $\frac{1}{6} \varepsilon^2$ (where ε/is essentially proportional to the pahse change of the wave while passing through the conductor), provided that the amplitude and phase of the second wave relative to the first one are correctly chosen. Furthermore, one can increase the relative phase by a factor of order ε/α without changing the order of magnitude of the energy absorption.

ELIMINATE MIRRORS (USE WAVES INSTEAD)

Figure 2c shows the design principle for a laser template accelerator in which no mirrors are used. Now there is no material object anywhere near the particle trajectory, so that the fields can reach any intensity without destruction. Instead of mirrors, the particle periodically encounters intense electromagnetic waves. These waves will first accelerate the particle in some direciton, then (as always happens when a wave interacts with a particle in vacuum if reradiation is negligible) overtakes it and decelerates it again. The net acceleration is zero. But the net displacement is not, and we choose it to be just large enough to remove the particle from the axis to avoid deceleration by the main accelerating electromagnetic field (i.e. the one which will accelerate the particle along the axis). After an appropriate time interval,a second wave moves the particle back to the axis just as the main wave is ready to accelerate it again. The displacing waves, of course, can be generated by the same template which generates the main accelerating field.

I am unable to report to you here complete details on how this device will work, because a computer calculation is still in progress.

Nevertheless, one can easily see that in principle it will work. Indeed, if the extra waves (drawn as wiggly arrows in Fig. 2c) are very intense compared to the main accelerating field (which accelerates the particles along z), then while an intense wave is acting on the particle, the trajectory will be approximately that which would be generated by that intense wave alone. That trajectory can be calculated and one finds that, typically, a displacement of the particle by some chosen distance, Δr, from the z axis can be accomplished with a given wave intensity during a time, $\delta t \sim \gamma^2$. After the intense wave is turned off, the particle moves in the main accelerating field alone, and its motion can again be easily calculated analytically. Thus, at least in this special case, the trajectory can be calculated to arbitrary accuracy and the device can be shown to work. Another special case of interest is one which can be integrated directly: the intense wave travels opposite to the particle. Again, it is clear that the device will work in principle. By contrast, the more general cases can usually not be solved analytically. One will have to await the results of computer calculations for these more general cases.

RESONANCE (R), RTL ACCELERATOR

One can use a resonance phenomenon to increase the field intensity in the accelerator. The resultant Resonance Laser Template (RLT) Accelerator is schematically shown in Fig. 8a. The inside of the template is covered by a partially reflecting layer, which transmits only a fraction, f, of the incoming radiation. The radius R is approximatley an integer times λ_r, and also λ_o, which allows the field to resonate inside this layer. The phenomenon can again be understood by considering a scalar field interacting with a pair of potential barriers. Each barrier has a transfer matrix as given in Eq. (8). If both barriers are equal, then both have the same transfer matrix. The solution is sketched in Fig. 8b. The field intensity just inside the resonator will be $\sqrt{f}$ times that what it is just outside it (f can be 10^2 or more). This design will help, if the laser puts out long pulses: what the device really does is a form of "photon stacking" inside the resonator.

DOUBLE RESONANCE (R2), R2LT ACCELERATOR

One can further increase the accelerating fields, using a second resonance phenomenon. The design is outlines in Fig. 9. Along a line, slightly off the z axis, a regular sequence of small objects (as e.g. in a grating) is placed. The intense electric field near the axis will oscillate the electrons in these pieces of material.

If the electric feld points, say, downward, then the electron cloud is accelerated upward, some of the electrons leave the object, as a result, the object becomes somewht positively charged. The charge tends to pull the electrons backward, so an oscillation is set up. If the frequency of the oscillating electric field and the frequency of the electron oscillation are in resonance, the electron oscillation amplitude can grow somewhat (but not too much because it is soon destroyed) and the momentary net charge of the small objects will increase. If, further, the accelerating particles are timed to arrive in the vicinity of the object just when the net electrostatic field of the object is able to accelerate the particles, a very powerful accelerator would result. The small objects would, of course be burned up after each shot of laser power. But while burning up, the acceleration would take place. I am unable to go into great detail now, because this m itself is a large subject, and, at any rate, we have not finished our calculations yet.[4] Nevertheless, I feel that this is an interesting enough possibility, to call it to your attention. I do not want to go on record quoting numbrs, but we have some preliminary results, if you are interested, I can show them to you privately.

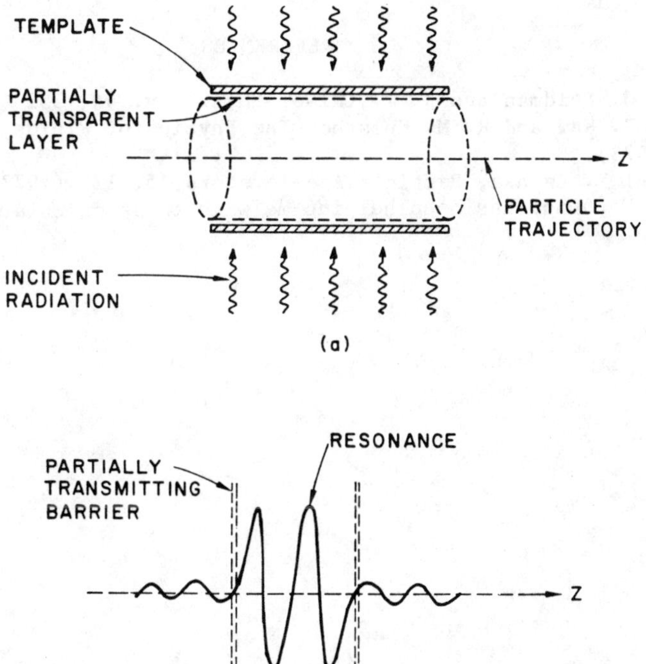

Fig. 8 a) Sketch of a Resonance Laser Template Accelerator.
b) Field distribution inside and outside the accelerating cavity.

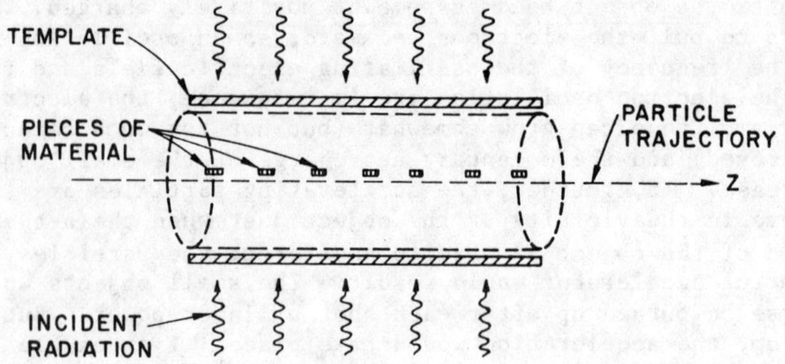

Fig. 9 Sketch of a Double Resonance Laser Template Accelerator.

REFERENCES

1. M. J. Feldman and R. Y. Chiao, Phys. Rev. 4A, 352 (1971).
2. P. K. Kaw and R. M. Kulsrud, The Physics of Fluids, 16, 321 (1973).
3. Paul L. Csonka, Particle Accelerators, 5, 129 (1973).
4. H. Watanabe has been helping me with these calculations.

SENSITIVITY OF A LASER DRIVEN GRATING LINAC TO GRATING ERRORS*

Norman M. Kroll
Stanford Linear Accelerator Center
Stanford University, Stanford, California 94305
and
University of California, San Diego†
San Diego, California 92093

ABSTRACT

The effect of grating errors on transverse beam stability is analyzed. We characterize grating errors by random groove displacements and find that transverse displacements due to such errors approach limiting values of the same order as the grating displacements themselves. It therefore appears that transverse stability requirements will not impose unusually stringent precision requirements on the grating structure.

INTRODUCTION

As described by Palmer,[1] the grating structure for a laser driven grating linac requires shaped groove spacings of the order of one half the laser wavelength and an overall length of several hundred meters. Random errors in the grooves are surely inevitable, and in view of the vast number of grooves, the effect of such errors upon beam stability must be assessed. We provide here an estimate of the relation between the magnitude of these errors and that of the mean deviation from the nominal orbit which these errors induce.

FORMULATION OF THE PROBLEM

We begin with a brief description of the strong focussing design discussed elsewhere in these proceedings.[1,2] The field components in synchronism with the electrons are written

---

*Work supported in part by the Department of Energy, contracts DE-AC03 76SF00515 (SLAC) and DE-AT03-81ER40029 (UCSD).

†Permanent address.

$$\vec{E} = E_o \cos py\, e^{-px} \left(\hat{z} \cos \phi - \hat{x} \frac{k}{p} \sin \phi\right) \tag{1}$$

$$\vec{B} = E_o\, e^{-px} \left[\hat{z} \sin py \cos \phi - \hat{x} \frac{p}{k} \sin py \sin \phi \right. \tag{2}$$

$$\left. + \hat{y}\left(\frac{p}{k} - \frac{k}{p}\right) \cos py \sin \phi \right] + B_o\, Sq(z)\, \hat{y} \quad .$$

Sq(z) is a square wave of amplitude ±1 and slowly increasing period L(z). The electron phase ϕ is given by

$$\phi = \phi_o - \frac{1}{2} \Delta\, Sq(z) \tag{3}$$

where here z is the coordinate of a particular electron. Over a grating section of length L/2 the fields vary as expi(kz - ωt) with $k \approx \omega/c$. The electron motion is ultra relativistic so that the variation of this quantity over a section is negligible. The value of (kz - ωt) for a particular electron is designated by ϕ. The phase shift, ±Δ, which occurs between sections is brought about by a shift of magnitude Δ/k of the grooves from section to section. The instantaneous shifts in the above formulas are, of course, an idealisation of shifts which occur in a distance small compared to L, and the modifications in the fields at these junctures which are required by Maxwell's equations are neglected in the analysis.

The transverse equations of motion of the electron are (taking $z \approx ct$ as the independent variable)

$$\frac{d}{dz}\, m\gamma\, \frac{dx}{dz} = -\frac{eE_o}{c^2}\left[\frac{p}{k} \cos py\, e^{-px} \sin \phi + \frac{B_o}{E_o}\, Sq(z)\right] \tag{4}$$

$$\frac{d}{dz}\, m\gamma\, \frac{dy}{dz} = -\frac{eE_o}{c^2}\frac{p}{k} \sin py\, e^{-px} \sin \phi \quad . \tag{5}$$

As described in Refs. 1 and 2, due to the operation of the strong focusing principle these equations have a stable straight line orbit with $\phi_o = 0$, y = 0, and x_o determined by

$$\frac{p}{k}\, e^{-px_o} \sin \frac{\Delta}{2} = \frac{B_o}{E_o} \quad . \tag{6}$$

There is also a family of more complicated stable orbits with slowly varying period L whose ϕ_o values lie in a relatively narrow band about zero. Here, however, we shall confine our attention to the simple $\phi_o = 0$ orbit and discuss the effect which grating errors have upon it alone. Setting $x_1 = x - x_o$ and assuming both x_1 and y small we have

$$\frac{1}{\gamma}\frac{d}{dz}\gamma\frac{dx_1}{dz} + K^2\, Sq(z)\, x_1 = 0 \tag{7}$$

$$\frac{1}{\gamma}\frac{d}{dz}\gamma\frac{dy}{dz} - K^2\, Sq(z)\, y = 0 \tag{8}$$

with

$$K^2 = \frac{|eE_o|}{\gamma m c^2}\, e^{-px_o} \sin\frac{\Delta}{2}\frac{p^2}{k} \tag{9}$$

$$= \frac{1}{\gamma}\frac{d\gamma}{dz}\tan\frac{\Delta}{2}\frac{p^2}{k}$$

where the second line of (9) follows from Eq. (1) and the obvious relation between E_z and the acceleration rate. Following Ref. 2 we choose the z variation of L so that $KL(z) \equiv \psi$ is constant.

It is apparent that the x_1 and y motion present identical problems so we confine our attention to the x_1 motion in the following. To take account of grating errors we introduce a driving force F(z) to the RHS of (7)

$$\frac{1}{\gamma}\frac{d}{dz}\gamma\frac{dx_1}{dz} + K^2\, Sq\, x_1 = \frac{F(z)}{\gamma m c^2} \equiv f(z) \quad , \tag{10}$$

where F(z) represents corrections to the RHS of (4) which arise from these errors. The general form of the displacements induced by f is given by

$$x_1(z) = \int_0^z dz_1\, g(z,z_1)\, f(z_1) \tag{11}$$

where $g(z,z_1)$ is the solution of Eq. (7) satisfying the boundary conditions

$$g(z,z_1) = 0$$
$$\left.\frac{\partial g}{\partial z}\right|_{z=z_1} = 1 \quad . \tag{12}$$

The quantity of interest is, of course, the expectation value of x_1^2 induced by the fluctuating force. It is given by

$$\langle x_1^2\rangle = \int_0^z dz_1 \int_0^z dz_2\, g(z,z_1)\, g(z,z_2)\, \langle f(z_1)\, f(z_2)\rangle \tag{13}$$

EVALUATION OF $\langle x_1^2 \rangle$

In order to evaluate Eq. (13) we need an expression for the force correlation function $\langle f(z_1)\, f(z_2)\rangle$ and for the Green's function $g(z,z_i)$.

Let $\gamma m\, f_L$ represent the x component of force in Eq. (4) due to the laser, that is

$$f_L = -\frac{eE_o}{\gamma m\, c^2}\frac{p}{k}\, e^{-px_o} \sin\frac{\Delta}{2}$$
$$= -\frac{1}{\gamma}\frac{d\gamma}{dz}\frac{p}{k}\tan\frac{\Delta}{2} \quad . \tag{14}$$

Taking account of the fact that the grating spacing s, the wavelength, and x_o are all of the same order of magnitude, we estimate that the value of $f(z_2)$ due to a displacement δs of a single groove at z_1 may be written

$$f(z_2) \approx \frac{\delta s}{s} f_L(z_1)\, C(z_1 - z_2) \tag{15}$$

where $C(0) = 1$. C may be expected to fall off on a scale of the order of s as z_2 recedes from z_1. Thus

$$f(z_1)\, f(z_2) \approx \left(\frac{\delta s}{s}\right)^2 f_L^2(z_1)\, C(z_1 - z_2) \quad . \tag{16}$$

We next estimate that the principal effect on $\langle f(z_1)\, f(z_2)\rangle$ may be attributed entirely to the displacement of the nearest groove, whence

$$\langle f(z_1)\, f(z_2)\rangle = \frac{\langle \delta s^2\rangle}{s^2} f_L^2(z_1)\, C(z_1 - z_2) \quad . \tag{17}$$

Finally, as will be clear below, the scale over which f_L^2 and $g(z,z_i)$ vary is enormous compared to s. Hence for insertion into (13) we write $C(z_1 - z_2) \approx s\delta(z_1 - z_2)$ yielding finally

$$\langle f(z_1)\, f(z_2)\rangle = \frac{\langle \delta s^2\rangle}{s}\, \delta(z_1 - z_2)\, f_L^2(z_1) \quad . \tag{18}$$

It seems clear that Eq. (18) has the correct dependence (for the purposes of Eq. (13)) on z_1, z_2, the laser field strength and $\langle \delta s^2\rangle$. The effect of the crude approximations which we have made can only be to replace s by a quantity of the same order of magnitude. Therefore, we shall henceforth think of s as a quantity of the order of the grating spacing rather than as the grating spacing itself.

The determination of g by solving (7) subject to the boundary conditions (12) is simple in principle, especially if one takes advantage of the fact that γ, and hence K^2 and L vary very slowly with z. In the interest of obtaining our final result in a simple form, however, we prefer to proceed in a more approximate manner. In particular we wish to replace (7) by

$$\frac{1}{\gamma}\frac{d}{dz}\gamma\frac{dg}{dz} + Q^2(z)\; g = 0 \tag{19}$$

where

$$\cos QL = \cos\frac{\psi}{2}\cosh\frac{\psi}{2} \quad . \tag{20}$$

This has the effect of replacing the strongly focused betatron oscillations of wave number Q described by (7) by simple harmonic oscillations with the same wave number. As shown in the appendix, this procedure is well justified when $\psi/2$ is small. ($\psi/2 < \pi/4$ is small enough.) Continuing then with (19) and assuming γ slowly varying we obtain by inspection

$$g(z,z_1) = \left[\frac{\gamma(z_1)}{\gamma(z)\; Q(z)\; Q(z_1)}\right]^{1/2} \sin\int_{z_1}^{z} Q(z')dz' \quad . \tag{21}$$

Substitution of (21), (18), and (14) into (13) yields

$$\begin{aligned}\langle x_1^2\rangle &= \frac{K^2}{Q^2}\,\frac{\langle\delta s^2\rangle}{s}\,\frac{\tan^2\frac{\Delta}{2}}{\gamma(z)\; K(z)}\,\frac{p^2}{k^2} \\ &\cdot\left(\frac{d\gamma}{dz}\right)^2\int_0^z \frac{dz_1}{\gamma(z_1)\; K(z_1)}\sin^2\int_{z_1}^{z} Q(z')dz' \quad .\end{aligned} \tag{22}$$

In writing (22) we have made use of the fact that K/Q is z independent and have, in addition, assumed $d\gamma/dz$ to be constant. To complete the evaluation we replace the $\sin^2$ factor by its average value and make use of (8) to obtain

$$\begin{aligned}\langle x_1^2\rangle &= \frac{\langle\delta s^2\rangle}{2sk}\,\frac{K^2}{Q^2}\,\frac{1}{\sqrt{\gamma(z)}}\int_0^z \frac{1}{\sqrt{\gamma(z_1)}}\,\frac{d\gamma}{dz_1}\,dz_1 \\ &= \frac{\langle\delta s^2\rangle}{sk}\,\frac{K^2}{Q^2}\left[1-\left(\frac{\gamma(0)}{\gamma(z)}\right)^{1/2}\right]\end{aligned} \tag{23}$$

Equation (23) is our main result. It shows that while the beam fluctuations grow initially, the effect quickly saturates on account of the "adiabatic damping" effect in (21). For $\psi < \pi/2$ Eq. (20) implies $Q^2/K^2 \approx \psi^2/48$.

As shown by the discussion of Ref. 2, the allowable ϕ_o range becomes small as ψ becomes small and $\psi < \pi/2$ is probably a smaller value than one would want to use. The specific design proposed in Ref. 2 has $\psi = \pi$, a value which is outside the established region of validity of the replacement of Eq. (7) by Eq. (18). The discussion in the appendix does, however, encourage one to believe that the qualitative behavior implied by (23) continues to hold, and for a preliminary estimate even to risk a quantitative application. Then we have $K/Q = 2$ and taking $s \approx 1/2\ \lambda$ we obtain

$$\langle \delta s^2 \rangle \approx \frac{\pi}{4} \langle x_1^2 \rangle \tag{24}$$

so that the groove displacement induces a particle displacement of the same order of magnitude.

Before concluding we recall that the above analysis has been confined to $\phi_o = 0$. Depending upon the sign, either the horizontal or vertical betatron wave number is reduced as ϕ_o is varied from zero, and approaches zero as the limit of stability is reached.[2] Since the betatron wave number Q^2 appears in the denominator of Eq. (23) it seems very likely that the $\langle x_1^2 \rangle$ induced by a specified $\langle \delta s^2 \rangle$ will increase and diverge as the stability limit of the ϕ_o range is approached. A further investigation to determine the effect which limitation on the attainable precision of grating ruling has upon the ϕ_o range would be desirable.

ACKNOWLEDGEMENTS

I am grateful to the other workshop participants for stimulation and helpful discussion, especially Paul Channell and John Lawson, who raised the issue discussed here, and Robert Palmer, who suggested that adiabatic damping might be important.

APPENDIX

Following Bruck[3] we note that Eq. (10) can be formally transformed to (we neglect the z dependence of γ here, and hence of K^2 and L also)

$$\frac{1}{\gamma}\frac{d}{d\bar{z}}\gamma\frac{d}{d\bar{z}}\bar{x} + Q^2\,\bar{x} = \bar{f}\big(z(\bar{z})\big) \tag{A1}$$

where

$$\bar{x} = \frac{1}{\sqrt{Q\beta}}\,x_1 \tag{A2}$$

$$\bar{z} = \int^{z} \frac{dz'}{Q\beta(z')} \tag{A3}$$

$$\bar{f} = (Q\beta)^{3/2}\, f \tag{A4}$$

and the Twiss matrix function $\beta(z)$ is the periodic solution of

$$\frac{1}{2}\,\beta\,\frac{d^2\beta}{dz^2} - \frac{1}{4}\left(\frac{d\beta}{dz}\right)^2 + K^2 Sq(z)\,\beta^2 = 1 \quad . \tag{A5}$$

We find that

$$\begin{aligned} Q\beta(z) &= \frac{Q}{K \sin QL}\left[\sin\frac{\psi}{2}\cosh\frac{\psi}{2} + \cos\left(\frac{\psi}{2} - 2Kz\right)\sinh\frac{\psi}{2}\right] \\ &\qquad\qquad 0 < z < \frac{L}{2} \\ &= \frac{Q}{K \sin QL}\left[\sinh\frac{\psi}{2}\cos\frac{\psi}{2} + \cosh\left(\frac{3}{2}\psi - 2Kz\right)\sin\frac{\psi}{2}\right] \\ &\qquad\qquad \frac{L}{2} < z < L \end{aligned} \tag{A6}$$

with $\beta(z+L) = \beta(z)$. One can readily verify that (A6) satisfies (A5) and continuity conditions on β and $d\beta/dz$ at $z = 1/2\,L$ and $z = 0, L$ (periodicity condition). In the small ψ limit (A6) becomes

$$\begin{aligned} Q\beta(z) &\approx 1 + \psi^2\left(\frac{1}{2}-\frac{z}{L}\right)\frac{z}{L} \quad ; \qquad 0 < z < \frac{L}{2} \\ &\approx 1 - \psi^2\left(\frac{3}{2}-\frac{z}{L}\right)\frac{z}{L} \quad ; \qquad \frac{L}{2} < z < L \end{aligned} \tag{A7}$$

so that $Q\beta$ oscillates about 1 with an amplitude $\psi^2/16$. The treatment given in the main text amounts to the neglect of the difference between the barred and unbarred quantities and in view of (A2), (A3), (A4), and (A7) appears to be well justified for $\psi < \pi/2$. For the $\psi = \pi$ case, inspection of (A6) indicates similar behavior for $Q\beta$ but with a maximum value of 2.41, minimum value of 0.5 and average value of 1.36. Hence the application of (23) to this case is not likely to be grossly in error.

REFERENCES

1. R. B. Palmer;

   Contribution to this conference.
2. Kwang Je Kim and Norman M. Kroll; Some Effects of the Transverse Stability Requirement on the Design of a Grating Linac; Contribution to this conference.
3. Henri Bruck; Circular Particle Accelerators, pp. 112-113; LA-TR-72-10 Rev., Los Alamos National Laboratory.

ARE GRATINGS INVISIBLE?

by

Paul J. Channell, AT-6, MS-H818
Los Alamos National Laboratory
Los Alamos, New Mexico 87545

ABSTRACT

I show that laser grating accelerators may encounter serious difficulties in operating near or even below the damage threshold caused by the formation of a dense plasma above the grating, which obscures the grating shape.

INTRODUCTION

The possibility of accelerating particles in the near field of laser light incident on a grating has been widely discussed.[1] The potentially high gradients and small volume of field make this scheme attractive both from the viewpoint of total accelerator length and from the viewpoint of stored energy. Two modes of operation have been envisaged: in one mode the incident field is kept below the damage threshold, and the grating is used for many laser pulses; in the other mode the grating is destroyed and then replaced after each laser pulse. I want to point out in this paper that, depending on laser pulse length, grating material, and grating purity, there may be significant limitations on operating near, or even below the damage threshold.

DISCUSSION

First let us observe that if a plasma of critical density, N_c, exists above the grating then the grating is effectively invisible. The critical density is that at which the plasma frequency equals the laser frequency and is given by

$$N_c = \frac{M_e \omega^2}{4\pi e^2} , \qquad (1)$$

where M_e is the electron mass, ω is 2π times the laser frequency, and e is the electron charge. I will show that under certain conditions such a plasma does exist above gratings.

Certainly near, and possibly below, the damage threshold significant amounts of material will be boiled or sublimed off the grating surface. To form a plasma of critical density for a grating appropriate to a CO_2 laser (10.6 μm) requires only ∿10 Å of material to be evaporated. I will assume that this material exists as a plasma just above the grating, that is, that the ionization time is very short.

Once the material is ionized it can no longer stream freely into the whole volume; the individual particles oscillate in the electric field with very small radial excursions. A plasma confined very near the grating surface would have little effect on accelerator operation.

If the plasma density were high enough and the temperature low enough, the electron collision time would be at the subpicosecond level; one might imagine that collisions would spread out the plasma. However, the development of strong space charge forces would freeze the electrons onto the ions and reduce the diffusion rate to that of the ions; this would be a significant process only on the microsecond time scale.

I want to point out, however, that a purely collisionless process exists that will rapidly spread the plasma uniformly over the grating. A crude picture of the grating geometry and field lines is shown in Fig. 1. There are two important points to be noted from this figure. First, the plasma need only spread a distance $\lambda/8$, where λ is the laser wavelength, to be uniform. Second, the electric field strength is not spatially uniform; indeed, for purposes of acceleration, it could not be.

When a particle is in a rapidly oscillating nonuniform field it "sees" an effective potential given by[2]

$$U_{eff} = \frac{e^2 E_0^2}{4m\omega^2} , \qquad (2)$$

where E_0 is the field strength and m is the particle mass. Let us note that the effective force on the ions is much smaller than the force on the electrons; we will approximate the effective force on the ions by zero.

The electrons will be pulled toward the low-field region, but will be retarded once again by the space-charge force. To see what

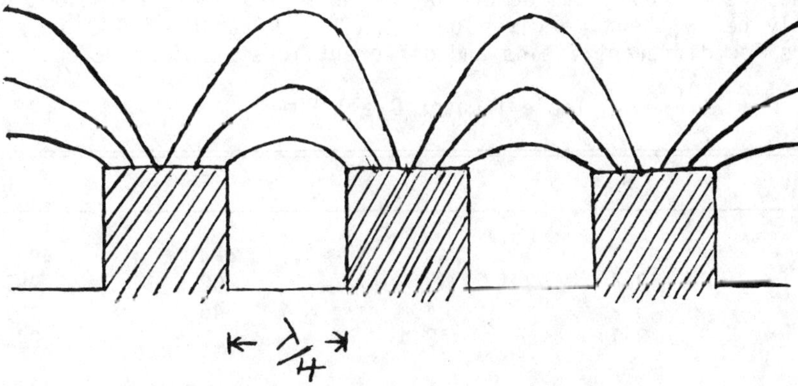

Fig. 1. Schematic of a grating accelerator with field lines shown.

will happen, let us write down the force equations for ions and electrons. These are

$$\frac{\partial \vec{V}_i}{\partial t} + \vec{V}_i \cdot \vec{\nabla}\vec{V}_i = \frac{e\vec{E}}{M_i} \quad , \tag{3}$$

$$\frac{\partial \vec{V}_e}{\partial t} + \vec{V}_e \cdot \vec{\nabla}\vec{V}_e = - \frac{e\vec{E}}{M_e} - \frac{\vec{\nabla}U_{eff}}{M_e} , \tag{4}$$

where $\vec{E}$ is the space-charge field.

Though the initial dynamics is very complicated, the plasma will quickly approach a steady state flow. To avoid steady space-charge build-up in this steady state, the right-hand sides of Eqs. (3) and (4) must be equal. We thus find that

$$e\vec{E} = - \frac{M_i}{M_i + M_e} \vec{\nabla}U_{eff} \simeq -\vec{\nabla}U_{eff} \quad . \tag{5}$$

Thus, space charge communicates the electron force to the ions virtually unchanged.

The approximate time for uniform spreading is

$$T \simeq \sqrt{\frac{M_i \lambda^2}{24\ U_{eff}}} \quad . \tag{6}$$

Using Eq. (2), Eq. (6) can be rewritten as

$$T \simeq \frac{2\pi}{|eE_0|c} \sqrt{\frac{M_i C^2 M_e C^2}{6}} \quad . \tag{7}$$

The laser dwell time before particle arrival must be held significantly below the time given by Eq. (7). A table of T/3 in picoseconds for different fields and different ions is given below.

Table I Laser Dwell Times

E_0 (GeV/m)	H	O	Fe	Cu
1	62	248	464	492
5	12.4	49.6	92.8	98.4
10	6.2	24.8	46.4	49.2
20	3.1	12.4	23.2	24.6

Note that the accelerating gradient will be about $eE_0/2$.

CONCLUSIONS

To avoid significant field modifications the laser dwell time should be held to less than T/3. Because very high-power lasers with short, accurately timed pulses will require significant technological advances, the numbers given in the above table represent fairly stringent restrictions on the operation of a grating accelerator. In particular, we can draw the following conclusions.

- Gratings should be made of as heavy a material as possible.
- Surface contamination by light elements must be eliminated.
- Operation above the damage threshold is probably not possible with foreseeable technology.

REFERENCES

1. R. B. Palmer, "Laser Driven Electron Accelerators," Proc. 1981 Particle Accelerator Conf., Washington, DC, March 11-13, 1981, IEEE Trans. Nucl. Sci. 28, p. 3370 (1981) and references contained therein.

2. L. D. Landau and E. M. Lifshitz, Mechanics, (Addison-Wesley Publishing Company, Inc., Reading, Massachusetts, 1960), p. 94.

IMPROVED DAMAGE THRESHOLD FOR OPTICAL ELEMENTS PLACED AT MINIMA (i.e., SHADOW) OF A RADIATION FIELD

Paul L. Csonka
Department of Physics and Institute of Theoretical Science
University of Oregon, Eugene, Oregon 97403
and
Argonne National Laboratory, Argonne, Illinois 60439*

ABSTRACT

It is shown that optical elements placed at interference minima of a radiation field will damage less. For example, when a 10^3 Å (or $3.15 \cdot 10^3$ Å) thick metallic foil is placed at the node of a standing wave pattern produced by 30 μm wavelength radiation polarized parallel to the foil surface, the damage due to Joule heating is calculated to decrease by a factor $6.8 \cdot 10^3$ ($6.8 \cdot 10^4$). When the angle of incidence is $\Theta = 5.7^0$ and the wave is polarized parallel to the plane of incidence, the factor is $5.8 \cdot 10^2$ ($1.8 \cdot 10^3$). Thus, higher fields are possible near mirrors, without damage. Applying this phenomenon to laser template accelerator design allows ≈ 1 GeV/m acceleration at any (including very high) particle energy. Experiments could determine how much of the calculated increase can be realized and where and if other damage mechanisms may take over.

INTRODUCTION

The development of intense electromagnetic beam optics is limited by the intensity which the various optical elements can sustain without damage.

I pointed out some time ago[1] that it is possible to improve the damage threshold of an optical element placed in the path of an intense electromagntic wave, by locating the element at intensity minima of a standing wave pattern.

Some authors reacted with scepticism to this suggestion and concluded apparently that it would be in violation of fundamental physical laws.[2] It seems, therefore, appropriate to discuss the phenomenon in more detail; that is the purpose of this note.

In particular, I claim the following:

1. Assume first that one intense electromagnetic wave incident on a thin film from one side will damage that film. Next, in addition to that wave, irradiate the film, simultaneously, from the opposite side, with a wave whose intensity is the same as that of the first one. If the phase of this second wave is appropriately chosen, then the two waves together

*Work supported by the U. S. Department of Energy.
0094-243X/82/910248-18$3.00

will not only not damage the film twice as much as one wave did, but in fact they can prevent radiation damage from occurring altogether.

2. For this phenomenon to be observable, the film must be at a location where the radiation field has an interference minimum in the presence of the film. For example, it would be wrong in general to place a thin plate in a resonant cavity at a location where an intense radiation field has a minimum without the plate, and then expect the plate not to burn up. Introduction of the plate will alter the field distribution of course, and the new distribution need not have a minimum where the old one had one.
3. The phenomenon will be most pronounced when the film thickness is small compared to $1/|k_z|$, where k_z is a component of the complex wave number, k, of the radiation inside the film, measured along the z direction; i.e., along the surface normal of a planar film.
4. Uncontrollable displacements of the interference minimum inside a thin film, away from or toward either surface, must be small compared to $1/|k_z|$. For this reason, the following simple experiment is not expected to given a spectacular demonstration of this effect: Place a thin dielectric film of thickness d on top of the plane surface of another dielectric with high refractive index. Illuminate the assembly with optical light at normal incidence. Reflection from the interface will cause a standing wave pattern to emerge. One might expect the damage threshold to depend on d. However, inhomogeneities in both dielectrics will cause random disturbances of the interference pattern. These inhomogeneities can have dimensions comparable to an optical wavelength, largely washing out the expected effect. To see the effect more clearly, the dielectric film should be far enough from the reflecting surface so that the effect of local fluctuations in the mirror will be averaged out at the location of the film. For example, thin films located of node planes and far from the walls, inside a resonant cavity, should show the effect.
5. The phenomenon can be considerd as "channelling" of electromagnetic radiation between thin plates.

In the following, I intend to discuss simple examples to illustrate the phenomenon. One example will show an application to laser accelerator design. Before coming to that, however, let me state what this paper is not intended to do.

First of all, I will not give an optimal design for a laser template accelerator for any set of parameters. Rather, the examples are selected to clearly illustrate the cancellation phenomenon. Second, in the examples to be presented, simple electromagnetic field configurations will be used; namely, those containing relatively few Fourier components as a function of position measured along the accelerator axis, in spite of the fact that more Fourier components could make an accelerator more effective.[(1),(3)]. Fourth, I will concentrate attention on the standing wave pattern as a function of only one dimension; namely, the dimension along the z axis. Generalization to other dimensions is straightforward.

Radiation source, positioning, materials, etc.

Questions related to radiation generation, positioning materials, machining, etc., are beyond the scope of this paper. Instead of studying these, it will be simply assumed that we have at our disposal

1. A source of coherent electromagnetic radiation capable of generating an appropriately polarized (to be specified later) wave train of frequency ω and time duration T.
2. Homogeneous isotropic materials which can be manufactured in any shape and size and whose complex index of refraction at the chosen frequency ω, is $n(\omega) = n_r + i\, n_i$ with n_i and n_r both real, and both to be specified later.

In what follows, we will be dealing with thin solid foils of typical thickness 10^3 Å, located with comparable accuracy in a radiation field. Such accuracies can be achieved by existing methods. Nevertheless, it would be desirable to produce a design where the solid foils would be positioned appropriately by the radiation field itself. That, of course, would be a reversal of the traditional state of affairs, where solid optical elements determine the radiation field. The economy inherent in such a design is a consequence of the high accuracy which is routinely achieved in the course of setting up optical interference patterns. Indeed, optical patterns are the most sensitive position monitors known. It appears that such a design is possible. A detailed discussion would be out of place here, but will be given elsewhere. The fact, however, that such a design is available, is an added incentive to study phenomena related to thin solid foils.

Although a discussion of material properties is beyond the scope of this paper, a brief survey is given in the Appendix.

Energy absorption in an optical barrier (or well).

Consider a layer of material which is bounded by two infinite planes and has thickness d. Outside the layer, there is vacuum on both sides.

A plane electromagnetic wave is incident on this layer from the left (See Fig. 1.) Its wave vector, vector $\bar{k}_1$, sublends an angle Θ_1 with the surface normal. The electric field amplitude is E_1. A "reflected" electromagntic wave travels away from the layer to the left. Its amplitude is E_{1r}, and the angle between its wave vector, $\bar{k}_{1r}$, and the surface normal is also Θ_1. Another electromagnetic wave is incident on the layer from

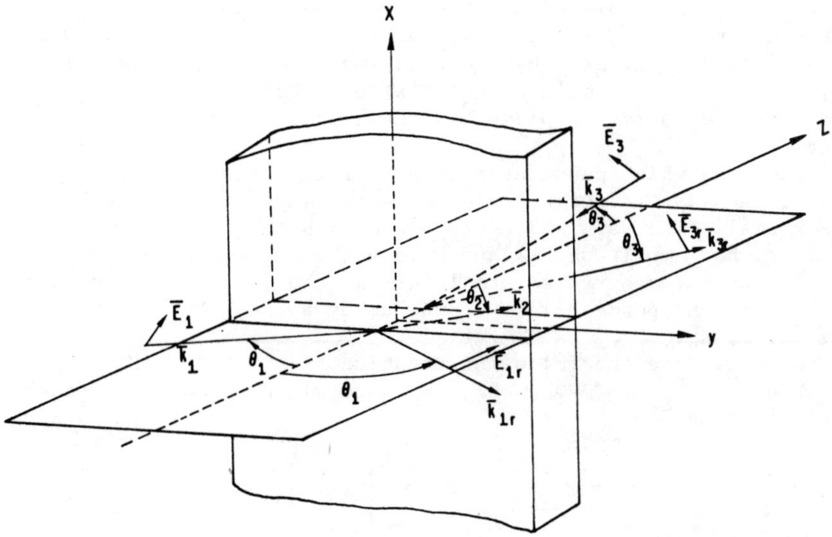

Figure 1

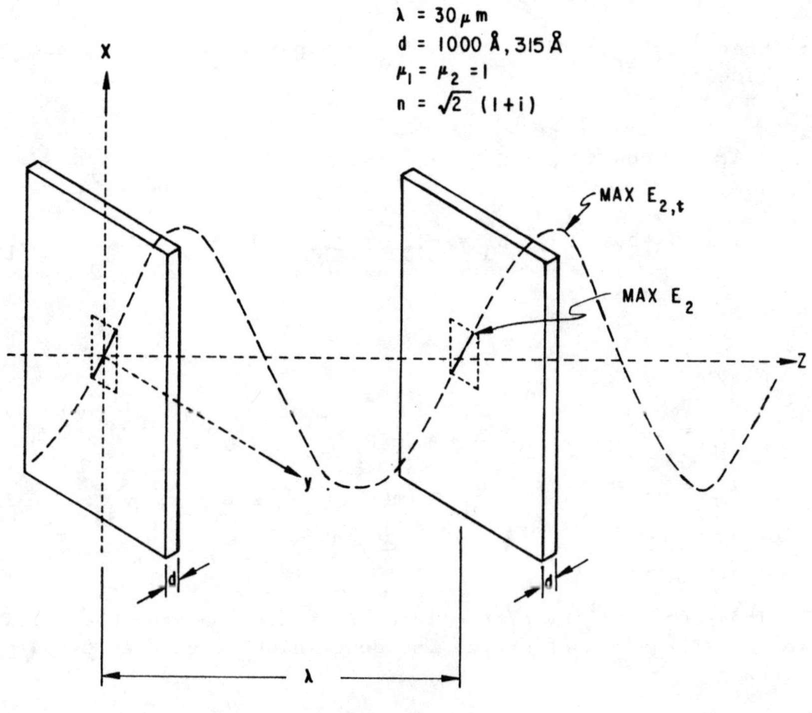

Figure 2

the right, the angle of incidence is Θ_3. The wave vector is $\bar{k}_3$, the electric amplitude E_3. A wave travels away from the layer to the right, its wave vector $\bar{k}_{3r}$ also subtends an angle Θ_3 with the surface normal, and the electric amplitude of this wave is E_{ir}. The surface normals as well as the vectors $\bar{k}_1$, $\bar{k}_{ir}$, $\bar{k}_3$, $\bar{k}_{3r}$ all lie in a single plane, which we call the plane of incidence.

The wave which penetrates the layer from the left has amplitude E_2 and wave vector $\bar{k}_2$ which makes an angle Θ_2 with the surface normal. The wave which propagates inside the layer toward the left, has amplitude E_{2r} and wave vector $\bar{k}_{2r}$ which makes an angle Θ_2 with the surface normal (k_2, k_{2r}, Θ_2, Θ_{2r} may be complex).

Denote the permeability of vacuum by μ_1 and that of the layer material by μ_2.

We introduce the following notation.

When $\bar{E}_1$, $\bar{E}_{ir}$, and $\bar{E}_3$, $\bar{E}_{3r}$ are all perpendicular to the plane of incidence:

$$A_1 = E_1 \ , \quad B_1 = E_{1\imath} \ , \quad P_1 = \frac{k_1}{\mu_1} \cos\theta_1$$

$$A_3 = E_{3r} \ , \quad B_3 = E_3 \ , \quad P_2 = \frac{k_2}{\mu_2} \cos\theta_2 \qquad (1a)$$

$$f = \mu_2 d \ , \quad B_3 = E_3 \ , \quad P_2 \frac{k_2}{\mu_2} \cos\theta_2$$

When $\bar{E}_1$, $\bar{E}_{ir}$, and $\bar{E}_3$, $\bar{E}_{3r}$ are parallel to the plane of incidence:

$$A_1 = E_1 \cos\theta_1 \ , \quad B_1 = -E_{1r}\cos\theta_1 \ , \quad P_1 = \frac{k_1}{\mu_1 \cos\theta_1}$$

$$A_3 = E_{3r}\cos\theta_3 \ , \quad B_3 = -E_3 \cos\theta_3 \ , \quad P_2 = \frac{k_2}{\mu_2 \cos\theta_2} \qquad (1b)$$

$$f = \mu_2 d \cos^2\theta_2$$

Also:

$$\alpha_r = \mathrm{Re}(P_1/P_2) \ , \quad \alpha_i = \mathrm{Im}(P_1/P_2) \ , \quad \alpha = \alpha_r + i\alpha_i$$

$$\varepsilon_r = \mathrm{Re}(fP_2) \ , \quad \varepsilon_i = \mathrm{Im}(fP_2) \ , \quad \varepsilon = \varepsilon_r + i\varepsilon_i \qquad (2)$$

$$C = \frac{1}{2}\left(e^{i\varepsilon} + e^{-i\varepsilon}\right) \ , \quad D = \frac{1}{2}\left(e^{i\varepsilon} - e^{-i\varepsilon}\right)$$

The penetrability of the layer to radiation from the left (i.e., the fraction of energy incident which penetrates the layer) is then

$$P = \frac{C^2 + D^2 - (\alpha + 1/\alpha)DC}{[C - (1/2)(\alpha + 1/\alpha)D]^2} \tag{3}$$

Consider the following special case:

$$A_1 = - B_3 \tag{4a}$$

$$\Theta_1 = \Theta_3 \tag{4b}$$

Arrange the phases of the two incident waves so that the electric field component parallel to either boundary surface has a node plane inside the layer, halfway between the two boundaries.

The energy flux absorbed by the layer per unit surface (counting both surfaces) per unit time will be (assume units in which $\mu_1 = 1)^{(3)}$

$$F = \frac{|E_1|^2 \cos\theta_1}{4\pi} \frac{8c}{|1+\alpha|^2} \left[- \alpha_r \operatorname{sh}\varepsilon_i - \alpha_i \sin\varepsilon_r \right], \tag{5}$$

where c is the velocity of light in vacuum. Recall that P_1 is real.

In the limit when $\varepsilon_r, \varepsilon_i \to 0$

$$F \to \left\{ \frac{|E_i|^2 \cos\theta_1}{4\pi} \frac{8c}{|1+\alpha|^2} \alpha_i 2\varepsilon_r \right\} \frac{1}{2} \left\{ \left(-1 - \frac{\varepsilon_i}{\varepsilon_r}\frac{\alpha_r}{\alpha_i} \right) + \frac{\varepsilon_r^2}{6} \left[1 - \left(\frac{\varepsilon_i}{\varepsilon_r}\right)^3 \frac{\alpha_r}{\alpha_i} \right] \right\} \tag{6a}$$

When the electric fields are perpendicular to the plane of incidence, then $\varepsilon_i/\varepsilon_r = - \alpha_i/\alpha_r$. In that case, Eq. (6a) reduces to

$$F \to \left\{ \frac{|E_1|^2 \cos\theta_1}{4\pi} \frac{8c}{|1+\alpha|^2} \alpha_i 2\varepsilon_r \right\} \frac{1}{2} \left[1 + \left(\frac{\alpha_i}{\alpha_r} \right)^2 \right] \frac{\varepsilon_r^2}{6} \tag{6b}$$

The simpler form (6b) is a good approximation, even when the electric fields are parallel to the plane of incidence, provided that

$$\mathrm{Im}(\cos\theta_2)/\mathrm{Re}(\cos\theta_2) \ll 1 . \quad (7)$$

Furthermore, even if condition (7) is violated, Eq. (6b) may nevertheless be a good approximation, provided that

$$d \ll \ell , \quad (8)$$

where ℓ is the effective oscillation amplitude of conduction electrons in the layer, moving parallel to the surface normal. (The physical reason is, that if condition (8) is satisfied, the conduction electrons are essentially immobilized as far as motion parallel to the surface normal is concerned. They are unable to absorb energy from field components along the surface normal.(4)

Consider now an electric field component, say the one parallel to the surface, $E_{\parallel}$. Near a node plane, this field intensity will be only a fraction of its value farther away. Indeed, if $\varepsilon \ll 1$, the field intensity will reach its maximum value near the layer boundaries. There its value is max $E_{2\parallel}$, which will be about a factor ε times less than max $E_{2\parallel}$ (thick) where max $E_{2\parallel}$ (thick) is what the maximum value would be in a thick layer

$$\max E_{2\parallel} \approx \varepsilon \max E_{2\parallel} \text{ (thick)} . \quad (9)$$

The phase and amplitude of an electromagnetic wave shifts as the wave travels across the layer. Write the waves which travel--say--toward the right as $\phi_1(\bar{r},t) = A_1 \exp[\bar{k}_1 \bar{r} - \omega t]$ and $\phi'(\bar{r},t) = A_3 \exp[+\bar{k}_{3r}\bar{r} - \omega^1 t]$, the former one referring to the wave on the left side of the layer, while the latter expression refers to the wave on the right side of the layer. Choose the z axis to be the surface normal pointing to the right and assume that the origin of the coordinate frame is located halfway between the two surfaces of the layer (see Fig. 1). Set y = 0. Then the shift just alluded to, measured along z, will be:

$$\Delta\phi = \phi_1'(z = d/2,t) - (z = -d/2,t)e^{i(k_1\cos\theta_1)d} \quad (10)$$

One can show that (3)

$$\frac{\Delta\phi}{|\phi_1|} \xrightarrow[\varepsilon \to 0]{} -i2\alpha\varepsilon \quad (11)$$

Thus, if one reduces the thickness d of a thin film, located near an intensity minimum, then ε is reduced, and the phase shift across the film is also reduced proportionally to ε and d. But the energy absorbed by a unit volume of the film according to

Eq. (6b), is reduced $\sim d^2$, and the total energy absorbed by the film will diminish $\sim d^3$. Clearly, if d is smaller, then larger phase shift can be obtained at a cost of less energy absorbed in the material.

When Eq. (4a) does not hold, F may still be of the order given in Eq. (6), provided that the deviation of B_3 from $-A_1$ is small enough.(3) As a result, one can arrange in certain cases $\Delta\phi$ larger than would be given by Eq. (11) for the case when Eq. (4a) holds. That fact can be utilized to construct mirror assemblies which absorb relatively little radiation and, therefore, can tolerate relatively large incident fluxes.(6) Having said that much, we will not investigate here the case when Eq. (4a) is violated. Instead, we shall continue to assume that Eq. (4a) holds. In the following we intend to discuss illustrations and an application of that particular case.

EXAMPLES

(1) Consider two plane waves of equal intensity, both polarized along y, counterpropagating along the z axis (see Fig. 2.) Both waves have wavelength λ. The two waves set up a standing wave pattern, whose node points lie in planes which are parallel to both x and y. The distance between neighboring node planes is $\lambda/2$. A thin plane film of thickness d is placed at some of the node planes. (Two are indicated in Fig. 2.) The electric field can be considered to be polarized perpendicular to the surface normal. The amplitude of the electric field inside each film has the same maximum value; namely, max E_2, which is much smaller than the maximum amplitude of the field. The ratio max E_2/max E_2 (thick) is $\sim \varepsilon/6$, as given by Eq. (9). The radiation energy absorbed by the plate is, according to Eq. (6b), proportional to ε_r^3, and, therefore, the energy absorbed per unit volume of the film is proportional to ε_r^2.

The incident radiation energy will not damage a film, whenever the incident energy is less than the incident damage threshold energy Q_{th}. For metals, for a wide range of radiation parameters, Q_{th} is proportional to the absorbed energy per unit volume. (See Appendix.) From Eq. (6), therefore, in the special case under discussion, Q_{th} will be proportional to

$$Q_{th} \sim \left[\frac{1}{2} \left[1 + \left(\frac{\varepsilon_i}{\varepsilon_r} \right)^2 \right] \frac{1}{6} \quad \varepsilon_r^2 \right]^{-1} . \qquad (12)$$

There are several materials (including C) which can be produced(5) in films of thickness < 1000 Å. To be specific, choose the radiation wavelength $\lambda = 30$ μm, the plate thickness $d = 10^3$ Å, $\mu_1 = \mu_2 = 1$, and $n = \sqrt{2}\,(1 + i)$ (approximately as in pyrolytic graphite) as indicated in Fig. 2.

Let us denote by Q_{th} the damage threshold energy for these particular parameters. Let Q_{th} (thick) be the same quantity in the case when α, λ, μ_1, μ_2 are as above, but when the film is "thick," i.e., for a film for which $\varepsilon_r \gtrsim 1$. For the parameters assumed $\varepsilon_r = \frac{2\sqrt{2}}{3}\pi \cdot 10^{-2}$ and from Eq. (6).

$$Q_{th} \approx 6.8 \cdot 10^3 \; Q_{th} \text{ (thick)} . \tag{13}$$

If d = 315 Å (quite easy for carbon films[5]) and all other parameters are as before, then

$$Q_{th} \approx 6.8 \cdot 10^4 \; Q_{th} \text{ (thick)} \tag{13a}$$

Equations (13) demonstrate that a metal film of thickness 1000 Å (or 315 Å) placed at nodes of such a standing wave pattern can tolerate a factor $6.8 \cdot 10^3$ (or $6.8 \cdot 10^4$) more incident energy than a thick film. The increase factor is substantial enough to justify further study of this phenomenon.

(2) Two waves of equal amplitudes, both with wavelength λ, both polarized along y, set up a standing wave pattern, as shown in Fig. 3. The waves propagate along $\bar{k}_1$ and $\bar{k}_3$, respectively. As a result of reflection from plates (see below) two other waves emerge, they travel along $\bar{k}_{1r}$ and $\bar{k}_{3r}$, respectively. All four k-vectors subtend on angle θ_1 with the z axis. Nodes of the electric field lie in planes which are parallel to the (x, y) plane. Neighboring node planes are $\lambda/2 \cos\theta_1$ distance from each other.

Thin plane films of thickness d are located at certain node planes.

Let us assume λ, d, μ_1, μ_2 and n as in Example 1. Assume further $\theta_1 = 60.166°$, so that $\sin\theta_1 = 0.8675$, $\cos\theta_2 = 0.4337 \cdot (1-i)/\sqrt{2}$, and $\varepsilon_r = \frac{2\sqrt{2}}{3}\pi \; 0.9059 \cdot 10^{-2}$. Then from Eq. (6b), we have

$$Q_{th} \approx Q_{th} \text{ (thick)} \cdot \begin{cases} 8.3 \cdot 10^3 & \text{, if } d = 1000 \text{ Å} \\ 8.3 \cdot 10^4 & \text{, if } d = 315 \text{ Å} \end{cases} \tag{14}$$

(3) Start with the same two waves which were described in Example 2. Again, thin plates of thickness d are located at certain chosen node planes.

This time, however, the films are not uniform (Fig. 4). The thickness d(x) of the plates may be allowed to be a function of x. Alternatively, the thickness may be uniform, but then the index of refraction of the plates is allowed to vary as a function of x. It is not necessary for us to specify here exactly which of these two possibilities (or a combination of them) we intend to consider. The important point is that the transmittance and the reflectance of the plates is now a function of x.

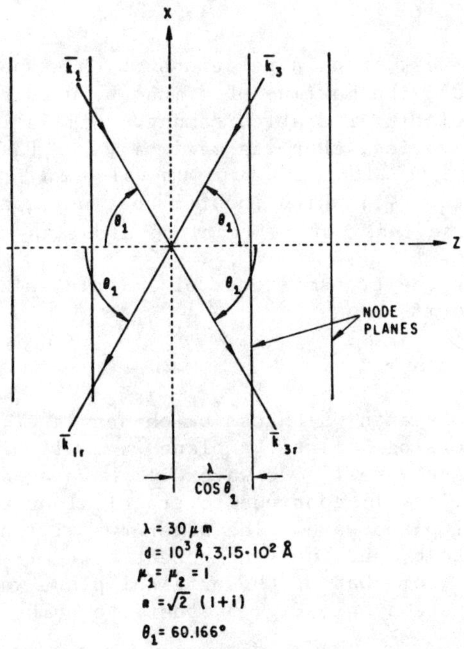

Figure 3

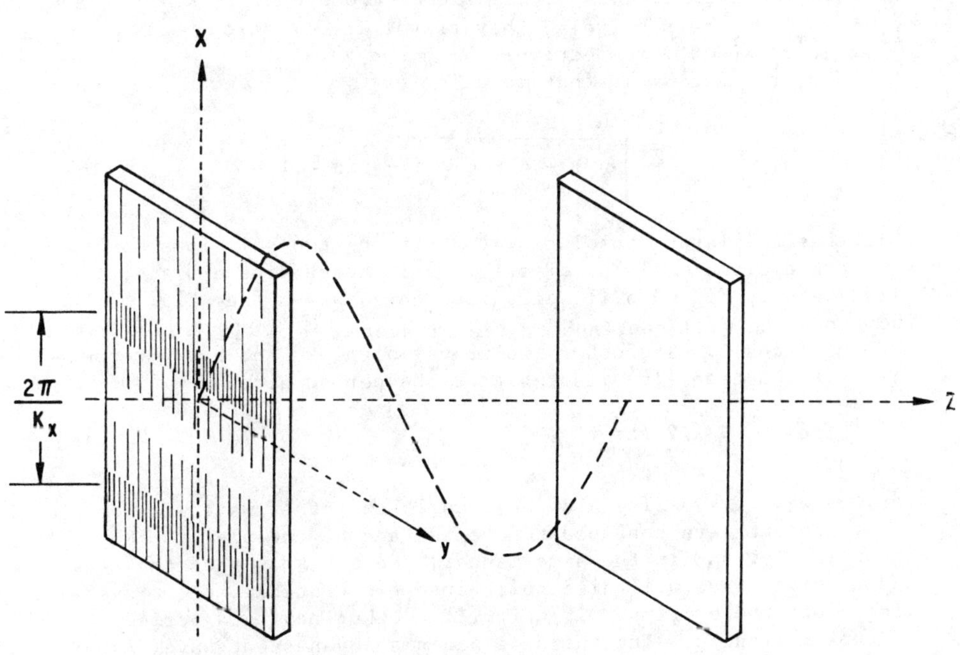

Figure 4

If the thickness varies, then in relations of the type of Eqs. (5), (6), or (8), the maximum of d must be used to be on the safe side when calculating radiation damage. Similarly, if the index of refraction varies, then its maximum value has to be used in Eqs. (5) and (6), but its minimum value in Eq. (7). Continue to assume $k_2 d \ll 1$, which insures that one can neglect variations of d or the index of refraction, along any ray while it traverses a plate.

We specify that the transmittance of all plates has the following functional form:

$$T = a + b \cos K_x x \quad , \tag{15}$$

where a and b are constants which can be chosen later. Usually $b \ll a$. While traversing a plate, a plane wave will in general split into three plane waves. One wave will have an amplitude proportional to a. It will continue to travel along the direction of the original plane wave. The other two will have amplitudes proportional to b, and the x component of their wave vector will differ by $\pm K_x$ from that of the original plane wave. For example, for the first wave in Fig. 3, the wave number along x is $k_{1,x} = \frac{2\pi}{\lambda} \sin \Theta_1$. As a result of traversing a plate, this wave will split into three waves, moving along directions $\bar{k}_1$, $\bar{k}_1'$, and $\bar{k}_1''$. The x components of the latter two are $k_{1x}' = k_{1x} + K_x$ and $k_{1x}'' = k_{1x} - K_x$ respectively, they travel in directions which make an angle θ_1' and θ_1'', respectively with the z axis.

Require that K_x be that solution of

$$K_x = \frac{2\pi}{\lambda} \left[\sqrt{1 - 4 \cos^2 \theta_1} - \sin \theta_1 \right] \tag{16}$$

which has the larger absolute value (if the two solutions of Eq. (16) are distinct). Then, clearly, while traversing a plate, the first wave in Fig. 3 will split into three waves (see Fig. 5). The first one will continue to travel along $\bar{k}_1$, its period along z being $\lambda/\cos \theta_1$. The other two travel along $\bar{k}_1'$ and $\bar{k}_1''$, respectively. Equation (16) insures that the period along z will be

$$\lambda/\cos \theta_1' = \lambda/2 \cos \theta_1 \tag{16a}$$

for the wave travelling along $\bar{k}_1'$, while the other wave will be an evanescent wave confined to the surface of the plate. (Not shown in Fig. 5.) In the same manner, one can show that the second original wave will also split into two branches, one travelling along the original direction, the other having a period $\lambda/2 \cos \theta$ along z. The third is again an evanescent wave. As a result of introducing the plates, the system will consist of altogether four waves propagating between the plates. It is important to realize that node planes of the first pair of waves are

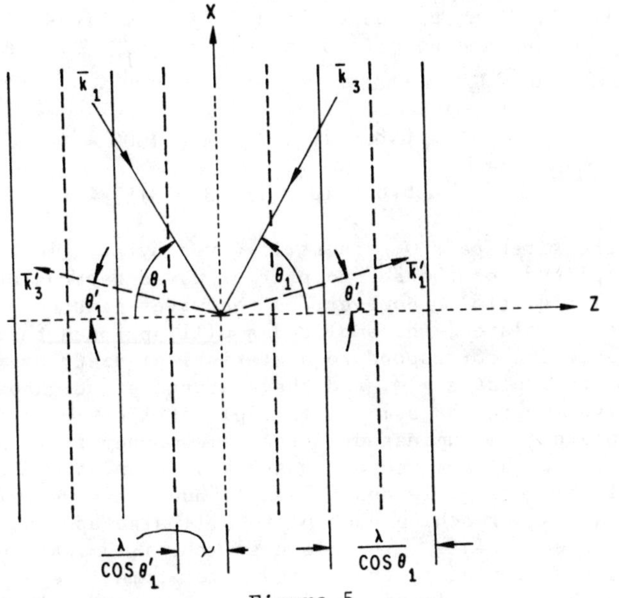

Figure 5

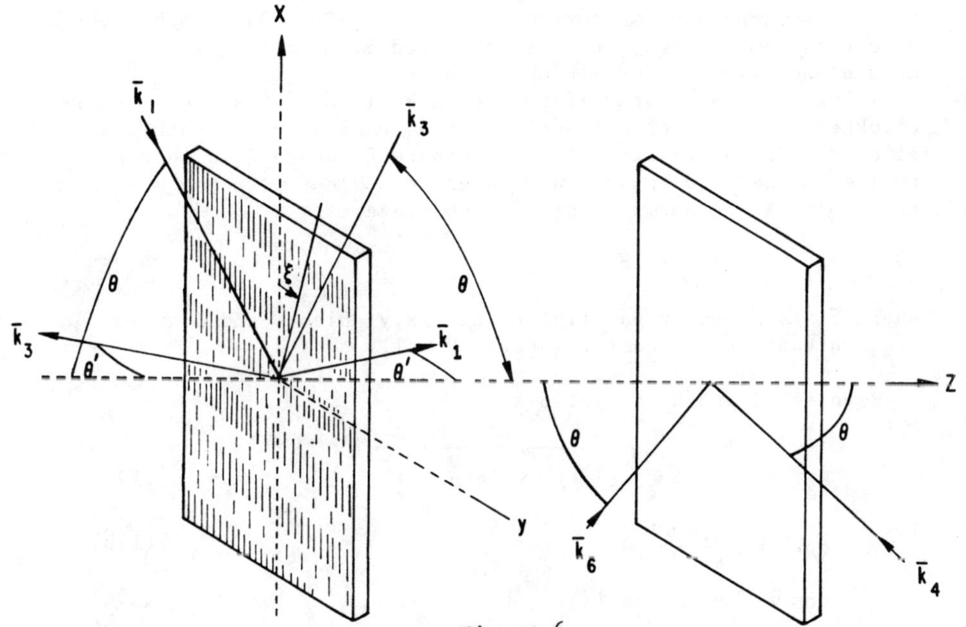

Figure 6

also node planes of the second pair, plates located here are at nodes of the total standing wave pattern.

Choosing λ, d, μ_1, μ_2 and θ_1 as before, one finds from Eq. (16b) that for the second pair of waves, $\sin \theta_1' = 0.1$, and $\varepsilon_r = \frac{2\sqrt{2}}{3}(1-3.2 \cdot 10^{-6})$, so that

$$Q_{th} = Q_{th}(\text{thick}) \begin{cases} 6.8 \cdot 10^{3}, \text{ if } d = 1000 \text{ Å} \\ 6.8 \cdot 10^{4}, \text{ if } d = 315 \text{ Å} \end{cases} \tag{17}$$

while for the first pair Q_{th} is still given by Eq. (14).

The amplitude of the second pair of waves depends on the constant b in Eq. (15). However, as the waves pass through a large number of plates, the amplitudes will approach the asymptotic values which correspond to a statistical distribution. If the number of plates $\to \infty$, and the distance of neighboring plates is everywhere the same (a multiple of $\lambda/\cos \theta_1$), if further each plate is illuminated in the same manner from the outside (by waves with wave vector $\bar{k}_1$ and $\bar{k}_3$), then edge effects along z will be negligible and the amplitudes of waves with wave vectors $\bar{k}_1'$ and $\bar{k}_3'$, reaching each plate will also approach a common value for each plate. In this case, the illumination must be periodic function of z, the period times some integer being the distance between neighboring plates. Such periodic illumination can be achieved with a suitable template.[1],[3]

(4) This case is similar to Example 3. Again, start with the two waves whose wave vectors are $\bar{k}_1$ and $\bar{k}_3$ (Fig. 3). Both waves have equal amplitudes, and are polarized along y. Again, place thin plane films at certain chosen node planes.

The films are not uniform. As described in Example 3, their thickness or index of refraction (or a combination of both) is allowed to be a function of a coordinate $\bar{q}$, where $\bar{q}$ is chosen to lie in the (x,y) plane, and makes an angle ξ with the x axis (Fig. 6). The transmittance of each plate is

$$T = a + b \cos K_q \, \bar{q} \cdot \bar{r}_{\perp} \, , \tag{18}$$

where $\bar{r}_{\perp}$ is a vector parallel to the (x,y) plane, and a and b are constants to be chosen later. Usually $b \ll a$.

We set ξ and K_q as follows

$$K_q = \frac{2\pi}{\lambda} \sqrt{(\sin \theta)^2 + (\sin \theta')^2} \tag{19a}$$

$$\text{tg}\xi = \frac{\sin \theta'}{\sin \theta} \tag{19b}$$

$$\cos \theta' = 2 \cos \theta \tag{19c}$$

Now, when a wave with wave vector $\bar{k}_1$ traverses a plate, it will split into three branches. One branch will have amplitude proportional to a, and will continue to travel along $\bar{k}_1$. The second branch will have a wave vector $\bar{k}_1'$, which lies in the (y,z) plane, and makes an angle θ_1' with z (Fig. 5). The third wave is evanescent. Similarly, the wave travelling along $\bar{k}_3$ will split into three branches. One with wave vectors $\bar{k}_3$, the other $\bar{k}_3'$, and the third wave is evanescent. The $\bar{k}_3'$ lies in the (y,z) plane, and makes an angle θ_1' with z.

In addition to the original two waves, illuminate the system with two more waves, waves No. 4 and 6. They travel along $\bar{k}_4 = -\bar{k}_1$ and $\bar{k}_6 = -\bar{k}_3$, as shown in Fig. 6. While traversing a plate, these waves will also split into two real waves each, with wave vectors $\bar{k}_4$, $\bar{k}_4'$ and $\bar{k}_6$, $\bar{k}_6'$, respectively. The $\bar{k}_4'$ and $\bar{k}_6'$ also lie in the (y,z) plane, and they too make an angle θ_1 with z. The waves with wave vectors $\bar{k}_1$, $\bar{k}_3$, $\bar{k}_4$ and $\bar{k}_6$ will set up a standing wave pattern. Nodes of the y component of the electric field, E_y, lie along planes parallel to the (x,y) plane, the distance between neighboring node planes is according to Eq. (19c) $\lambda/2 \cos\theta_1$. Choose the phases so that a node plane of this pattern coincides with a node plane of the first two waves, and also with one node plane of waves No. 4 and 6. Then, any node of the first two waves will also be a node of waves No. 4 and 6, and also of E_y in the waves with wave vectors $\bar{k}_1'$, $\bar{k}_3'$, $\bar{k}_4'$, $\bar{k}_6'$. The thin plane films are placed in these common nodes of the fields mentioned. Note that waves with $\bar{k}_1'$, $\bar{k}_3'$, $\bar{k}_4'$, and $\bar{k}_6'$ are polarized in the plane of incidence as they reach these films.

When evaluating Q_{th}, we distinguish two cases:

a. Condition (8) holds. Then Q_{th} will depend strongly on the microscopic properties of the specific material chosen. This case will not be discussed further.

b. Condition (8) is violated. Then one has to consider Eq. (6a). Assume λ, d, μ_1, and μ_2 have the same values as before. Equations (19) determine θ such that $\sin\theta_1' = 0.1$. Choose $n = 3.5\,(1 + i)$. Then for the waves whose wave vector is any one of $\bar{k}_1'$, $\bar{k}_3'$, $\bar{k}_4'$, or $\bar{k}_6'$, one finds[7] that $1/2\,(-1 - \varepsilon_i\alpha_r/\varepsilon_r\alpha_i)^3 \simeq 8.17 \cdot 10^{-4}$, and $(1/2)(1/6)\varepsilon_r^2(1 - \alpha_r\varepsilon_i/\varepsilon_r\alpha_i^3) = 9.0 \cdot 10^{-4}$.

Therefore,

$$Q_{th} \approx Q_{th}\ (\text{thick}) \cdot 5.8 \cdot 10^2 \quad . \tag{20}$$

On the other hand, if all parameters are as before, but d = 315 Å and n = 6.2 (1 + i), then

$$Q_{th} \approx Q_{th}\ (\text{thick}) \cdot 1.8 \cdot 10^3 \quad . \tag{20a}$$

The longitudinal electric field is non-zero for the waves we studied. Assuming conservatively that thick films of the same material can withstand without damage 1 Joule/cm² incident energy in 30 psec pulses, Eq. (20) shows that the averaged (over time) longitudinal (i.e., parallel to the surface normal) electric accelerating forces acting on an electron at a mirror surface can reach 0.7 GeV/m (1.2 GeV/m) for the parameters chosen above. These fields can be used to accelerate particles. For other parameters, the accelerating field component may be higher, or lower. At any rate, acceleration of order 1 GeV/m is possible without destroying the structure, and this value is independent of the energy of the accelerated particle; it is valid even if γ is high. Such high gradients could not be obtained if the films were thick, or if they would be located away from the nodes of the standing wave pattern generated by the electromagnetic waves.

SUMMARY

When a thin foil is located at an interference node of a standing wave pattern generated by electromagnetic waves, it can withstand without damage more intense incident waves than when the foil is thick, or when a thin foil is located away from a node. The resistance to damage increases as the foil thickness decreases.

The optical effectiveness (e.g., phase and amplitude change caused by the foil) decreases as the thickness of the foil is decreased. ($\sim$ d). This decrease, however, is slower than the increase in resistance to damage caused by Joule heating ($\sim d^2$). The ratio (optical effectiveness)/(Joule heat damage) increases with decreasing foil thickness. Therefore there is advantage in employing a sequence of thin foils, rather than a single thick plate, as an optical element (e.g., as a mirror).

The improved resistance results from the fact that inside such a film the maximum field intensity (as well as maximum Joule heating) is much lower than outside the film. One can say that the electromagnetic radiation is "channelled" between such foils.

The increased resistance to damage makes it possible to have higher electromagnetic fields near optical elements; e.g., mirrors. That, in turn, would allow the contruction of a template laser accelerator which will accelerate electrons (or any particle with one electron charge) at a rate of ≈ 1 GeV/m, no matter how heavy or how energetic the particle may be; i.e., even for high γ.

The calculated decrease due to Joule heating can be several orders of magnitude for realistic cases. Since other damage mechanisms are also proportional to some power of the electric field inside the foil, one expects damage caused by most mechanisms to decrease in a thin foil near an interference node. The decrease for some mechanisms may be even more dramatic than for Joule heat damage; for other mechanisms, it may be less pronounced. At any rate, it is plausible that total damage resistance will not improve as much as calculated here for Joule heat damage alone. Nevertheless, since that factor is so large, a significant increase in total damage resistance is expected.

Many mechanisms exist which can cause electromagnetic radiation damage to surfaces. Most of these mechanisms cannot be calculated accurately. Experimental work will be needed to determine how pronounced this phenomenon really is for various cases of practical importance.

Acknowledgements

I am grateful to P. Channell, D. Milam, R. Martin, J. McClure, A. Saxman, and B. Veal for stimulating discussions.

APPENDIX

Assume that a material surface under study will survive without noticeable damage when a radiation pulse of time duration T, average frequency ω and averaged (over a time $2\pi/\omega$) intensity I(t) as a function of time t, is incident on it, whenever

$$\int_0^T I(t)\, dt < Q_{th} \quad . \tag{1A-1}$$

The Q_{th} is then called the incident damage threshold energy. (Clearly $I(t) = 0$ for $t < 0$ and for $t > T$).

Experimental evidence indicates[8] that for thick (compared to the heat diffusion length) metals, when 10^{-11} sec $< T < 10^{-6}$ sec, the Q_{th} is proportional to $\sqrt{T}$

$$Q_{th} \sim T^{\frac{1}{2}} \quad . \tag{A-2}$$

The heat diffusion length is $(\pi\alpha T/4)^{\frac{1}{2}}$, where α is the thermal diffusivity. The value of α for metals is typically 1 or 2 cm^2/sec. Thus, for 10 psec pulses the scaling law (A-2) is expected to hold down to a thickness of a few hundred Angstroms.

If the heat diffusion length $(\pi\alpha\, T/4)^{\frac{1}{2}}$ exceeds d, then damage will be controlled by heat conduction sideways through the film; and by heat radiation from the film surfaces. For very thin films, the latter process will dominate. In this case, Q_{th} will no longer be proportional to $\sqrt{T}$,for long enough T it will be proportional to T.

For dielectrics generally Eq. (A-2) is observed not to hold. Damage commences at the location of impurities, cracks, and probably at the surface of microcrystals. Q_{th} is determined by the local composition and mechanical condition of the material. Depending on which damage mechanism is dominant, Q_{th} may be proportional to E_{in}, the incident electric field, or E_{in}^2 or, generally to E_{in}^q, where q is an empirical exponent which is expected to be real and positive. For metals q = 2, since, as we saw, damage depends on absorbed energy, and that is proportional to E_{in}^2. High q is an advantage when protecting thin films in the manner described in this paper.

Clearly, the damage behavior of metals is much easier to understand and predict than those for dielectrics. Furthermore, for metals q is relatively high. On the other hand, dielectrics have certain properties which make them more desirable as foil material. This leads one to explore the possibility of using "hybrid" materials; e.g., semiconductors. In particular, one looks for a material from which thin foils can be manufactured (e.g., C,Si). A material with lower $|n|$ may be preferred, so that $kd \ll 1$ even for larger layer thickness d. High heat conductivity (as close to metallic as possible) will probably be a requirement, in order to decrease local overheating generated by impurities, imperfections, and inhomogeneities; i.e., to approach a metallic damage mechanism.

REFERENCES, FOOTNOTES

1. Paul L. Csonka, Particle Accelerators 5, 129 (1973).

2. See, for example:

   Robert B. Palmer, Particle Accelerator Conference, March 11-13, 1981, Washington, D.C.

   John D. Lawson, "A Survey of Some Ideas for Accelerators Using Laser Light." Paper presented at the Chinese National Accelerator Conference, Shanghai, October 1980.

3. Paul L. Csonka, "Near Field" Laser Accelerators, Lecture presented at the Workshop on the Laser Acceleration of Particles, Los Alamos, New Mexico, February 1982.

4. Unless the field is so strong as to remove electrons from the layer.

5. Carbon films can be bought off the shelf with an area of a few square inches, for a few 10^2 dollars per box of 25. Thickness is determined to an accuracy of about 100 Å for

thin films. Films are available down to nominal thickness of 30 Å (but ± 100 Å).

6. This will be discussed elsewhere.

7. $|n|$ was specified so that for the chosen parameters and $n_r = n_i$ the two terms in the second curly bracket in Eq. (6a) should be approximately equal.

8. T. T. Saito, D. Milam, P. Baker, and G. Murphy, "1.06 μm 150 psec Laser Damage Study of Diamond Turned, Diamond Turned/Polished and Polished Metal Mirrors," 7th Annual Conference on Laser Induced Damage in Optical Materials, July 29-31, 1975, Boulder, CO.

ACCURATE POSITIONING OF OPTICAL ELEMENTS BY A RADIATION FIELD

Paul L. Csonka
Institute of Theoretical Science, and Department of Physics
University of Oregon, Eugene, OR 97403

and

Argonne National Laboratory, Argonne, IL 60439

ABSTRACT

It is shown that thin optical elements such as foils can be accurately positioned directly by a radiation field (e.g. optical) standing wave pattern. An application: to protect optical elements from radiation damage by locating them at intensity minima of a standing wave pattern.

***

This short note concerns itself with the question of how to position thin optical elements with high accuracy in a radiation field.

Assume that we wish to protect some optical element of thickness d, from damage caused by intense electromagnetic radiation of wavelength λ incident on the surface of the element. To achieve protection, we set up a standing wave pattern generated by the incident radiation and locate the optical element at an intensity minimum of this pattern.[1,2] The protection against damage is most pronounced if $d \ll \lambda$ and if the element is positioned with an accuracy $\ll \lambda$, preferably with accuracy $\approx d$.

To be specific, let us consider a class of simple optical elements, namely planar foils of uniform thickness. As d gets smaller, the foil's resistance to radiation damage increases.[2] At the same time, the optical effectiveness (relectivity, phase shift across foil, etc.) decreases. But for a wide range of parameters this decrease is slower than the increase in resistance to radiation damage, so that one gains by using a sequence of thin foils rather than a single thick one. That fact may be applied to advantage when one designs optical elements exposed to intense electromagnetic radiation, such as mirrors, or - of interest here - a laser template accelerator[2] and so on.

In both of the applications just mentioned, one would have to construct a sequence of several thin plates, each located at intensity minima of a standing wave pattern within a tolerance $\ll\lambda$. If λ is in the infrared region, conventional methods of construction would reach the required accuracy. At shorter λ however, the difficulties encountered will rapidly increase. Even where conventional methods are adequate, a simpler approach would be desirable.

It is therefore useful, and also amusing, to realize that one can use an electromagnetic standing wave pattern to insure accurate positioning of thin optical elements, e.g. of thin planar foils. Since optical interference patterns are among the most sensitive

0094-243X/82/910266-05$3.00

position monitors known, it is advantageous to have the pattern itself directly arrange the position of elements.

The standing wave pattern need <u>not</u> be generated by a laser. The radiation which produces it need <u>not</u> be monochromatic. The method is <u>not</u> sensitive to the detailed spatial indensity distribution within the interference pattern. What <u>is</u> important though, is that the minima of the pattern be located correctly within an accuracy $\ll \lambda$, preferably $\approx d$.

The physical reason why the method to be described is relatively insensitive to radiation characteristics, can be understood as follows. The method consists of allowing each plate to settle down to an equilibrium position at or near an interference minimum. The history of settling down will, of course, depend on the details of the radiation field. But the final equilibrium positions depend duly on where the nodes are located.

In interference optics, standing wave patterns are invariably generated by appropriately positioned solid optical elements, such as mirrors, lenses and so on. The arrangement to be described here is a reversal of that usual state of affairs: now solid optical elements, such as foils, are to be positioned through feedback by an appropriate interference pattern. The point to be noticed is simply that contemporary technology has reached the level where foils can routinely be made thin enough, and sources produce radiation intense enough to physically move those foils to appropriate positions.

The idea is straightforward. It will be outlined for the simple special case when uniform rigid planar foils located in vacuum are to be positioned.

Let ρ be the density of the foils. Then ρd is the mass per unit surface for each foil. Assume that a radiated power $I(\Theta)$ per unit surface is incident on the foil at an angle Θ (measured from the surface normal). The radiation is partly transmitted by the foil, but a fraction $R(\Theta)$ of the intensity is specularly reflected.

Then the radiation pressure exerted per unit foil surface is

$$P(\Theta)=2(\sin\Theta)R(\Theta)I(\Theta)/c, \qquad (1)$$

where c is the velocity of light in vacuum. If the radiation arrives at the foil from various directions, then the total pressure will be the integral $\int d\theta P(\theta)$.

If no other force acts on the foil, it will accelerate at a rate of

$$a = \frac{P}{\rho d} = \frac{2}{c\rho d}\int d\theta (\sin\theta)\, I(\theta)\, R(\theta)\, . \qquad (2)$$

Assume now that the foil is parallel to a fixed solid reflecting wall, at a distance ℓ from it as shown in Fig. 1. Assume further that $I(\Theta)$ is independent of ℓ. If the foil is placed further away from the wall, but still parallel to it, at a distance ℓ', then the number of times a light beam bounces back and forth between the foil and the wall will decrease with increasing ℓ'. Furthermore, if the foil is tilted by an angle $1 \gg \phi' > 0$ as defined in Fig. 1, so that the radiation arrives at an angle of incidence $\Theta'=\Theta+\phi'$, then again the radiation pressure force on the foil will decrease, because in

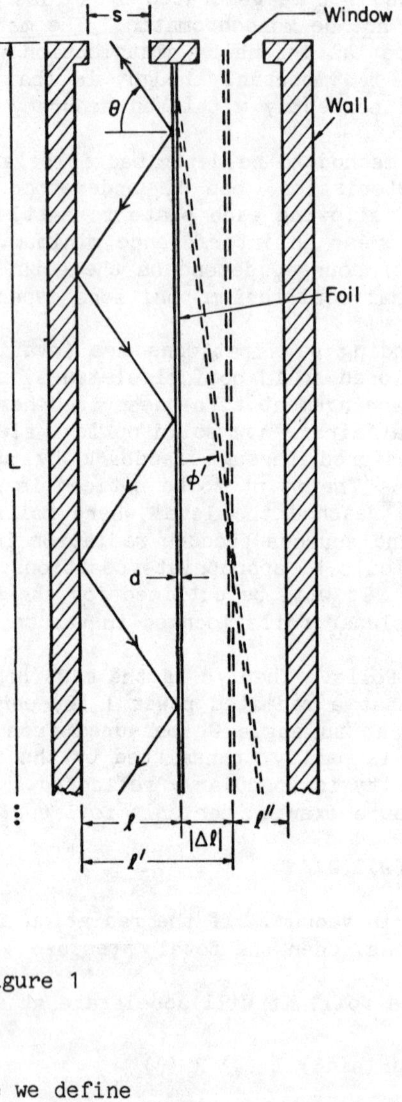

Figure 1

expressions of type (2) $\sin\Theta$ is replaced by $\sin\Theta' < \sin\Theta$ and at the same time the number of beam reflections between the wall and the foil decreases.

If the foil is placed between two parallel walls, as in Fig. 1, illuminated symmetrically from both sides through windows s wide, then it will tend to locate itself halfway between the two walls, and align itself with them. To see this in more detail, consider the case when the length of the foil, $L \gg \ell \,\mathrm{tg}\Theta$ i.e. when the beam is bounced many times between the foil and the wall. Then the number of bounces is approximately proportional to $s/2\ell'$. Let the foil be parallel to the wall, i.e. $\phi'=0$. Then the total pressure force on the foil is the difference of the total force acting from the left side, minus that acting from the right side:

$$F_t = F_\ell - F_r \approx \frac{2}{c}\int d\theta . \tag{3}$$

$$(\sin\theta) I(\theta) R(\theta) \frac{s}{2}\left(\frac{1}{\ell'} - \frac{1}{\ell''}\right) .$$

Denoting $\ell' = \ell + \Delta\ell$, $\ell'' = \ell - \Delta\ell$, and assuming $\Delta\ell \ll \ell$, one obtains

$$F_t \approx -d\rho\,\omega_o^2\,\Delta\ell , \tag{4a}$$

where we define

$$\omega_o^2 = \frac{1}{\rho d}\,\frac{2}{c}\int d(\theta)(\sin\theta) I(\theta) R(\theta) \frac{s}{\ell^2} . \tag{4b}$$

The force F_t given in Eq. (4) will cause the foil to describe a harmonic oscillation with circular frequency ω_o around its equilibrium position: the plane halfway between the two walls. The oscillation will presist indefinitely, since, by assumption, no other force acts on the foil. In practice, of course, the oscillations will be damped and the foil will settle down into its equilibrium position. For typical values, see Table I.

One source of damping is the atmosphere. If the system is to be used in vacuum, and yet, past damping is desired, one may allow the plates to settle down before full evacuation.

Of special interest is a system which consists of several equidistant plates between two fixed walls. Windows allow radiation to enter the space on both sides of each plate. This system can be treated as was done above: there will be an effective restoring force between each pair of neighboring plates. That will tend to settle down each plate into its own equilibrium position relative to its neighbors. Without damping, the plates will describe coupled oscillations which are characterized by a superposition of normal modes ("phonons"). In the presence of damping, all oscillation amplitudes will asymptotically approach zero, and the plates will stay near their respective equilibrium position.

Actually, foils are not perfectly rigid. Distortions will occur, and they can be described as a superposition of normal surface modes. If elasticity is important, these modes will in general travel along the surface, and have to damp out before equilibrium is reached. If, on the other hand, elasticity is negligible, then the foil will act as a chain of loosely tied composite foils. When such a foil is hung from approximately the correct position, then, provided that the radiation is intense enough, most of its bulk will nevertheless tend to settle down at its correct location, thus compensating for any error in initial positioning.

To insure appropriate radiation flux on both sides of each plate, a sequence of windows has to be opened on the outer walls of the device, through which the illuminating radiation centers. The window need not have uniform transmittance, indeed the transmittance may be taylored to achieve the desired $I(\Theta)$. That can be done by allowing the transmittance of each window to vary as a function of position. Such sequence of windows forms a template [1,2]. This same template may be used to set up the interference pattern at whose nodes the foils are to be located.

One may wonder if the radiation field which positions the foils may overheat and thus damage them. Actually the positioning radiation can arrive in long pulses, and may even be continuous. In that case, even in the limit $I(\Theta)$ 0, the plates will settle down at their equilibrium position, provided that no other force disturbs them. In practice, one can not wait indefinitely for equilibrium to be reached, and in any case there will be other forces perturbing the system. Therefore, in practice $I(\Theta)$ will have to be finite; nevertheless, it can be well below the threshold which would damage the foils. For example, if the parameters are as in Table I, (and e.g. the radiation is polarized perpendicularly to the plane of incidence) then the equivalent normal incident radiation flux is orders of magnitude below damage threshold for usual materials.

REFERENCES, FOOTNOTES

1. Paul L. Csonka, Particle Accelerators, 5, 129 (1973).
2. Paul L. Csonka, "Near Field Laser Accelerators", lecture delivered at the Workshop on The Laser Acceleration of

Particles, Los Alamos, New Mexico February 1982, and "Improved Damage Threshold for Optical Elements Placed at Minima (i.e. Shadow) of a Radiation Field", Proceedings of the Workshop on The Laser Acceleration of Particles, Los Alamos, New Mexico, February 1982.

3. If it is not, but increases away from the equilibrium position of the foil, then the effect to be described will be even more pronounced.

Table I

$\lambda = \mu m$	
$R = 0.1$	
$\theta = 10^{-1}$	
$I(\theta)\sin\theta = 10^{-2}\ W/_{cm^2}$	
$\delta = 2\ g/cm^3$	
$d = 10^3\ \mathring{A}$	
$\ell = 300\ \mu m$	
$s = 150\ \mu m$	
$R = 0.1$	1
$\omega_o = 2.4\ sec^{-1}$	$7.75\ sec^{-1}$

SUMMARY OF WORKSHOP

J D Lawson
Rutherford Appleton Laboratory, Chilton, Oxon. UK.

1. INTRODUCTION

The possibility of accelerating charged particles by exploiting the high fields associated with laser beams has been considered at intervals over the past 20 years. A number of proposals, not all of which have been sound, have been discussed in the literature.

The purpose of the present workshop has been to examine the physical basis of these proposals and their limitations, to make exploratory studies of parameters which might be achieved (such as energy, current, repetition rate and pulselength), and, assuming a favourable outlook, to suggest further studies or experiments which might advance the subject.

Conventional accelerators are efficient and versatile, and it is evident that merely to demonstrate that acceleration is possible is not enough; laser accelerators must show promise of performance which is superior, or costs which are low.

The application that excites most interest is undoubtedly to high energy physics. Energy limitations of existing machine types are now well understood, and any radically new approach that might seem to give promise of greatly increased energy or reduced cost is clearly worthy of close scrutiny. It was appropriate, therefore, that the first lecture should be devoted to 'High Energy Physics and New Accelerators'. Limitations to existing types of accelerator, and future performance characteristics which would be considered as interesting have been studied at two recent workshops, and these were presented in outline. Other papers presented at the beginning of the meeting surveyed the four classes of laser accelerator which had previously been identified. Two of those involve the use fields in free space, and two make use of dielectric or plasma media.

2. CLASSES OF ACCELERATOR SELECTED FOR STUDY

In this section the basic principles of the main classes of accelerator selected for study are briefly presented, together with their possibilities and limitations as perceived at the beginning of the workshop. In section 3 the findings of the workshop in the different areas will be discussed.

2.(i) Near Field Accelerators

Continuous interaction with a traveling wave possessing an electric field in the direction of propagation requires that the particle beam should be within a wavelength of a periodic structure or dielectric surface. This immediately places constraints of beam size

0094-243X/82/910271-05$3.00 Copyright 1982 American Institute of Physics and emittance. Furthermore, since the accelerating field is of the same order as the fields in the conductor or dielectric, the accelerating field is limited by breakdown in a manner analogous to that in conventional accelerators. Contrary to some earlier claims, it can be shown to be impossible to set up the required type of field configuration remote from a material boundary.

2.(ii) Far Field Accelerators

A particle which moves in a straight line cannot receive continuous acceleration from a plane wave or gaussian laser beam. If it moves parallel to the wave vector it experiences oppositely directed and nearly equal transverse electric and magnetic forces oscillating in time with a frequency equal to the optical frequency multiplied by $1-\beta$. If now the orbit (with $\theta=0$) is modulated to become a helix or sinewave, with wavelength $\Lambda=\lambda_o(\beta_z^{-1})$ where λ_o is the optical wavelength, then for particles of the correct phase, the alternating transverse electric field from the lightwave always has a component in phase with the transverse motion of the charge, so that continuous acceleration can occur. Phase oscillations about the 'stable phase' occur as in conventional accelerators. The practical requirement that $\Lambda>>\lambda_o$ requires that $1-\beta_z$ be small, so that the transverse velocity is small, and the acceleration a 'second order' effect. This is the inverse mechanism of the 'ubitron' and free electron laser, suggested by Gorn in a patent dated 1947. The large ratio between Λ and λ_o for relativistic particles was first exploited in Motz' undulator in 1951. At high energies it becomes difficult to produce sufficient transverse acceleration to maintain a compact beam with adequate transverse velocity, and the rate of acceleration decreases.

2.(iii) Inverse Cherenkov Accelerators

In a medium with refractive index n the phase velocity of a plane wave is slowed down to give a longitudinal component of accelerating field traveling at the particle velocity of $c\sec\theta=n\beta c$. (This is the principle of Shimoda's first suggestion in 1962, recently tested by experiments at Stanford). The density, and hence the value of n, is limited in practice by gas breakdown or scattering. This implies a small angle θ, so that for a plane wave or gaussian beam the longitudinal z-component of field is considerably less than the transverse component.

2.(iv) Collective Accelerators

The interaction of intense laser light with dense plasma can give rise to a wide range of phenomena, especially if non-linear processes are included. A particular mechanism, that has been studied by simulation and experiment, is the trapping of particles in the wake of traveling longitudinal plasma oscillations. Such wakes can be set up by a single laser pulse, or perhaps more effectively by making use of simultaneous pulses of frequencies ω_1 and ω_2 such that $\omega_1 - \omega_2$ is the plasma frequency of the medium.

3. PROGRESS MADE AT THE WORKSHOP

Participants were divided into three study groups to look at the three categories 'near field', 'far field' and 'media', the latter category comprising both the collective and inverse Cherenkov schemes. The aim was to examine the physical mechanisms, and, if possible, sketch out some parameters which might be achievable in a practical device. Reports of the groups appear earlier in this volume. These are summarized below, with additional comments.

3.(i) Near Field Accelerators

The starting point for this group was the grating accelerator already proposed by Palmer. Two lines of work were followed; in the first this concept was examined in more detail, the problem of transverse focusing was identified, and solved in principle. Speculative parameter ranges for possible accelerators were presented, with emphasis on high acceleration rate. The limits to the acceleration rate were extensively discussed. At high power densities breakdown by field emission occurs, followed by thermal damage to the grating surface if the pulse is long enough. If these limits are avoided, it seems difficult to reach 1 GeV per metre. The question was raised whether one could work above these limits, using very short pulses or, as an extreme measure, allowing the grating to be destroyed after each shot. In this way figures of several GeV per metre, albeit with small currents, might be achievable.

The second line of enquiry was to note that the grating represents an unconventional form of linac, and to investigate the scaling of various features of linac operation over the wavelength range 10μ to 10 cm. What is the 'optimum' wavelength? This must depend on the ultimate use, but figures of merit can be devised representing a suitable compromise between overall energy consumption of the accelerator, accelerating gradient and accelerated current. It was felt that the optimum perhaps lies between 1 mm and 1 cm, though this may be in a region where adequate power sources are not readily available.

3.(ii) Far Field Accelerators

This group studied both the universe FEL mechanism, using a magnetic undulator, and also the two-wave proposal by Pantell and Smith in which the transverse motion is provided by a forward microwave field traveling at an angle to the axis.

Parallel analysis of these two schemes showed their essential equivalence. Even though the acceleration is a second order effect, the very high fields in laser beams enable accelerating rates of the order of hundreds of MeV per meter to be obtained at least over limited distances. The inevitable spreading of finite gaussian beams, together with the difficulty of producing adequate modulation at high γ, suggest that energies would be limited to a few GeV. If the optical beams can be repeatedly reformed, this range might be extended somewhat.

3.(iii) Inverse Cherenkov Accelerators

The limitations of this type of accelerator were outlined in 2 (iii). A design was presented using a 'cone' of waves in which the electric field is radial. Near the axis of the cone in the region where the waves overlap to give a field on the axis, the radial field distribution follows the well-known Bessel function form, and the electric field is in the z-direction. This is limited by gas breakdown in the region of high transverse fields about $\lambda_0/2$ from the axis. Using a 70 terawatt laser beam to give an accelerating rate of 500 Mev/metre the acceleration length in this design is limited by scattering to 50 metres, giving an energy gain of order 25 GeV.

3.(iv) Collective Accelerators

In this class the most tantalizing results were presented. Unlike the other concepts, enormous fields in the direction of propagation seem attainable, though it is not yet evident how to control these to give coherent acceleration over the long distances required to attain very high energies. Several methods of setting up these high fields were suggested, and experiments which appear to support the relevant theory were described. The most promising concept is probably the 'beat wave' accelerator suggest by Tajima and Dawson. Intense laser beams, with frequencies whose different is the plasma frequency of a plasma into which they are fired, produce strong non-linear bunching. The associated large fields can accelerate plasma particles, or particles in a beam injected into the plasma. The process may be regarded as the nonlinear saturated limit of Raman scattering, in which the plasma waves produced move forward at about the speed of light. Fields well in excess of 1 GeV/metre might be expected from this mechanism; indeed simulation by Sullivan showed gradients as high as 50 GeV/m. Experiments described by Joshi had yielded 1 MeV electrons, in good agreement with theory.

4. CONCLUSIONS AND OUTLOOK

That particles can be accelerated to high energies using laser light is not in doubt. Whether useful accelerators using this technique will, in fact, be constructed is an open question on which opinions vary. It was agreed by participants at the workshop that useful progress had been made in providing a clearer framework for the subject, and that there was a case for support of further work to address some of the many questions still unresolved.

Since the most interesting characteristic of laser beams is the very high energy density, and hence electric field, which can be obtained, attention was naturally focused on the attainment of a high rate of acceleration. Other important aspects of accelerators such as energy spread, pulse length, accelerated current and repetition rate received less attention, though it was evident with the grating linac that the short wavelength and pulselength implied rather few particles per pulse and low mean currents. It was disappointing that none of the laser concepts showed promise of meeting the criteria for interesting high energy physics machines laid down in the first lecture. What did emerge was that some wavelength between laser wavelengths and the 10 cm wavelength of existing linacs might be an 'optimum' for some applications. There is clearly a need for thorough analysis of this problem, perhaps supported by well chosen experiments, particularly on field emission breakdown as a function of pulselength and wavelength.

Although we had the benefit of excellent educational lectures by Loree and Saxman of LANL, there were too few experienced laser physicists in the working groups; for this reason the laser aspects of some of the proposals remained somewhat sketchy. For example, the cost and feasibility of reasonable repetition rate combined with high power, was one of the relevant questions for not specifically discussed.

Further study will yield better understanding and clearer appreciation of the constraints. In the vacuum far and near field accelerators the nature of the fundamental constraints is clear, and there would appear to be little scope for further invention. The most hopeful development is probably a highly efficient and economical high power source at a wavelength substantially longer than 10μ. The limitations to the inverse Cherenkov scheme was likewise fundamental and can be clearly perceived.

On collective accelerators, the field is still 'wide open'; the basic physics is interesting, and further work must produce new knowledge of plasma behaviour at extreme conditions. Whether this knowledge will point out the way to a useful device is a question to which the answer is awaited with interest.

In conclusion, there is clearly scope for further work to confirm the general conclusion of the meeting, and to address the outstanding problems which have been noted. The initiative for most of the investigations so far has come from accelerator physicists, and it is now essential that detailed participation of laser experts should be encouraged, so that the limitations and possibilities, both fundamental and techical, of the laser configurations that might be used are clearly perceived. For accelerator physicists the problem of creating beams of suitable quality for injection into the accelerators deserves further study.

Two years for reflections on the contributions to the present workshop, and for further studies along the lines indicated, should enable a more definitive assessment to be made at the next.

LIST OF PARTICIPANTS

Roger Bangerter, LANL
Mario Bosco, UCSB
Winston Bostick, Corrales, NM
Paul Channell, LANL
Alexander Chao, SLAC
Frank Cole, FNAL
Richard Cooper, LANL
Paul Csonka, U. Oregon
Jorge Fontana, UCSB
David Forslund, LANL
Roger Freedman, UCSB
Dennis Gill, LANL
Terry Godlove, DOE
Louis Hand, Cornell
Heino Henke, CERN
Heinrich Hora, U. of New South Wales
F. R. Huson, FNAL
Chan Joshi, UCLA
Dennis Keefe, LBL
Tat Khoe, Argonne
Kwang-Je Kim, LBL
Ed Knapp, LANL
Norman Kroll, UCSD
John Lawson, Rutherford
Chris Levey, U. of Wisconsin
Greg Loew, SLAC
Alfredo Luccio, BNL
Siva Mani, Schafer Associates, Inc.
B. D. McDaniel, Cornell
Melvin Month, DOE
Phil Morton, SLAC
H. Motz, Oxford
Vittorio Nardi, Stevens Institute
David Neuffer, FNAL
Craig Olson, Sandia
Robert Palmer, BNL
Richard Pantell, Stanford
Claudio Pellegrini, BNL
Melvin Piestrup, Stanford
Donald Prosnitz, LLNL
Alberto Renieri, Frascati
Leonard Rivkin, UCLA
Stephen Rockwood, LANL
Robert Rossmanith, DESY
Allessandro Ruggiero, FNAL
Ronald Ruth, LBL
H. A. Schwettman, Stanford
Andrew Sessler, LBL
Sidney Singer, LANL
John Slater, MSNW
Todd Smith, Stanford
S. Solimeno, University of Napoli
Phillip Sprangle, NRL
Ravi Sudan, Cornell
Donald Sullivan, Mission Research
David Sutter, DOE
Toshi Tajima, University of Texas
Cha-Mei Tang, NRL
Sergio Tazzari, INFN
Lee Teng, FNAL
Maury Tigner, Cornell
Tom Wangler, LANL
Thomas Weiland, DESY
William Willis, CERN
Perry Wilson, SLAC

AIP Conference Proceedings

		L.C. Number	ISBN
No.1	Feedback and Dynamic Control of Plasmas	70-141596	0-88318-100-2
No.2	Particles and Fields - 1971 (Rochester)	71-184662	0-88318-101-0
No.3	Thermal Expansion - 1971 (Corning)	72-76970	0-88318-102-9
No.4	Superconductivity in d-and f-Band Metals (Rochester, 1971)	74-18879	0-88318-103-7
No.5	Magnetism and Magnetic Materials - 1971 (2 parts) (Chicago)	59-2468	0-88318-104-5
No.6	Particle Physics (Irvine, 1971)	72-81239	0-88318-105-3
No.7	Exploring the History of Nuclear Physics	72-81883	0-88318-106-1
No.8	Experimental Meson Spectroscopy - 1972	72-88226	0-88318-107-X
No.9	Cyclotrons - 1972 (Vancouver)	72-92798	0-88318-108-8
No.10	Magnetism and Magnetic Materials - 1972	72-623469	0-88318-109-6
No.11	Transport Phenomena - 1973 (Brown University Conference)	73-80682	0-88318-110-X
No.12	Experiments on High Energy Particle Collisions - 1973 (Vanderbilt Conference)	73-81705	0-88318-111-8
No.13	π-π Scattering - 1973 (Tallahassee Conference)	73-81704	0-88318-112-6
No.14	Particles and Fields - 1973 (APS/DPF Berkeley)	73-91923	0-88318-113-4
No.15	High Energy Collisions - 1973 (Stony Brook)	73-92324	0-88318-114-2
No.16	Causality and Physical Theories (Wayne State University, 1973)	73-93420	0-88318-115-0
No.17	Thermal Expansion - 1973 (lake of the Ozarks)	73-94415	0-88318-116-9
No.18	Magnetism and Magnetic Materials - 1973 (2 parts) (Boston)	59-2468	0-88318-117-7
No.19	Physics and the Energy Problem - 1974 (APS Chicago)	73-94416	0-88318-118-5
No.20	Tetrahedrally Bonded Amorphous Semiconductors (Yorktown Heights, 1974)	74-80145	0-88318-119-3
No.21	Experimental Meson Spectroscopy - 1974 (Boston)	74-82628	0-88318-120-7
No.22	Neutrinos - 1974 (Philadelphia)	74-82413	0-88318-121-5
No.23	Particles and Fields - 1974 (APS/DPF Williamsburg)	74-27575	0-88318-122-3
No.24	Magnetism and Magnetic Materials - 1974 (20th Annual Conference, San Francisco)	75-2647	0-88318-123-1
No.25	Efficient Use of Energy (The APS Studies on the Technical Aspects of the More Efficient Use of Energy)	75-18227	0-88318-124-X

No.26	High-Energy Physics and Nuclear Structure - 1975 (Santa Fe and Los Alamos)	75-26411	0-88318-125-8
No.27	Topics in Statistical Mechanics and Biophysics: A Memorial to Julius L. Jackson (Wayne State University, 1975)	75-36309	0-88318-126-6
No.28	Physics and Our World: A Symposium in Honor of Victor F. Weisskopf (M.I.T., 1974)	76-7207	0-88318-127-4
No.29	Magnetism and Magnetic Materials - 1975 (21st Annual Conference, Philadelphia)	76-10931	0-88318-128-2
No.30	Particle Searches and Discoveries - 1976 (Vanderbilt Conference)	76-19949	0-88318-129-0
No.31	Structure and Excitations of Amorphous Solids (Williamsburg, VA., 1976)	76-22279	0-88318-130-4
No.32	Materials Technology - 1976 (APS New York Meeting)	76-27967	0-88318-131-2
No.33	Meson-Nuclear Physics - 1976 (Carnegie-Mellon Conference)	76-26811	0-88318-132-0
No.34	Magnetism and Magnetic Materials - 1976 (Joint MMM-Intermag Conference, Pittsburgh)	76-47106	0-88318-133-9
No.35	High Energy Physics with Polarized Beams and Targets (Argonne, 1976)	76-50181	0-88318-134-7
No.36	Momentum Wave Functions - 1976 (Indiana University)	77-82145	0-88318-135-5
No.37	Weak Interaction Physics - 1977 (Indiana University)	77-83344	0-88318-136-3
No.38	Workshop on New Directions in Mossbauer Spectroscopy (Argonne, 1977)	77-90635	0-88318-137-1
No.39	Physics Careers, Employment and Education (Penn State, 1977)	77-94053	0-88318-138-X
No.40	Electrical Transport and Optical Properties of Inhomogeneous Media (Ohio State University, 1977)	78-54319	0-88318-139-8
No.41	Nucleon-Nucleon Interactions - 1977 (Vancouver)	78-54249	0-88318-140-1
No.42	Higher Energy Polarized Proton Beams (Ann Arbor, 1977)	78-55682	0-88318-141-X
No.43	Particles and Fields - 1977 (APS/DPF, Argonne)	78-55683	0-88318-142-8
No.44	Future Trends in Superconductive Electronics (Charlottesville, 1978)	77-9240	0-88318-143-6
No.45	New Results in High Energy Physics - 1978 (Vanderbilt Conference)	78-67196	0-88318-144-4
No.46	Topics in Nonlinear Dynamics (La Jolla Institute)	78-057870	0-88318-145-2
No.47	Clustering Aspects of Nuclear Structure and Nuclear Reactions (Winnepeg, 1978)	78-64942	0-88318-146-0
No.48	Current Trends in the Theory of Fields (Tallahassee, 1978)	78-72948	0-88318-147-9
No.49	Cosmic Rays and Particle Physics - 1978 (Bartol Conference)	79-50489	0-88318-148-7

AIP Conference Proceedings

No.50	Laser-Solid Interactions and Laser Processing - 1978 (Boston)	79-51564	0-88318-149-5
No.51	High Energy Physics with Polarized Beams and Polarized Targets (Argonne, 1978)	79-64565	0-88318-150-9
No.52	Long-Distance Neutrino Detection - 1978 (C.L. Cowan Memorial Symposium)	79-52078	0-88318-151-7
No.53	Modulated Structures - 1979 (Kailua Kona, Hawaii)	79-53846	0-88318-152-5
No.54	Meson-Nuclear Physics - 1979 (Houston)	79-53978	0-88318-153-3
No.55	Quantum Chromodynamics (La Jolla, 1978)	79-54969	0-88318-154-1
No.56	Particle Acceleration Mechanisms in Astrophysics (La Jolla, 1979)	79-55844	0-88318-155-X
No. 57	Nonlinear Dynamics and the Beam-Beam Interaction (Brookhaven, 1979)	79-57341	0-88318-156-8
No. 58	Inhomogeneous Superconductors - 1979 (Berkeley Springs, W.V.)	79-57620	0-88318-157-6
No. 59	Particles and Fields - 1979 (APS/DPF Montreal)	80-66631	0-88318-158-4
No. 60	History of the ZGS (Argonne, 1979)	80-67694	0-88318-159-2
No. 61	Aspects of the Kinetics and Dynamics of Surface Reactions (La Jolla Institute, 1979)	80-68004	0-88318-160-6
No. 62	High Energy e^+e^- Interactions (Vanderbilt , 1980)	80-53377	0-88318-161-4
No. 63	Supernovae Spectra (La Jolla, 1980)	80-70019	0-88318-162-2
No. 64	Laboratory EXAFS Facilities - 1980 (Univ. of Washington)	80-70579	0-88318-163-0
No. 65	Optics in Four Dimensions - 1980 (ICO, Ensenada)	80-70771	0-88318-164-9
No. 66	Physics in the Automotive Industry - 1980 (APS/AAPT Topical Conference)	80-70987	0-88318-165-7
No. 67	Experimental Meson Spectroscopy - 1980 (Sixth International Conference , Brookhaven)	80-71123	0-88318-166-5
No. 68	High Energy Physics - 1980 (XX International Conference, Madison)	81-65032	0-88318-167-3
No. 69	Polarization Phenomena in Nuclear Physics - 1980 (Fifth International Symposium, Santa Fe)	81-65107	0-88318-168-1
No. 70	Chemistry and Physics of Coal Utilization - 1980 (APS, Morgantown)	81-65106	0-88318-169-X
No. 71	Group Theory and its Applications in Physics - 1980 (Latin American School of Physics, Mexico City)	81-66132	0-88318-170-3
No. 72	Weak Interactions as a Probe of Unification (Virginia Polytechnic Institute - 1980)	81-67184	0-88318-171-1
No. 73	Tetrahedrally Bonded Amorphous Semiconductors (Carefree, Arizona, 1981)	81-67419	0-88318-172-X
No.74	Perturbative Quantum Chromodynamics (Tallahassee, 1981)	81-70372	0-88318-173-8

No. 75	Low Energy X-ray Diagnostics-1981 (Monterey)	81-69841	0-88318-174-6
No. 76	Nonlinear Properties of Internal Waves (La Jolla Institute, 1981)	81-71062	0-88318-175-4
No. 77	Gamma Ray Transients and Related Astrophysical Phenomena (La Jolla Institute, 1981)	81-71543	0-88318-176-2
No. 78	Shock Waves in Condensed Matter - 1981 (Menlo Park)	82-70014	0-88318-177-0
No. 79	Pion Production and Absorption in Nuclei - 1981 (Indiana University Cyclotron Facility)	82-70678	0-88318-178-9
No. 80	Polarized Proton Ion Sources (Ann Arbor, 1981)	82-71025	0-88318-179-7
No. 81	Particles and Fields - 1981: Testing the Standard Model (APS/DPF, Santa Cruz)	82-71156	0-88318-180-0
No. 82	Interpretation of Climate and Photochemical Models, Ozone and Temperature Measurements (La Jolla Institute, 1981)	82-071345	0-88318-181-9
No. 83	The Galactic Center (Cal. Inst. of Tech., 1982)	82-071635	0-88318-182-7
No. 84	Physics in the Steel Industry (APS.AISI, Lehigh University, 1981)	82-072033	0-88318-183-5
No. 85	Proton-Antiproton Collider Physics - 1981 (Madison, Wisconsin)	82-072141	0-88318-184-3
No. 86	Momentum Wave Functions - 1982 (Adelaide, Australia)	82-072375	0-88318-185-1
No. 87	Physics of High Energy Particle Accelerators (Fermilab Summer School, 1981)	82-072421	0-88318-186-X
No. 88	Mathematical Methods in Hydrodynamics and Integrability in Dynamical Systems (La Jolla Institute, 1981)	82-072462	0-88318-187-8
No. 89	Neutron Scattering - 1981 (Argonne National Laboratory)	82-073094	0-88318-188-6
No. 90	Laser Techniques for Extreme Ultraviolt Spectroscopy (Boulder, 1982)	82-073205	0-88318-189-4
No. 91	Laser Acceleration of Particles (Los Alamos, 1982)	82-073361	0-88318-190-8